Tunnelling Contracts
and Site Investigation

Other books on tunnelling
and geotechnical engineering
from E & FN Spon

Tunnelling Contracts and Site Investigation

P.B. Attewell

Spon Press
an imprint of Taylor & Francis
LONDON AND NEW YORK

First published by E & FN Spon, an imprint of Chapman & Hall
2-6 Boundary Row, London SE1 8HN, UK

This edition published 2011 by Spon Press
2 Park Square, Milton Park, Abingdon, Oxon, OX14 4RN

Simultaneously published in the USA and Canada by Routledge
711 Third Avenue, New York, NY 10017

ISBN 978 0 419 19140 7

A catalogue record for this book is available from the British Library

Library of Congress Catalog Card Number: 95-70021

CONTENTS

PREFACE

In recognition of the importance of site investigation for ground engineering works, and the need for site (ground) investigation specialists to be aware of the contractual significance of their results and interpretations, the writer began work in the early 1980s on two areas of the subject: first, a survey of UK tunnelling contracts and associated site investigation costs, and, second, a review of the place of site investigation in tunnelling contracts. The impetus for the work stemmed largely from the major tunnelling activity in the north-east of England and particularly on the Tyneside Sewerage Scheme which is now substantially completed. The former work was reported* and circulated to those companies that had responded to the request for information, and was also made available to other enquirers. A significant feature of the analyses was the manner in which all the cost data on tunnelling contracts and their site investigations were recalculated to a common base date, so allowing realistic conclusions to be drawn as to the cost effectiveness of the investigations. Statistical packages were used to resolve trends between geological, geotechnical, geometrical and cost data. Several earlier, quite short, and unpublished versions of the latter review were produced as a private initiative and circulated in a limited manner within the industry. The project was then put on the 'back burner' until in late 1989 when, realising that many site investigations for tunnelling works were being conducted without a firm understanding of the contractual problems which could ensue, the writer decided to revise and considerably expand the text.

This present book is divided into two parts: Part One is the core text and Part Two is a compilation of 'Supplementary Information' items, each expanding a particular element of the core text. Most of these Part Two special subject entries are of a 'stand alone' nature, some of them are fairly substantial, and some are much more peripheral to the core of the subject than others.

The book generally relates to the relatively 'routine' small diameter (up to about 3m) bored tunnels. Shafts are mentioned occasionally in the text. Not discussed at all are immersed tube tunnels, as completed across the River Medway in Kent, as under construction for the cross-harbour Hong Kong Airport Railway, or as planned for the Peveza-Aktion crossing at the entrance to the Ambracian Gulf in north-west Greece, and also tunnels formed between diaphragm walls and a pre-cast roof slab, or tunnels built by reinforced caissons, both methods of construction as used, for example, for rail tunnels in Amsterdam. The book has been written primarily from a practical tunnelling point of view for people who have some, but not necessarily an extensive, knowledge of tunnelling. For this reason it has not been considered necessary to include drawings of tunnels and tunnelling systems. However, the book could also be of some interest to recent graduates with a geological background who have a little experience in the ground engineering industry and who wish to gain a broader understanding of how their ground investigation activities might interact with tunnelling construction and its contractual framework.

Having indicated the scope of the book, it is also necessary to say that it is *not* a book dedicated to tunnel construction, which would necessarily include the theory and practice of excavation and ground support, *nor* is it a book covering the fundamentals of site investigation. Some specialist elements of ground investigation are given a fairly detailed treatment for the reasons that they may be well-known to tunnelling engineers, and the information is less readily retrievable. An example of this treatment on the ground investigation side of the subject area is the analysis of clay minerals in a soil, and on the

---

*Attewell, P.B. and Norgrove, W.B. (1984) *Survey of United Kingdom Tunnel Contract and Site Investigation Costs*, Construction Industry Research and Information Association RP 324, 154pp.

(perimeter of the) construction side is ground vibration and noise. Some of the other environmental issues that could impinge on tunnelling operations are also covered in Part Two, and an attempt has been made throughout the book to acknowledge the importance of health and safety issues.

It is sensible to identify some of the elements of site (including ground) investigation that are considered only very briefly or not at all in the book. Although it is important to choose the correct techniques for drilling and sampling, these subjects are generally beyond the scope of the intentions in the present work. In any case there is suitable coverage in other books and readily accessible major reports dedicated exclusively to operational site investigation. Geologists and engineers responsible for site investigation soon become aware, for example, of what core barrels are appropriate for particular types of ground – double tube, triple tube, retractable triple tube, wire line, and so on. The same comment applies to the selection of core bits for drilling in rock. At the root of ground investigation is, of course, the testing of soil, rock and water, both in the laboratory on samples retrieved from boreholes, trial/inspection/exploratory pits/trenches and suitable exposures where such exist and also in the ground. Attention in this book is directed only to those tests which are regularly performed for the purposes of tunnelling contracts and to add comments when thought to be appropriate. Some of the tests recommended for and adopted in site investigation more generally are again considered only very briefly or are not considered at all. These include such tests as the field vane, static cone penetration, pressuremeter, plate loading and *in situ* density. Although the subject of soil and rock description and classification is covered only in a relatively cursory manner, the importance of this element of site investigation is fully recognised. It is also noted that over the last few years there have been substantial developments in the area of knowledge based systems applied to site investigation, and, for these systems to be valid, the soil and rock descriptions and classifications need be very rigidly formulated.

The writer has benefited from his consulting experience over the years with a number of contractors, consultants, public and local authorities, notably with Northumbrian Water, and now with Exploration Associates Limited (Managing Director: Mr Ken Marsh), a subsidiary company of Northumbrian Water Plc. Mr Barry Tate, Management Consultant, commented on very early versions of the work, and knowledge stemming from discussions with Mr Bob McMillan of Northumbrian Water Limited and Mr Neil Hayes of Kennedy Construction Limited has been incorporated in the document. Dr Alan Common and Mr Fred Fountain of Exploration Associates Limited have contributed to the Supplementary Information entry on quality assurance. Needless to say, the errors, and indeed the gaps, that inevitably remain are the responsibility of the writer alone.

This text has been expanded and revised over a period of several years. With an accelerating recognition of the need for quality in site investigation and geotechnical testing, while maintaining a competitive price structure, there is a growing awareness in the industry of the need to remove, or at the very least modify, the adversarial character of civil engineering contract forms and procedures and move towards what could be termed a 'partnership in contractual responsibility'. It is necessary also to point out that some of the contract forms, such as the ICE Conditions of Contract for Ground Investigation and the New Engineering Contract, that are discussed in the book have been in the process of revision at the time of writing. It is most likely, therefore, that portions of the text – indeed any text – on these subjects will be out-of-date by publication day.

ACKNOWLEDGEMENTS

Extracts from the ICE Conditions of Contract and the New Engineering Contract are reproduced by permission of Thomas Telford Services Ltd, Thomas Telford House, 1 Heron Quay, London E14 4JD.

Extracts from British Standards are reproduced with permission of BSI. Complete copies of the Standards can be obtained by post from BSI Publications, Linford Wood, Milton Keynes MK14 6LE.

ABBREVIATIONS

AC	Adjudicator's Contract	MAC	maximum admissible concentration
ADR	alternative dispute resolution	MBA	methylene-bis-acrylamide
AGIS	Association of Ground Investigation Specialists	MICL	maximum intact core length
		MQB	Mining Qualification Board
BCS	British Calibration Service	NACCB	National Accreditation and Certification of Certification Bodies
BGS	British Geological Survey		
BOD	biochemical oxygen demand	NAMAS	National Measurement Accreditation Service
BPEO	best practical environmental option		
BSE	bovine spongiform encepalopathy	NATLAS	National Accreditation of Testing Laboratories
CAR	contractors' all-risks (insurance)		
CCS	Confederation of Construction Specialists	NATM	New Austrian Tunnelling Method
		NEC	New Engineering Contract
CCSJC	Conditons of Contract Standing Joint Committee	NMT	Norwegian Method of Tunnelling
		NPL	National Physical Laboratory
CCT	compulsory competitive tendering	NRA	National Rivers Authority
CEDR	Centre for Dispute Resolution	OCR	overconsolidation ratio
CIRIA	Construction Industry Research and Information Association	OEL	occupational exposure limit
		OFS	overload factor (simple)
COSHH	Control of Substances Hazardous to Health (Regulations)	Ofwat	Office of Water Services
		OJ	Official Journal (of the EU)
CPT	cone penetration test	PAH	polynucleararomatic hydrocarbon
CSC	Consumer Service Committee	pfa	pulverised fuel ash
D&C	design and construct	PHH	polynuclear heterocyclic hydrocarbon
DMAPN	dimethylaminoproprionitrile	PIN	Periodic Indicative Notice
EN	Euro Norm	PSC	Professional Services Contract
EPB	earth pressure balance (machine)	PUSWA	Public Utilities and Street Works Act 1950
EQS	Environmental Quality Standards		
eVDV	estimated vibration dose value	PVC	polyvinylchloride
FCEC	Federation of Civil Engineering Contractors	QA	quality assurance
		RMR	Ross Mass Rating
FR	friction ratio	RQD	Rock Quality Designation
GRP	glass fibre reinforced pipe	RSR	Rock Structure Rating
HDPE	high density polyethylene	SPT	standard penetration test
HMIP	Her Majesty's Inspectorate of Pollution	SWQO	statutory water quality objectives
		TBM	tunnel boring machine
IPC	integrated pollution control	TP	third party (insurance)
JCPDS	Joint Committee on Powder Diffraction Standards	TQM	total quality management
		UCS	unconfined compressive strength
JCT	Joint Contracts Tribunal	UES	Uniform Emission Standards
LEL	lower explosive limit	WSC	Water Services Company
LQD	lithological quality designation	WWM	welded wire mesh

PART ONE

TUNNELLING CONTRACTS
AND
SITE INVESTIGATION

1

INTRODUCTION

1.1   GENERAL

Tunnelling is a high-risk business. Its operatives are exposed to severe physical risks and the contracting parties undertake major financial risks. These risks are increasing with the increased use of mechanised tunnelling. The main constraint on the success of a tunnelling operation is usually the nature of the ground through which the tunnel is driven. The cost of a thorough site investigation is usually only a small fraction of the cost of the construction work (Attewell and Norgrove, 1984b). However, if the site investigation is skimped, the client may be involved in additional expenses resulting from an over-conservative design, from the work running into difficulties during construction (NEDC, 1964), or for claims for extra payment on the grounds that the information provided for tendering (contract bidding) purposes was inadequate or inaccurate and thereby misleading.

Site investigation, which needs generally to abide by European Community legislation, comprises a preliminary review of information relating to the construction site, observation of surface features at the site and the drawing of conclusions therefrom, and the specification and implementation of an in-ground investigation from which parameters necessary for the geotechnical design of the works can be drawn. Because of the importance of ground investigation (Attewell and Norgrove, 1984a, b; Norgrove and Attewell, 1984), it is essential that the work be carried out by experts under full professional control (Uff and Clayton, 1986). Descriptions of site investigation procedures sometimes accompany descriptions in the literature of civil engineering works. However, information on investigations that were deficient, and which may have led to contractual claims, is often not given, any comments on shortcomings usually being revealed in discussions on professional and research papers. Yet it is such comments that are perhaps of the greatest value.

Site investigation funds are limited. It needs to be stated that the ground investigation proper provides a most unfocused view of the ground interior, with only a small fraction of 1% normally being exposed. What is seen before construction cannot therefore be regarded as 'representative' of what is actually there to be tunnelled. This means that there must always be a question mark, *ab initio*, against any ground investigation results.

The aim of this current work is to outline some of the experience gained from involvement in tunnelling contracts, mainly in the north-east of England. The review is directed primarily at practising engineers who are involved in the procurement and management of site investigation and tunnelling contracts. It is not intended to be comprehensive, nor to comprise a manual on site investigation practice or tunnelling methods. For these, the reader is referred to standard reference works (Dumbleton and West, 1976; British Standards Institution, 1981; Weltman and Head; 1983). Reference should also be made to the four publications under the heading of 'Site Investigation in Construction' produced by the Site Investigation Steering Group (1993) under the auspices of the Institution of Civil Engineers. It is also noted that the new European Codes will eventually be operational in the UK. (Part 1 - General Rules - of Eurocode 7 (Commission of the European Communities, 1989) related to ground engineering was scheduled for publication towards the end of 1994, Parts 2 and 3 covering laboratory testing and field testing and sampling, respectively, are not expected to appear until 1998, and Part 4, intended to cover specific foundation types and geotechnical processes, had not, at the time of writing, been programmed. Reference may be made to the article by Parker, 1994.) This current review stems mainly from the author's experience of 3 metres or less diameter tunnels for sewerage and main drainage schemes in urban areas. Most of these tunnels have been hand-driven in soil with shield protection, and both with and without compressed air, and so the focus in the book is substantially on this system of tunnelling and support. However, the author has experience of slurry shields used in weak tills below the water table, man-entry pipe jacked tunnels, mini-tunnels, pipe jacked microtunnels, cutter boom shield tunnels, rock tunnels excavated with TBMs, rock tunnels with boom cutters in short shields and rock tunnels excavated with tracked roadheaders. It is useful at this early stage to comment briefly on these systems in order to 'set the scene' for what follows later in the book.

1.2   TUNNELLING SHIELDS

Most tunnel driving systems incorporate some form of tunnelling shield. A tunnelling shield serves two general purposes:

* in unstable ground to provide temporary protection for men, machine and equipment until such time as permanent ground support can be erected;
* to provide housing and support for tunnelling boring plant and permanent support erection facilities.

The most commonly used systems are as follows.

1.2.1 No shield

In this case the rock or firm soil to be tunnelled is sufficiently competent to stand without any temporary support until permanent support is constructed. Soil may be excavated by hand tools, such as clay spades, or by tracked or wheeled plant containing diggers or back-acters if the tunnel is big enough to admit it. Rock may be excavated by boom cutter mounted on a tracked or wheeled vehicle, or by explosives in drilled holes. There are no real restrictions as to tunnel cross-sectional shape. Primary support may be by legs and girders for a square or rectangular tunnel, by colliery arches, (grouted) anchorages with mesh and spray concrete for a horseshoe-shape tunnel, and by (grouted) anchorages with mesh and spray concrete or pre-cast concrete segments (bolted or wedge-block expanded, as appropriate) for a circular tunnel. Permanent support will often take the form of

cast-in-situ or slip-formed concrete. The New Austrian Tunnelling Method (NATM) of anchorages, mesh and shotcrete for primary support operates without a shield, but with substantial (radial) ground movement and (usually) porewater pressure monitoring (*see* Section 6.8). Also included under this heading is the Italian infilaggi method which involves the installation of overlapping umbrellas of micropiles to support weak rock (the equivalent of forepoling), particularly faulted ground when the faults are filled with clay gouge and water. Further support is offered by shotcrete, and, where necessary in very weak rock, steel rings are inserted at intervals of about 1m. A PVC layer, protected by a geotextile, may be installed before a permanent concrete lining is cast.

1.2.2 Simple hand shield

This is an open-face shield in which manual excavation takes place, the soil being loaded into tubs or onto a belt or chain conveyor. The tunnel is of circular cross-section and a cast iron segmental lining or, more usually, a bolted pre-cast concrete segmental lining (Lyons, 1978) is usually erected within the protection of a tailskin to the shield. There will also be pea gravel/Lytag/cement-pfa injection behind the extrados of the lining in order to limit ground loss.

1.2.3 Cutter boom shield

This will usually take the form of an open-face shield in which a cutter boom is mounted for excavation purposes in rock and 'hard' soil. The tunnel will be of circular cross-section and the same comments related to ground support apply as for the hand shield.

1.2.4 Back-acter shield

This is again an open-face shield in which a back-acter is mounted for the excavation of soil and of 'soft' rock of sufficiently low strength not to require picks for its removal, the action of the bucket being sufficient.

1.2.5 Full-face tunnel boring machine (TBM)

This is a shield containing a rotating cutting head

or heads, the cutters (picks, discs or buttons) being set on the head in cruciform configuration. The face may be separated from the rest of the shield by a bulkhead, this being particularly the case with the 'drum digger' for soil. The term 'TBM' is often reserved for tunnelling in rock in which this plant would be used only for significant tunnel lengths. Squeeze in deeper rock would suggest the need for some form of retractable facility across the diameter. All these shields allow excavated ground to pass through the bulkhead at a controlled rate. This type of machine cuts a circular cross-section, but some continuous miners can now cut rectangular sections. One of these is the Robbins Mobile Miner which carries a large wheel with peripheral disc cutters. The wheel rotates about a lateral axis and swings from side to side on a hydraulic boom. Development has also reached the stage of a triple-head machine for simultaneously excavating three tunnels, say two for a double track railway and for a station. A Japanese Hitachi Zosen TBM has been built to construct tunnels 7.8m in diameter and 17.3m wide for a new Osaka business park station on subway line 7, construction having been scheduled to begin early in 1995.

1.2.6 Slurry shield

This type of machine is used for tunnelling in weak ground in which high water pressures (up to 30m head) are present. The slurry takes the form of either the natural ground and water, slurried up together through the rotary action of the excavator arms, or the natural soil can be supplemented with a bentonite/water mix injected under pressure to the ground side of a bulkhead. The bulkhead isolates the ground to be excavated at the face from the rear of the machine and the section of tunnel already cut and lined out. Induced pressure in the slurry serves to balance the groundwater pressure and also offer face support. Wirebrush tail seals (ideally treated with non-hydrocarbon grease, but sometimes hydrocarbon grease may be necessary) prevent the slurry passing from the face down the sides of the shield. Slurried soil is pumped to the surface, and any bentonite that might be present is separated out as far as possible from the soil by means of cyclones. This bentonite is then re-cycled in the system with

fresh bentonite. Preliminary separation of cuttings from the slurry may be by sieve, 6mm mesh, followed by hydrocyclones having a theoretical cut of $100\mu m$ or multicyclones having a cut of $30\mu m$. With the slurry shield system, the *earth pressures* must be balanced automatically by mechanically coordinating the speed of excavation, the cutter face pressure and the jacking force. The *groundwater pressure* must be balanced by adjusting the slurry pressure. A circular cross-section is cut and a pre-cast concrete segmental lining is normally erected within the protection of the shield. Herrenknecht and Iseki, for example, manufacture machines which operate on the slurry principle.

Slurry pressure can be more finely controlled by using a volume of compressed air behind an intermediate partial bulkhead. Double bulkhead machines of this type are known as 'hydroshields' or 'Mixshields'.

Generally, a slurry machine is good for water-saturated sand and sandy-gravel ground, but in coarse gravels if a bentonite cake does not form on the face a soil collapse can occur. Although slurry machines can be used for clayey ground, the slurry treatment takes longer and is more expensive.

Cutter-head design comprises combinations of disc cutters and teeth to cope with sand, gravel and cobbles. So that the cutters can be replaced from behind the head it is necessary for an airlock to be incorporated into the shield bulkhead so that access is possible under compressed air.

1.2.7 Earth pressure balance machine (EPB)

This is a full-face tunnel boring machine for use in soft ground, again with a bulkhead through which the excavated material is transported from the face, this time by a balanced screw auger or screw conveyor. The face is supported by excavated material held in front of the forward bulkhead at a pressure equal to or slightly higher than that exerted by the surrounding earth and groundwater. Pressure is controlled by the rate of passage of excavated material through the balanced screw auger, in which the pitch of the auger thread is very tight near to the face but increases further back down the line, or by valves on the screw conveyor. A experienced operator is needed to achieve this control. Earth pressure is normally reduced to about 80% of the face pressure at the point of entry of the screw, with a progressive reduction of pressure along the screw. The screw conveyor can be a welded structure, but this is less strong than a screw machined from a solid cylinder of steel. As in the case of the slurry shield, wire brush tail seals prevent the ground on the high pressure side of the bulkhead seeping along the sides of the shield.

Smaller EPBs use a central drive unit for the cutting wheel, but larger units use a peripheral drive so allowing higher cutting wheel torques to be applied for harder and generally more demanding ground conditions.

There needs to be provision, usually by means of airlocks incorporated into the bulkheads, for retraction of the forward section of the screw conveyor and for man-entry under compressed air into the excavation chamber so that any repairs can be effected and cutters replaced when necessary (refer to British Standards Institution, 1990a). Tool replacement might be quite frequent when cutting in strongly frictional ground. This operation of replacing a full chamber of fluidised spoil with compressed air can be more difficult in the case of an EPB machine but relatively easy (perhaps a matter of minutes) with a Mixshield (Herrenknecht, 1994).

Machines operating on this EPB principle, of which the Lovat machine is an example, are usually considered particularly suitable for cohesive or silty ground that does not contain cobble or boulder obstructions and in which the water head at the face is no more than about 3m, and although they are capable of working in sandy conditions they do tend to be less economic than slurry machines. The optimum ground material conditions for the operation of an earth pressure balance machine comprise homogeneous soils of fine to medium particles having a low permeability, with a permeability of 10^{-5}m/s being an empirical limit for EPB operation. If the ground is too permeable, the spoil will not flow smoothly along the screw, having a tendency to arch at the entrance to the screw, allowing groundwater will find its way into the excavation chamber and screw conveyor, and inducing collapse at the face. Because this ideal requirement for soil permeability rarely exists, successful excavation may depend on the choice and use of soil conditioning agents to improve the

plastic fluidity of the ground material - plain water, bentonite-based muds, chemical polymers, and foaming agents. When foaming agents, which comprise air bubbles encapsulated in a detergent-like fluid in order to keep the ground particles apart and thereby reduce friction, were initially introduced it was primarily to lubricate the flow of the spoil through the cutter head compartment and screw, and so reduce torque, to improve the permeability of the material in very wet ground in order to prevent the passage of water out through the discharge gate, and to improve the overall consistency of the spoil so that it could be more easily handled from the discharge gate to the point of final disposal. If the water content of the spoil is too high, and the particle sizes are predominantly medium to coarse (higher permeability) it becomes difficult to form the necessary plug of impermeable spoil in the screw. Furthermore, the wet, sloppy spoil cannot be handled easily at ground surface and the operation of disposal becomes inefficient because the weight of the spoil in the haulage skips and the storage hopper increases to reduce their holding capacity. On the other hand, if the spoil is too dry, it will not flow properly into the screw which will instead cut itself a hole in the dry, compacted material. It will also tend to form a cake inside the skips and hopper and will need to be cleaned out frequently by hand. Automatic systems are required for foam injection. The volume used depends on the advance rate, cutting wheel torque, and the pressure of the earth. It is difficult, however, to calculate the actual volume because 90-95% of the medium is air. Foam uses a quite low percentage of chemical additives which means that disposal of the spoil will usually have fewer adverse environmental effects. Alternatively, composite conditioner, such as a thick bentonite and a thinner chemical polymer, could be mixed at the surface and then pumped through a rotary coupling into the arms of the cutting head so that the soil conditioning could begin immediately on excavation.

The EPB technique of tunnelling was chosen for the two 8km long, 8.5m external diameter East Channel rail tunnels of the Storebælt link in Denmark where the site investigation indicated glacial till containing substantial granite boulders up to 3m diameter underlain by heavily fractured marl in which water flows through the fissures could reach 500 litres per minute. At the deepest point, about 80m below sea level, it was calculated that the water would exert a maximum pressure of 8 bars on the tunnelling machine. Pressures and potential water flows were reduced by a system of well pumping from the fissured marl, taking advantage of the partial aquiclude conditions afforded to the marl by the overlying till (Wallis, 1993).

Slurry and EPB systems were also used to tunnel through very wet fissured Upper Chalk on the French side of the Channel Tunnel. A 9.7m diameter Group Fives-Lille (FCB) EPB machine to a Kawasaki design is scheduled (at the time of writing) to operate in Portugal on the Lisbon Metro extension close to the River Tagus under only 5m of cover and will use grout and mortar soil injection as a back-up for the face support together with a front-end injection of bio-degradable polymer foam, the latter having been used successfully with the EPB machine at Lille Metro, France. A 9.5m outside diameter Lovat EPB machine has been used for the 1.8km St Clair river crossing - part of National Rail's main US-Canadian rail link between Toronto and Chicago - but has suffered clay contamination in the lubrication system. This was caused by damage to five of the sixteen concentric annular elastomeric seals which surround the inside and outside of the machine's peripheral head bearing.

In another example an EPB machine, ordered by a French consortium GIE Lyon Nord and manufactured by Mitsubishi Heavy Industries, has worked on the northern part of Lyon's ring road in France for client Société Concessionaire du Boulevard Périphérique Nord de Lyon. At 11m outside diameter, 13.6m long, and 2000 tonnes weight it is the largest shielded EPB machine in operation at the time of writing. The privately funded $500m contract was for driving two 3.2km one-way traffic tunnels under the Saône River north of Lyon. The tunnelled ground is of hard gneiss and soft alluvium and so the machine was designed to have variable excavation capabilities. A period of 22 months was allowed for excavation and lining out of the tunnels.

An 9.5m Lovat EPB machine was used to cut through soft glacial clay with only 4.5m of cover to the tunnel crown below the St. Clair River bed during construction of the rail tunnel between Canada and the USA, the largest 8.4m internal diameter and also 1824m long single track

subaqueous railway tunnel completed on 8 December 1994. Reinforcement in the concrete segmental lining (6 segments, 1.5m long and 400mm thick, plus a trapezoidal key, all fitted with continuous neoprene rubber gaskets and bolted connectors) was epoxy coated to safeguard against chloride attack.

There does seem to be a current trend away from slurry machines and towards EPB machines, not least because of the absence of the separation plant and pumping equipment that are necessary adjuncts to the slurry shield principle. Advances in techniques for treating the excavated ground with mud injection or other suitable conditioning agents mean that an EPB machine can now tunnel through a wider range of ground material than can be tunnelled by a slurry shield, but the additional surface works that are needed for this facility offset to some degree the advantage over slurry. Another major advantage is one of flexibility in that modern EPB machines can also be operated in 'open mode', that is, operated with the bulkhead doors open and without pressure at the face. This means that higher rates of progress can often be achieved in open mode when the ground conditions permit, but when the face becomes unstable the machine can then be quickly converted to closed mode operation. EPB machines such as the Lovat ME 202SE used on the London Underground Jubilee Line extension can change rapidly between the closed mode and the non-pressurised open (free-air) mode. Hydraulic flood doors may also be incorporated for sealing off the tunnel face in a matter of only seconds if unstable water-bearing ground is encountered.

1.2.8 Air pressurised shield

This TBM system, of which a Denys machine is an example, incorporates a bulkhead in which the groundwater pressure is balanced by air in the excavation chamber. As with the other machines noted above acting on similar principles, a circular cross-sectional tunnel is cut.

1.2.9 Microtunnelling machines

This is a remotely controlled (from ground surface) machine, less than 0.9m in diameter. The tunnel is non-man-entry, and the temporary support system at the tunnel face is either slurry or earth pressure balance. In a Herrenknecht machine, for example, pumped slurry pressures may be up to 8 bars. Forward progress of the shield and permanent ground support are achieved by a pipejack system from rams located in a shaft bottom, although there may be provision for interjack stations. The pipejack system is not limited to microtunnelling machines, but can also be used for larger diameter man-entry tunnels. In a relatively new development the Iseki 'Perimole' small diameter (200mm or 250mm) microtunnelling machine, incorporating a pipejacking system, advances through the ground by displacing and compacting the soil around the head rather than by excavating it. In addition to concrete, clay, grp, and steel pipes, the system is also able to install plastic pipelines. Reference may also be made to the paper by Washbourne (1993).

1.2.10 Thrust boring machines

This is a method of constructing small diameter, substantially horizontal pipelines by soil displacement.

1.2.11 Auger boring machine

This is a non-steerable machine used for constructing horizontal pipelines and making use of continuous flight augers for excavation and spoil removal. (Some machines have an antiquated steering system of rods outside the head.)

1.3   TUNNEL SIZE

1.3.1 Definitions

'Small diameter' refers to sizes up to 8 ft (2.4m) internal diameter. The transition from 'small' to 'large' occurs at 8 ft to 10 ft (3.05m). (8 ft internal driven at +300mm overcut provides a 9ft external diameter tunnel.) 'Microtunnelling' is less than 900mm diameter, and incorporates a pipejacking system. Discussions in British industry over the minimum man-entry size preceded publication of BS 6164: 1990 (British Standards Institution, 1990a). The British Standard states that 'for tunnels under construction the minimum size for man entry should be not less than 900mm in order to

facilitate rescue; tunnels smaller should be non man entry'. Engineers on the continent of Europe are tending to re-define man-entry as diameters greater than 1m. There also appears to be some pressure from the machine manufacturers to make the minimum size for man-entry somewhat larger - at least 1.2m.

1.3.2 Stones

Some bentonite and slurry tunnelling machines have a cone facility for crushing boulders having a size up to about 30% of the machine diameter and unconfined compressive strength up about 200MPa. Since much of the work that these machines are asked to perform is in glacial till deposits, this is a necessary facility. The Iseki Crunchingmole shield, for example, which is used for tunnelling up to 3500mm outside diameter, will crush large stones of up to 20% of the shield diameter down to a size that can be pumped away in the slurry. The Unclemole, for tunnelling up to 2100mm outside diameter, can crush stones of up to 30% of the outside diameter of the shield.

1.4   MICROTUNNELLING

This comprises the use of a steerable remote controlled tunnel boring machine to allow installation of pipelines up to DB 900 by pipejacking. Microtunnelling was first developed in Japan during the late 1970s, with a second generation of German machines following in the early 1980s. Because there is increasing interest in and use of the microtunnelling system, particularly for 'no-dig' installation of service pipes, it is relevant to make some further observations on the system.
- All the services (slurry, electrics, and so on) pass through the jacking station into pipes which are an integral part of the temporary pipes. The electric cable which powers the machine passes along a duct in the external circumference of the pipe so that one continuous cable can be used, no joints being required.
- The system can handle gradients up to 1 in 40.
- The machines often operate at very shallow depths and are therefore more likely to tunnel through contaminated land. Non-man-entry and remote operation mean that there is no

direct human exposure to toxic, corrosive or asphyxiating chemicals in soil and groundwater.

- The problems of microtunnelling machine breakdown between shafts should not be understated. A man-entry tunnel may have to be driven from the target shaft or a new shaft may have to be sunk to retrieve the machine, but this may not be too expensive a job if the tunnel is shallow. However, a necessary consideration at the planning stage is the presence of open land along the tunnel route for additional shaft sinking, since in a built-up area such a method of retrieval might not be possible. (Very shallow microtunnelling may be chosen in preference to open-cutting simply on the basis of environmental and ground surface disruption considerations - for example, in anticipation of a strong protest lobby against constructing a new gas or water main by open-cut across a golf course, a more expensive microtunnelling option may be adopted.)

- The slurry action at the tunnel face can cause fine particles to get into the disc bearings on a rock machine. This tends to be the problem that limits a microtunnelling design distance in rock to between about 130m and 150m. Slurry pumping efficiencies may also be a factor that limits the tunnelling distance. Such short distances, and a consequential need for a higher density of shafts along the route, raise the cost of the operation.

- If the machine sticks at the face, but has not broken down, an attempt can be made to 'blow it out' by retracting the rams at the shaft and at any interjack stations, and then pressurising the slurry at the machine head.

- Unusually, suppose that a microtunnelling machine cannot, for whatever reason, be driven forward to a shaft for dismantling, but must still be retrieved. In rare cases special provision may be made for retraction of the machine back to the thrust shaft. A double skinned machine can be used - at the end of the drive the outer skin of the machine can be disconnected from the body of the cutting head by retracting short-stroke hydraulic cylinders within the machine. The cutting head can then be withdrawn from the outer skin and pulled back to the start shaft within the pipe already installed in the ground.

- If a slurry tunnelling machine, such as a Herrenknecht, is in clay, then it will be usual to 'stand off' the head of the machine. If it is in silty ground the machine will tend to be forced into the face to give fast progress, but this could cause ground heave if the tunnel face is close to ground surface. These two effects need to be balanced by a good driver at the remote control console near the top of the shaft from which the lining pipes are jacked.

- When tunnelling in a mudrock, granular particles, which may be natural or may be the result of crushing of boulders at the tunnel face, assist the mudrock slurrying action and prevent the creation of a 'clay ball' which could clog up a microtunnelling machine ports and so not allow the slurry water to wash through. (In the case, for example, of a Herrenknecht machine the ports are only about 1.5 inches (38mm) diameter.) Without such preventative action a Herrenknecht microtunnelling machine may be able to cut 13 to 14m/day in sandstone but only 3 to 4m/day in a mudstone. Another answer to the mudstone/mudrock problem is to incorporate a high pressure jetting system at the machine head. Another design feature, as in the Iseki Unclemole, which helps to eliminate or reduce 'balling-up' involves the use of a cone crusher at the tunnel face instead of a pressure door. The eccentric movement will then tend to pelletise or extrude sticky, mouldable clay.

- There seem to be no actual problems when microtunnelling in rock having an unconfined compressive strength up to 100MPa (with the exception of certain types of sandstone in which the problems seem to be unrelated to strength), and even a very strong rock for short periods of time, but there can be a slurry loss through fissures. This slurry may not only need to be substantially replaced as a direct result of loss but there will also be a need for continuous replacement as it thickens due to collection of fines not separated out by cyclones. (A suitable disposal point for the waste slurry will also need to be found.)

- Mixed faces can create a problem because the machine head has a tendency to travel towards the weaker material and so lose alignment. The cutters can also 'jump' against boulders and hard bands of rock, and although the cutting head is designed to penetrate through boulders and flints without difficulty, the small diameter of a micro-tunnel face does mean that cobbles take on the same significance as boulders in the context of excavatability.

- When microtunnelling in rock there will need to be sufficient forward thrust applied to the face without stalling the rotation. This is one of the many problems that need to be continuously addressed by the remote control operator. There is an advantage in using disks on the cutting head at all times, even when tunnelling in soil, if the ground investigation suggests that there is any possibility of encountering rock in stratified and cobble/boulder form. The Herrenknecht tunnelling machine accommodates this facility.

- Several special uses of the microtunnelling technique have been reported in the literature. One, for example, involves drilling around the outside diameter of a large-diameter tunnel in, say, water-bearing sands and gravels for the purposes of installing freezing pipes which will form an ice wall around the tunnel, so allowing the main tunnel to be driven in the dry. (Care is needed during the subsequent freezing process to ensure that a soft core is left in the main tunnel in order to facilitate excavation.)

- Remote control, non-man-entry operation of microtunnelling/pipejack systems in Coal Measures rocks (for example, Kennedy Construction's Herrenknecht machine on the River Tyne South Bank Western Interceptor Sewer) or in contaminated ground (for example, Donelon's Lovat M60 machine on the Burslem to Strongford Link Sewer project for Severn Trent Water) offers health and safety advantages. When it is necessary to tunnel through ground which contains potentially flammable gases such as methane, gas monitoring equipment in the form of a gas sensing head will be required and the machine will need to be driven by flameproof equipment.

- Top works (pumps, bentonite separation, pipework) are expensive to install - £15k to £20k (1992). These capital costs have to be recovered as a cost per metre tunnelled, often from a limited tunnelling distance.

- It is more difficult for a contractor to sustain a Clause 12 claim (ICE Conditions of Contract, 6th Edition, 1991) on the basis of unforeseen ground conditions because with the remote tunnelling operation the ground at the location of the problem cannot be seen. However, good site investigation is particularly important in the case of microtunnelling for selection of the correct type of plant for the ground and groundwater conditions, and for determining the need or otherwise for ground improvement. This latter may take the form of groundwater lowering by well point dewatering, deep sump or deep well pumping, or by freezing, or by grout (particulate, such as cement or clay, or chemical, such as silicate or resin) injection.

1.5   PIPEJACKING

Pipejacking is about 100 years old, being first used in the United States by the Northern Pacific Railroad between the years 1896 and 1900 to jack pre-cast pipelines into place. It was later applied to sewer construction, again using concrete pipes. It was first used in the UK in the late 1950s, and by the early 1970s was being increasingly used for sewer construction. The relevant British Standard for concrete pipes is BS 5911: 1989 (British Standards Institution, 1989) and there is also reference to pipejacking in BS 6164: 1990 (British Standards Institution, 1990a). For man-entry tunnels (900-3000mm diameter) the double-concentric reinforced concrete pipes, in lengths of between 1.2 and 2.5m, may be either rebated at the joints or butt-jointed incorporating steel or glass reinforced plastic collar bands. Both of these joint types incorporate an elastomeric sealing ring consisting of a circular cord or specially profiled sections if water pressures are likely to be high. Specially rebated pipes may be produced for insertion into the jacking shield and similarly rebated pipes are produced for the trailing pipes at any intermediate jacking stations. Concrete jacking pipes, 1.0 to 2.5m long, for microtunnelling (300-875mm diameter) are similar in design to the larger diameter man-entry pipes, but are not necessarily, although usually, reinforced, and are generally of steel banded form in order to sustain the jacking stresses for powering the machine forward. There are also vitrified clay pipes, of greater wall thickness and in lengths of from 1.0 to 1.2m, for microtunnelling. The joints of these pipes are normally of double spigot type incorporating a loose joint band of steel or, in the case of the smaller diameters, of rubber-coated polypropylene. The internal diameter of this type of pipe is usually from 150 - 600mm, but larger diameters up to 1m are available.

There are several quite significant *benefits* offered by the pipejacking technique, and there seems to be little doubt that these will lead to its increased popularity as the demand for microtunnelling grows.

- Fewer joints (circumferential only) compared with those in a segmentally lined tunnel and therefore the reasonable assurance of a greater degree of watertightness.
- A one-pass method - no requirement for a secondary lining, the pipe being the permanent lining, and so money is saved.
- Smooth internal factory finish gives excellent fluid flow characteristics.
- Minimal ground loss and so little attributable surface settlement. (If the ground at the tunnel face is water-bearing and under pressure, then pressurised slurry, as noted above, may have to be used in order to prevent ground loss.)
- For alignment, although established laser guidance methods are less effective in small, non-man-entry tunnels, new systems have been developed using gyroscopes and computer software (the Iseki Unclemole, for example, uses an 'AS' automatic guidance system controlled through fuzzy logic software). A laser is mounted three to four pipes behind the face, the drive control software 'remembering' where the machine was before and so adjusting for line and level. It is possible, using packings, to pipejack around bends. Atkin (1991) describes the use of a marine giro compass mounted in the shield to monitor horizontal alignment, with readings relayed back to the control cabin monitor by means of a small television camera.

There may also be *problems* associated with pipejacking:

- Pipes may need to be thicker for tunnelling in rock in order to carry the greater thrust forces. Typical maximum jacking rig force might be about 350 tonnes at 500 bars pressure using double-action jacks; a ratchet mechanism might be used. Another system could use six 200 tonne double-acting hydraulic rams with a stroke a little greater than an individual pipe length. Steel-banded pipes may be needed (260 tonnes maximum jacking force, but for microtunnelling Spun Concrete Ltd produces a pipe, 675mm internal diameter, 875mm outside diameter, capable of taking a 400 tonne end load). Standard steel collars are normally used on Spun Concrete sewer pipes but stainless steel collars may need to be used (at greater cost).
- Although there is no excessive friction for jacking in rock there could be problems in boulder clays, sands and silts. Interjacks would then be necessary, and it might mean an increase in pipe size for man-entry in order to recover the cylinders. This could also require an increase in access shaft diameter in order to get the larger diameter pipes down. Guide-line frictional forces tend to lie between 0.5 and 2.5 tonnes/m^2 of external circumferential area, dependent of course upon the ground conditions and the type of excavation.
- There is a problem relating to the sealing of the overcut void that needs to be addressed. In water-bearing ground the seal is mainly necessary to limit ground loss around the break-out area. In rock the seal has to effectively block one end of the overcut void to (a) prevent the face slurry from washing out bentonite lubricant that is usually injected around the pipe, and (b) to prevent fines from the excavation face filling the void and so seizing up the pipeline. In rock, the frictional effect only really occurs over the bottom quarter of the lining, but it is especially important to keep the overcut void stuffed full of bentonite. Bentonite injected through the ports in the tail shield at a maximum pressure of 100 bars will be needed for this purpose

even when the machine is being operated in earth pressure balance mode. The bentonite will need to be prepared in a high-shear colloidal mixer for rapid hydration.

- In very weak ground, where there can be a problem with the weight of a TBM, use of a pipejack with a closed face having an extrusion facility and with bentonite pumped into the extrados void is a possibility. However, maintaining line and level under these ground conditions can be very difficult. Compressed air may be needed, and if the tunnel crown is close to ground surface or under water there could be a danger of a blowout (*see* Section 5.6). Furthermore, it might not be possible to apply the bentonite at the normal three-times overburden pressure.

- Problems may be experienced when jacking from the assembly pit into ground having a high water head. This is a situation that arose on Northumbrian Water's River Tyne South Bank Interceptor Sewer construction at Derwenthaugh. Sheet piled coffer dams were used for the thrust pits and a line of secant piles was constructed against the forward face of the thrust pits before breaking out. An entrance ring for the slurry machine was welded to the piling and a hole was cut by thermic lance before the shield was launched. As the slurry shield advanced there was a need to chain the rearmost pipe to the cofferdam (shaft) wall to prevent hydrostatic pressure from reversing the machine when the rams were retracted in order to fit the next pipe into the drive. This constraint could be relaxed only when there were sufficient pipes in the ground to achieve the necessary pipe-ground friction, and on one occasion 21 pipes were needed before the machine was able to 'hold its ground'.

- Swelling clay pressures acting on the pipes will require increased jacking pressures, or, if foreseen before the contract is priced, will lead to a higher density of shafts along the route and smaller jacking distances.

- In chemically aggressive ground a resistive outer lining must be specified for the jacked pipes. Retention of this outer lining against the friction of the ground during jacking can obviously be a problem. For Northumbrian

Water's South Bank Interceptor Sewer's drive at Derwenthaugh under the River Derwent, where the ground pH was as low as 2.3, contractor Lilley used an Iseki Crunching Mole and, with Spun Concrete, developed a 6mm thick composite glass and polyester reinforced plastic coating for Spun Concrete's jacking pipes to give the necessary abrasion-resistant chemical protection to the pipes. For pipes subjected to internal acidic sewage conditions, as in Dubai (Darling, 1994), it may be necessary to incur added complexity by using, for example, glass fibre reinforced pipes (GRP) internally coated with a layer of varnish, the GRP then having a concrete surround.

- There are problems associated with larger diameter pipejack tunnels (man-entry, above about 1.2m diameter):
  - Transportation of large pipes (say over 8ft is more difficult and expensive than the transportation of equivalent pre-cast tunnel ring segments. (It may be possible to use half-pipes to help transportation, but the pipe costs increase, as do the number of joints and the potential for loss of seal.)
  - A large access shaft to accommodate the pipejack equipment, especially if the shaft is deep, will be expensive. Alternatively, it might be possible to excavate an adit at the shaft bottom to house the larger jacking equipment, but this again would be expensive.
  - Interjacks are more easy to operate, so allowing for longer tunnels between shafts, but there are correspondingly greater requirements for ventilation under man-entry conditions.

Jacking pipes incorporating flexible joints to comply with BS 5911: 1989 are offered in several diameter sizes specified in millimetres, ranging from 900mm to 3000mm, although manufacturers may be prepared to produces sizes outside this range. The specific sizes of jacking pipes on offer are shown in Table 1.1. For further, but earlier, information, and for an insight into the research on this subject, reference should be made to Craig (1983) and to Norris (1992), respectively.

Table 1.1 Diameters of jacking pipes in millimetres

900	1050	1200	1350	1500	1650	1800	1950	2100	2250	2400	2550	2700	2850	3000

1.6   GENERAL COMMENT

The expression 'site investigation' in geotechnical engineering and engineering geology has traditionally been used to encompass all the elements of the investigation - the so-called 'desk study' and 'ground walk', the in-ground investigation and sample retrieval, laboratory sample testing, and then report writing. Increasingly, geotechnical engineers and engineering geologists are using the expression 'ground investigation' to denote all these facets of the exploration and not simply those relating to the actual in-ground works - the boring and/or trial pitting, any *in-situ* testing, sampling and laboratory testing. However, after some deliberation the decision was taken to retain the 'traditional' expression for the purposes of this book.

This contents of the book provide some guidance on site investigation methods and practices, including examples relating to deficiencies in investigations and problems with tunnelling contracts. Some information is also provided on the presentation of site investigation information in tunnelling contracts. There are examples of problems that might arise during the course of a tunnelling contract, with further examples illustrating the value of good quality site investigations. The opportunity has also been taken to describe some more general experiences in tunnelling contracts, not all of those experiences being closely related to site investigation problems. Although an integral part of tunnelling, there is no discussion on shaft sinking although it is recognised that the chosen methods of shaft excavation and of lining construction, be it by underpinning, by caisson, by secant piling or by jet grouting, are dependent to varying degrees on the geological and geotechnical conditions of the ground.

It is not possible in this book to consider in any detail the considerable legislation that impinges on work in the tunnelling industry, although some of the legislation is mentioned from time to time in the text. Most is directed towards health and safety issues, which assume great prominence in tunnelling. Some of the more prominent Regulations are given in Section 6.1.

Letting of contracts in the UK, where the estimated value of those contracts exceeds a threshold level, is subject to European Community Utilities Directive 90/531/EEC, which has been incorporated into UK statute through the Utilities Supply and Works Contract Regulations 1992. The requirements stemming from the Directive and Regulations are considered briefly in Chapter 5 of the book.

This review is substantially based on the 6th Edition of the ICE Conditions of Contract (1991) because although the 5th Edition is still used at the present time in the UK it is a natural expectation that the 6th Edition will succeed it. There are, however, four significant and in some cases related developments which are attracting support in the field of construction contracts. First, there are the ICE Design and Construct Conditions of Contract (1992) which more fully acknowledge, than is the case in the ICE 6th Edition, the increasing role of the contractor in the design of civil engineering works. Second, notwithstanding the ingrained adversarial nature of British political and legal 'life' there is recognition that contractual relations in construction are best conceived and conducted in a non-adversarial spirit. Third, given the current generally unsatisfactory level of pre-contract preparation (for whatever reasons, the lack of time usually being quoted as one reason), the 6th (and 5th) Edition of the ICE Conditions of Contract can then encourage contractual disharmony through a claims scenario and might therefore ultimately be considered as a candidate(s) for abandonment as representing statements of potentially unachievable goals. Fourth, and stemming from these points, there are new forms of civil engineering contract which include a recognition that designers are unwilling or incapable of designing to accommodate contract risk and that contractors for their part seem unwilling to accept any risk, seeing their role rather as the provider of the service to deliver the product. These new (to the civil engineering profession) forms, which are discussed in Chapter 2 of the book, and notably the IChemE form and the ICE's New Engineering Contract form, go a long way towards promoting what might be termed a *partnership in contract responsibility*, acknowledging the inherent risks in tunnelling work and invoking a professional and equitable attitude towards their definition and management. Although the short-term costs might in some instances be higher, the eventual benefits to the profession could be enormous.

2

CONTRACTS

2.1 INTRODUCTION

A contract is an agreement or bargain reached by the parties and which the courts will enforce. The parties to the contract are bound by it because they have agreed so to be bound. For testing whether or not an agreement exists, the courts will examine what the parties actually said and did and not what was in their minds. Bargains rather than promises are at the core of the law of contracts. Each side must give something of value in exchange for what they receive from the other. An agreement in which one party received something without giving anything in return is not enforceable, but there are types of gifts that can be made enforceable if the donor puts a promise in writing and seals it, so creating a deed (*see* Section 2.2.2 below). Reference may be made to the book by Eggleston (1994b) for a discourse on civil engineering contracts.

The form of contract most generally referred to in this book is the *ICE Conditions of Contract and Forms of Tender, Agreement and Bond for Use in Connection with Works of Civil Engineering Construction 6th Edition December 1991*, which was launched on 28 January 1991 to supersede the earlier 5th Edition (1973). As was the case with the 5th Edition, it aims to share the contractual risk more equitably between the contracting parties than did the earlier 4th Edition and, in fact, there is no significant change in the allocation of risk between the two later Editions. Compared with the 5th Edition, the text in the 6th Edition has been shortened and/or broken up to make it more readable, the language has been modernised (and this includes changes to some of the traditional wording), it reflects current management practices and some case law that has post-dated the issue of the 5th Edition in 1973, and it generally aims to simplify and clarify contractual procedures. Armstrong (1991) lists the clauses that

constitute the main changes between the 5th and 6th Editions. Of these clauses, those which could have a significant bearing on site investigation are:

- Contractor's design - Clauses 7(6), 8(2), 20 and 58(3);
- Employer's information obligations, particularly at the time of tender, on ground conditions - Clauses 11, 12;
- Risks - Clauses 7, 8, 11, 12;
- Conciliation and arbitration - Clause 66.

The underlying importance of Clause 12 in the 6th Edition is the fact that the contractor is not asked to price for a risk that he cannot foresee. The 6th Edition is particularly directed towards the admeasurement style of contract based on Engineer design (for the client/employer) of the Works (save only for a new provision for limited contractor design - *see* below) and a priced bill of quantities. Design by or for the client prior to tender is sometimes known as the 'conforming design', and it is upon this design that the bidding contractors price for the work, although alternative designs may be submitted by one or more of the tendering contractors. There is also a full 'design and construct' set of conditions of contract (noted below) based on the ICE 6th Edition. This 'design and construct' set forms part of a package of contracts which include the Minor Works form (which was issued in May 1988) and possibly also a future form covering the maintenance/refurbishment of existing works. Notwithstanding these developments, there seems to be increasing recognition by the industry that a new style of cost reimbursable contract for risk sharing should lower the incidence of dispute and litigation. It is possible that the New Engineering Contract (NEC), launched in draft by the Institution of Civil Engineers in January 1991 for discussion in the industry, and noted below, may achieve this aim by addressing alternative forms of contract such as those embodying a target cost formula, also as noted below.

2.2   ICE CONDITIONS OF CONTRACT 6TH EDITION 1991

Reference may be made to Eggleston (1993). The following clauses and sub-clauses are particularly relevant to the theme of this book:

<u>Clause 11(1)</u>   The Employer shall be deemed to have made available to the Contractor before submission of the Tender all information on

(a) the nature of the ground and sub-soil including hydrological conditions and

(b) pipes and cables in on or over the ground

obtained by or on behalf of the Employer from investigations undertaken relevant to the Works. (August 1993 corrigendum.)

<u>Clause 11(2)</u>   The Contractor shall be deemed to have inspected and examined the Site and its surroundings and information available in connection therewith and to have satisfied himself so far as is practicable and reasonable before submitting his Tender as to

(a) the form and nature thereof including the ground and sub-soil
(b) the extent and nature of work and materials necessary for constructing and completing the Works and
(c) the means of communication with and access to the Site and the accommodation he may require

and in general to have obtained for himself all necessary information as to risks contingencies and all other circumstances which may influence or affect his Tender.

<u>Clause 11(3)</u>   The Contractor shall be deemed to have

(a) based his Tender on the information made available by the Employer and on

his own inspection and examination all as aforementioned and

(b) satisfied himself before submitting his Tender as to the correctness and sufficiency of the rates and prices stated by him in the bill of quantities which shall (unless otherwise provided in the Contract) cover all his obligations under the Contract.

<u>Clause 12(1)</u>   If during the execution of the Works the Contractor shall encounter physical conditions (other than weather conditions or conditions due to weather conditions) or artificial obstructions which conditions or obstructions could not in his opinion reasonably have been foreseen by an experienced contractor the Contractor shall as early as practicable give written notice to the Engineer.

<u>Clause 12(2)</u>   If in addition the Contractor intends to make any claim for additional payment or extension of time arising from such condition or obstruction he shall at the same time or as soon thereafter as may be reasonable inform the Engineer in writing pursuant to Clause 52(4) and/or Clause 44(1) as may be appropriate specifying the condition or obstruction to which the claim relates.

<u>Clause 12(3)</u>   When giving notification in accordance with sub-clauses (1) and (2) of this Clause or as soon as possible thereafter the Contractor shall give details of any anticipated effects of the condition or obstruction the measures he has taken is taking or is proposing to take their estimated cost and the extent of the anticipated delay in or interference with the execution of the Works.

2.2.1 Comment

Clause 11(1)(b), an August 1993 addition to the ICE Conditions of Contract, 6th Edition, places firm responsibility on the Employer, usually via the Engineer, for doing the searches to provide the contractor with full information on the presence and location of buried pipes and cables. Insurance claims often arise from disruption caused by damage to pipes and cables during the progress of drilling and trial pitting, particularly on private land where records of buried services may be insufficiently clear or non-existent. At the time of letting and implementing the ground investigation there should be the same assumption of responsibility by the employer in respect of the ground investigation works, with no disclaimers entering into the conditions of contract.

Sub-clauses 1, 2 and 3 of Clause 12 above were all contained in Clause 12(1) of the earlier 5th Edition. Under Clause 12(1) *written* notice to the Engineer is now specified whereas *notice* was required under the 5th Edition. Under Clause 12(2) intention of making a claim is now stated as being pursuant to Clause 52(4) *and/or Clause 44(1)* (Extension of time for completion), whereas under the 5th Edition only Clause 52(4) was specified. Under Clause 12(3) the contractor is required to give details of the measures *he has taken* is taking or is proposing to take, whereas in the 5th Edition there was no express recognition that the contractor had already taken measures although it could be argued that in this case the present effectively incorporates the past.

In the context of Clause 12(1) above, the implications of the word 'foreseen' have been the subject of wide discussion in the literature. It seems reasonable to take 'foreseen' as meaning what can properly be provided for with the information available, not what can be imagined as occurring. The 6th Edition does not change the substance of the 5th Edition Clause 12 wording in respect of conditions which could not reasonably have been foreseen. Most of the serious and protracted disputes have in the past arisen under this clause and are likely to do so in the future. The clause is not immutable, however, and it is possible for the contracting parties by agreement to change the balance of risk by allowing the contractor only a stated (perhaps tendered) percentage of any claim. Another possibility is to require the Engineer's initial assessment, if favourable to the contractor, to be binding unless the contractor pursues the whole claim to arbitration in which case it could be clawed back.

The definition of *physical conditions* has come to the fore in an Appeal Court judgment in the case of *Humber Oil Terminals Trustee Ltd* v *Harbour and General Works (Stevin) Ltd*. This judgment involved the relationship between Clause 12 and Clause 8 of the Conditions of Contract - a relationship that is unlikely to have been examined previously. The problem involved a jack-up barge equipped with a 300 tonne crane. Although the legs of the barge had previously been loaded to 933 tonnes, in operation the barge collapsed under applied loads that were less than this. The contractor brought a claim under Clause 12 on the basis that the collapse was due to physical conditions which, as an experienced contractor, he could not have reasonably foreseen. The matter was referred to arbitration, and the employer's response was that the collapse was not due to physical conditions that could not have been so foreseen, and that even if it was, the contractor's claim was invalid because of the provisions under Clause 8(2) of the contract under which the contractor must take full responsibility for the adequacy, stability and safety of all site operations.

On the matter of the first Clause 12 claim, the arbitrator identified the reasons for the collapse as being an initial small settlement followed by a large settlement of a leg, and concluded that the collapse was itself unforeseen and that it was indeed unforeseeable. But although the soil conditions at the base of the collapsed leg were foreseeable and the geometry and other details of the leg were well understood, the basis of the Clause 12 claim was not whether the *collapse* was unforeseeable but whether the *physical condition* was unforeseeable. The arbitrator, in considering whether the soil strengths and the stresses to which it was subjected at the time of failure were a 'physical condition', concluded that Clause 12 contained no limitation on the meaning of 'physical condition' and that there was no reason why a combination of strength and stress should not fall within its terms. It was also judged that Clause 8(2) could not be held as applying to a case where inadequate, unsafe or unstable operations or methods of construction were brought about by the contractor having encountered physical conditions within the meaning of Clause 12.

Reference of the arbitrator's award to the Official Referee and thence to the Court of Appeal saw the original award upheld. The term 'physical conditions' must now be taken to include conditions of stress and strain in the ground that in the course of construction, could not reasonably have been foreseen by a contractor experienced in the particular type of work and which affected his progress towards completion of the works. Operation of Clause 12 in this sense is unaffected by the provision of Clause 8(2) and the contractor's responsibilities under that clause.

Brierley and Cavan (1987), writing from an American viewpoint, have noted two types of 'differing site conditions for tunnels'.

'The first type is when the ground really *is* different - clay instead of sand, rock instead of soil, or wet ground instead of dry. This type of change is relatively easy to identify during construction, but can have a widely varying impact on project cost and schedule. A geotechnical investigation done sufficiently well can avoid this type of change, yet they can and still do occur even on well run projects.

The second type of differing site condition is when the contractor says that the ground is not *behaving* as anticipated. In essence, the contractor is saying that progress with the selected equipment and procedures is not sufficient to complete the project on time and at a profit. This type of change is much more difficult to identify and requires a very sophisticated approach by the owner, based on good data, good contracts, and good intentions.'

Material differences, as noted above for the 'first type' are rarely as clear cut as in the examples. If they were, there would then really be little reason for dispute since the shortcomings of the prediction would be manifest. Very often, those differences which give rise to dispute involve changes within material types. An obvious example concerns rock having an unconfined compressive strength greater than was expected from the site investigation test programme. The contractual problem then arises because the contractor does not have suitable equipment to excavate the stronger material according to his UK ICE 6th Edition Clause 14(1) programme and his Clause 14(6) method statement if such is requested by the Engineer. The ground is not then behaving as expected, and so the problem merges into one of the 'second type' which can be dealt with on the basis of Clause 12, or Clause 51 (an ordered variation) with perhaps Clause 44 (an extension of time for completion).

The question of 'contractor incompetence' (Brierley and Cavan, 1987) also enters into the equation. If 'improper' equipment or 'improper' methods of construction are selected by the contractor, progress is likely to be delayed, and with liquidated damages on the horizon the contractor is left with little option but to attempt to recover some costs even when the ground is ultimately adjudged not to differ from that 'anticipated'.

A major feature of the 6th Edition of the ICE Conditions of Contract are the several new provisions for dealing with the contractor's design liability (*see*, for example, Cornes, 1994). Clause 8(2) in the old 5th Edition provided for contractor liability in the design or specification of the permanent works if this were 'expressly provided in the contract'. It was not clear whether this liability should be based on the usual professional requirement of 'reasonable skill and care' or on a higher level of 'fitness for purpose' (*see* Section 7.10.4), which would seem to be implied in a contract where the contractor undertakes both the design and the construction and which seems to have been the view taken by the courts. Procurement of insurance might be more difficult under the latter circumstances. The 6th Edition clarifies this responsibility in the new Clause 8(2) with the words, 'The contractor shall exercise all reasonable skill care and diligence in designing any part of the permanent works for which he is responsible'. It does not, however, address the several problems that could arise if the degree of permanent works design is substantial. In such a case it is likely that the contractor's proposals in outline at least should form part of the contract documents. If they do, then it is not clear under the 6th Edition how ambiguities and discrepancies that might arise will be dealt with, how these particular works are to be billed, who takes on responsibility for changes in quantities and any associated requests for re-ratings, and how variations are to be accommodated in the contract. Such matters are addressed in the New Engineering Contract (Institution of Civil Engineers, 1993; *see* Section 2.8 below). Other clauses in the 6th Edition suggest that a contractor's obligations may require more than skill, care and diligence, so a higher level of responsibility may be demanded with respect to other duties as defined in other contract documents such as the works specification. On the other hand, and in one sense of re-balancing the risk, the contractor's liability for the care of the works, and their insurance, from the date of issue of the (renamed) 'Certificate of Substantial Completion' has ended. The 5th Edition Clauses 59(a) and 59(b) dealing with nominated subcontractors are now merged in the 6th Edition into a single Clause 59.

A major change in the 6th Edition is the provision in Clauses 66(4) and 66(5) for disputes to be taken to conciliation prior to arbitration. The Institution of Civil Engineers is now required to compile and maintain a list of conciliators as well as its list of arbitrators. Interestingly there are no clauses related to safety matters, but the Health and Safety Executive has maintained that some 20 clauses in the 6th Edition are not compatible with the law in respect of safety. The relevant clauses relate to the contractor being responsible for safety on a working site, whereas under the 1974 Health and Safety at Work etc Act, in which under Sections 2 and 3 the responsibilities of the employer are stated in broad terms, everyone who has an involvement with the project, including the client and Engineer, has a firm legal responsibility for safety. Under the Management of Health and Safety at Work Regulations 1992 this duty of care is defined more explicitly as a duty of line management (for example, workers are to be adequately trained, the provision of information being insufficient), and risk assessment of work activities is made a legal requirement. In the Workplace (Health and Welfare) Regulations 1992, which were

introduced on 1 January 1993 and which expand the requirements beyond those specified by the 1974 legislation, emphasis is also placed on identifying the risk factor(s) in the work. Attention will also need to be paid to BS 6164:1990 (British Standards Institution, 1990a) which gives a comprehensive guide to the rules and regulations which need to be followed carefully in respect of fire precautions. Since criminal law takes precedence anyway over civil law, and the former changes quite frequently, it may be argued that the 6th Edition is quite correct in setting out a contractual responsibility between employer and contractor and not engaging in any exposition of criminal liabilities and law. The parties to a contract can be reminded of such legal issues in contract specifications. In summary, the contractor, being in charge of construction, has primary responsibility for safety, although both the employer and the Engineer have a duty towards their own staff in particular and to the works in general. With both national and European Community pressures in the area of safety increasing, it is possible that amendment clauses covering such matters may be needed in the future.

It can be claimed that the ICE 6th Edition form of contract is relatively inflexible compared with other forms in the matter of methods of working. Where, for example, support systems have to be rapidly matched to ground conditions, as with the use of the New Austrian Tunnelling Method (*see* Section 6.8 and Supplementary Information 17 in Part Two of the book), then sets of amendments must be formulated for the 6th Edition form to operate. The actual form to be used depends upon the amount of responsibility for and control of the contract that the client wishes to retain. NATM has been used, for example, on the Channel Tunnel (access adits, marshalling tunnels beneath Shakespeare Cliff, the UK undersea cross-over cavern, and a section of running tunnel beneath Castle Hill), on the London Heathrow Express tunnels, and on sections of the London Underground Jubilee Line extension. It has also been scheduled for use on the Heathrow baggage tunnel and on London Crossrail. However, a collapse of a NATM subway tunnel in Munich in September 1994, in which two people died, and the progressive collapse of station and running tunnels on the Heathrow Express project in October 1994 have led to questions being asked as to the appropriateness of the system for tunnelling in such weak materials such as the London Clay. As a result of the Heathrow collapse work on London Underground's Jubilee Line extension was immediately suspended pending results of investigations of the collapse.

2.2.2 'Signed Sealed and Delivered'

Under the ICE 6th Edition form of contract, there is provision for the Form of Agreement (page 51) and the Form of Bond (pages 53 and 54) to be 'Signed Sealed and Delivered'. The Companies Act 1989, which came into force on 1 July 1990, has introduced important changes to the requirement for the sealing of documents. These have been discussed in outline by Jerram (1990).

As a result of the case *Whittal Builders Co Ltd* v *Chester-le-Street District Council* (1987) intention has been held to be more important than the actual sealing of a document. Under the Law of Property (Miscellaneous Provisions) Act 1989 sealing is now no longer necessary provided that it is clear that the intention is for the document to be a deed, supported by the signature of the contracting party and which is witnessed. The case of *Venetian Glass* v *Next Properties* (1989) also established the fact that it is not necessary for the document to be actually handed to the other party for it to be 'delivered', since the document will become binding once it becomes known that the signature has been applied. (The company sent a signed and sealed deed to its solicitor and thereby showed its intention to be bound by it.) However, delivery could under specified conditions be conditional and not binding until those conditions had been fulfilled.

For a contract to be made there must be an agreement which the law views in terms of *offer* and *acceptance*. One party to the contract, the offeror, declares a readiness to take on an obligation subject to certain conditions. This offer is communicated to the other party, the offeree, who then has the opportunity to accept or reject it. If it is accepted, the acceptance must be complete and unconditional. Acceptance is then communicated to the offeror and the agreement is then complete. There must also be a *consideration*. The law will only enforce a promise when it has been brought by a person who wishes to enforce it. *Consideration* is the price of this promise. It is usually money, but it need not be, nor need it be a fair reward of the contract performance. There must further be an *intention* to create legal relations. .This is presumed by the court in the case of commercial agreements but can be rebutted if the context requires it.

There has to be a consideration for 'simple' contracts to be made orally or 'under hand' (in writing), and without it no contract exists. Contracts under seal (deed), which are of a substantial nature, do not have to be supported by a consideration. An example of a contract where a deed is necessary and in which there is no consideration between the contracting parties (the bondsman and the beneficiary) is a performance bond (which clients are increasingly likely to require in the future). Although such agreements used to be referred to as contracts under seal they should now be referred to as deeds. A deed may be employed as a device to extend the period of limitation.

'Limitation period' is the time within which an action for breach of contract can be brought from when the breach actually occurred. It is a subject that is considered at greater length in Section 6.10. Whereas the limitation period of a contract under hand or an oral contract in England and Wales is six years, it extends for twelve years for a contract made under deed (seal). (The period of limitation in tort is six years and for personal injury, three years.) This alone is a strong argument for a client to enter into a contract under deed (seal) rather than an oral contract or one under hand, but the pricing of the work by a contractor will reflect the extended period of liability.

Establishing the start of a limitation period is itself controversial. It seems reasonable that it should begin when the completion certificate has been issued by the Engineer, because before that time the contractor has an opportunity to correct any faults in his work. The breach of contract then occurs when the work is handed over and the employer takes possession of the defect(s).

2.3 ICE DESIGN AND CONSTRUCT CONDITIONS OF CONTRACT 1992

These (D & C) Conditions of Contract have been drawn up specifically to accommodate problems associated with the increasing incidence of contractor design, as noted in Section 2.2.1 above and in Section 6.10 on Insurance. They apply when the contractor is responsible for all aspects of design and construction, including any design by, and on behalf of the Engineer. The Form of Tender under these Conditions provides for payment on a lump sum basis, but does not preclude other forms of payment being used. Reference may be made to the book by Eggleston (1994a).

In substance, these Design and Construct Conditions of Contract are not a variant on the ICE 6th Edition, but are an entirely new set of Conditions based on very different premises, and although the layout and clause numbering generally follow those of the 6th Edition the implications of the clauses are in many cases quite different.

Under this form of contract, the contractor has an opportunity of defining the responsibilities of the employer's representative, who does not need to apply the impartiality required of the Engineer under Clause 2(8) of the ICE 6th Edition, need not be a professional engineer, and could be an accountant or a quantity surveyor. His (or her) name may not be known to the contractor at the award of the contract. He will, however, need to make important engineering judgements and decisions on difficult and contentious issues. Specific to the theme of this book, he must, for example, decide, under Clause 12(6), on reasonable foreseeability by an experienced contractor and also, under Clause 18(2), decide whether investigation is a necessary consequence of a Clause 12 situation. In addition to the employer's representative there will often be a need for an independent checker, and certificates will need to be submitted to confirm that the contractor's designs are compatible with the employer's requirements. The style of contract also lends itself to the application of Peck's observational method (Peck, 1969) which accommodates variations in design and construction in response to ground conditions as revealed by monitoring systems installed by the contractor and costed out in the bid for the construction. Rate of construction should be an integral part of the design when the observational method is to be implemented.

Atkinson (1995) discusses ambiguities in the definition of the word 'contract' under the D & C form. In the case of the ICE 6th Edition Clause 1(1)(e), the contents of the contract are identifiable documents, and are contained in the Form of Tender. In the case of the D & C form, Clause 1(1)(g) defines 'contract' as '.....the Conditions of Contract the Employer's Requirements the Contractor's Submission and the written acceptance thereof....' The 'Contractor's Submission', which is not accommodated in the Form of Tender, is defined in Clause 1(1)(f) as '.....the tender and all documents forming part of the Contractor's offer together with such modifications and additions thereto as may be agreed between the parties prior to the award of the Contract.'

Two important points are highlighted by Atkinson. First it follows that the contractor's submission is not always the same as the tender, so creating the possibility of uncertainty and ambiguity. The tender is an offer for work based on the documents listed in the Form of Tender. Should any revised offers be made during a period of negotiation it is then possible that other documents could be drawn into the contract. There would seem to be no mechanism within the contract form for identifying which offer provides the basis for the contract.

The second point is that although documents can be amended during negotiation, there is no requirement for such amendments to be confirmed in writing. There is also provision under Clause 1(1)(g) for the contract to include '.....such other documents as may be expressly agreed between the parties....' but there is no provision for identifying these 'other documents'. Atkinson states that all documents encompassed within the 'contract' should be listed at the end of negotiations.

Clause 11 of the D & C Conditions of Contract deals with site information: the information obtained by the employer from investigations relevant to the works, and inspection by the contractor of the site and its surroundings together with information available in connection therewith. Under Clause 11(1) the employer provides the contractor with information (including information on 'pipes and cables in on or over the ground') obtained before submission of his tender, there being a duty on the employer to make available all the relevant information that he has obtained and which is not otherwise readily available. The contractor is quite entitled to expect that this has been done. This requirement also raises the question as to how much ground and other factual information the employer should reasonably be expected to acquire, and so provide to the contractor, and how much assessment (interpretation) of that information he should undertake and provide.

The employer's information to the contractor is dated as at the day of tender, but under a D & C contract the Contractor's Submission is the key document. This Submission can differ from the tender and be finalised some time after the tender date. Negotiations between employer and contractor leading to the ultimate contractor design on submission of his tender may draw forward information relevant to the contract that the employer had not originally considered to be significant and which, being made available after the date of tender, might lead to dispute. Problems could be avoided by concluding a written agreement at the end of the negotiations which clearly identifies the contents of the contract and especially the Contractor's Submission. Atkinson states that the D & C form should be amended so that the date of 'knowledge' is the date of the Contractor's Submission.

The contractor may need to do further (ground) investigation (under Clause 18) after the award of the contract and for the purposes of detailed design. Under Clause 18(1) permission to do so is required from the employer's representative. However, the difficulty for the employer is that the contractor may proceed to complete a design which significantly increases the risk of encountering unforeseen physical conditions during the progress of the tunnelling works. The ground itself that is excavated and supported may be foreseeable and have been foreseen, but it may

behave differently under different tunnelling regimes which impose different stress conditions on it. It follows, therefore, that the employer would be wise to attempt, at the tender evaluation stage, to compare the design solutions proposed by the tenderers, assess their interaction with the ground information provided, and consider, from the whole range of designs, where changes might be made during negotiations and the implications of those changes in the context of the ground information already provided. The employer should require tenderers to describe not only the ground investigations to be carried out during the design phase but also to provide a schedule of laboratory tests and a demonstration of how the information will be used in the design of the tunnelling works.

Under Clause 18 the contractor must bear the cost of any investigation unless it is a consequence of a Clause 12 situation or a variation. An alternative strategy would be for him to do no investigation, expecting to recover costs under Clause 12 if a risk (to him) arises. However, against that position, under these conditions of contract the experienced contractor has become a contractor experienced in both design and construction and so his ability to claim on the basis of ground investigations and tests conducted during the design phase would appear to be limited.

Because the contractor now designs the works he should bear the responsibility for suggesting solutions for avoiding a Clause 12 situation. While this strategy is reflected in amendments to Clauses 12(3) and 12(4) in the ICE 6th Edition of the Conditions of Contract, Clause 12(4) in the ICE D & C form does not expressly allow the contractor to be ordered to investigate and report on alternative measures, nor to be instructed on how to remedy the situation. Under Clause 12(4) the employer's representative can only consent to the contractor's measures, or issue a variation or suspend the works. Clause 12 therefore has the effect of shifting responsibility for the action onto the contractor.

Clause 8(2)(a) makes it clear that the contractor's design responsibility extends only to the exercise of reasonable skill, care and diligence, as is the normal requirement of a

Table 2.1 Main types of civil engineering contract

Types of Contract	Comment
Cost plus a percentage fee	Wholly reimbursable to the contractor
Cost plus a fixed fee	Partially reimbursable to the contractor
Target (shared overrun)	Partially reimbursable to the contractor
Guaranteed maximum	Partially reimbursable to the contractor
Lump sum (i.e. wholly lump sum)	Not reimbursable to the contractor

consulting engineer, and although the contractor is not required to carry full professional indemnity insurance against negligence in the design he could still be liable to the employer for any failure on his or his servant's part to exercise reasonable skill, care and diligence that results in damage to the employer. Such failure may not be covered by insurance under Clauses 21 (Works) and 23 (third party), and so this places a responsibility on the contractor to set the specification for his level of design checks and the procedure for carrying them out. More generally, under Clause 8(3) the contractor may be required to set up a quality assurance system, to be approved by the employer, which applies to the contractor's system of checks on his design and setting out of the Works and for testing all materials and workmanship.

The employer may require the contractor to maintain his professional indemnity insurance in respect of design. These policies are renewable annually, but with no guarantee of renewal. They provide cover only for claims made in that particular year and not for those arising from negligence in that year which gives rise to a claim in a later year.

The Brighton stormwater interception scheme provides one example of a design and build contract. A £20 million, 132 000m³ capacity retention tank, 6m internal diameter and more than 5km long, was constructed parallel to the coastal interceptor sewer and up to 35m under the beach in order to comply with the EC Bathing Water Directives and offset flow onto the beaches from short outfalls at times of storm sewage overflows.

For certain tunnel functions it may be decided that the work should be let by a client on a design/build/operate basis for a prescribed number of years.

2.4   GENERAL STYLES OF CONTRACT

Apart from an admeasurement style of contract, the main forms of contract in civil engineering are given in Table 2.1.

The term 'reimbursable to the contractor' means reimbursement of costs actually incurred on the contract. In any partially reimbursable contract the provision of some goods and services is covered by a lump sum payment while others are paid for on a reimbursable basis.

2.5   TARGET COST CONTRACT

Under a target cost contract (and unlike a 6th Edition admeasurement type of contract) the bidding contractors do not price for risk (specifically ground conditions risk) in the tender. By excluding such unknowns the range of tender pricing should, in theory at the very least, be reduced and the disadvantage to the perceptive, responsible and experienced contractor who prices sensibly for such risks should be eliminated. Contractors compete to design and construct the work to a performance specification by submitting estimated target costs, including bonus percentages.

There is provision for the contractor to take an active part in the site investigation. The contractor is paid for all the work actually done, including measures taken to overcome difficult ground. The client decides whether the ground conditions were of an unexpected condition that could not reasonably have been foreseen by the experienced contractor at the time of tender. If they were unexpected, the tender target cost is then raised by the necessary amount, and the contractor's bonus is unaffected. If they could have been foreseen, then the original target cost remains. (The problem of deciding, or, more specifically, who decides, which conditions are 'unexpected', 'unforeseeable', 'latent' still remains.)

At the end of the contract, the contractor's hoped-for percentage bonus is reduced or increased depending upon whether the actual cost of the works exceeds the final target cost or not. Provision for liquidated damages will usually remain in the contract.

The underlying theme of a target cost contract, as currently exercised in the UK, is that over-expenditure and under-expenditure should be shared according to some formula between contractor and client. Such a form of contract can be defined at the present time in the UK by a number of special clauses added to the ICE Conditions of Contract.

A target need not simply embrace the matter of cost, but can also encompass the quality of the end product and the time to complete. Flexibility in this form of contract arises from the extent to which variations between the cost and the target are shared between the contracting parties.

This degree of sharing can be incorporated into a 'share formula' which is determined by the risks inherent in the particular type of work being carried out and the level to which the client is prepared to share the risks, both favourable and unfavourable, with the contractor. The final price to be paid by the client to the contractor may comprise a base sum to cover the net site expenditure plus or minus an incentive adjustment, the latter to be calculated from a comparison between the net site expenditure and the target value of the contract.

As an example of an incentive adjustment, Boyd and Stacey (1979) quote, for the Wyresdale water transfer rock tunnel in Lancashire, a not-to-be-exceeded figure of 3% of the target value, whether negative or positive. The target value would correspond to the contract price as for a normal admeasurement contract, but for it to be equivalent to the site expenditure the sums comprising the contract (tender) price should exclude the contractor fee element. Such a target value would then be subject to the same variations, including a contract price fluctuation, as on a normal admeasurement

contract. Bidding contractors would be required to specify a fee payable, usually as a percentage of target value, to cover profit and other costs.

There are various styles of incentive adjustment. One method would be for the client and contractor to share equally the cost difference between the site expenditure and the target value. If the latter exceeded the former, the contractor, in addition to his site expenditure and fee, would receive the benefit of half the difference between the two. On the other hand, if the former exceeded the latter, payment to the contractor would be the site expenditure and fee less half the cost overrun difference between the two.

This form of contract is most appropriate when the risks, through lack of prior knowledge, are high - for example, when there is little or no site investigation. In such circumstances the client will support most of the risk. For a particularly risky project (and tunnelling constitutes one of the most risky types of civil engineering construction) the client would wish to limit his share of any overspend up to a pre-stated level. Targeted cost would then be on a sliding scale, with a guaranteed maximum price to give the client assurance as to his upper limit cost.

There may be two particular reasons for not undertaking a site investigation. First, it may not be physically possible for drill rigs to reach position over deep tunnels in mountainous terrain. The geological settings for such tunnels may therefore be known only sketchily, and the possibility of unforeseen physical conditions occurring at the tunnel face will be correspondingly high. Ideally, therefore, a flexible tunnelling system would be specified at the outset, but there are examples of deep rock tunnels which have received little or no prior investigation but which have been driven using full-face machines rather than by drill and blast. If the ground is known to be geologically complex, there may then be a second reason for ignoring site investigation. It may be argued that very local information from discrete boreholes could prove to be misleadingly optimistic and that the cost of building up a reliable picture of the ground conditions would be unacceptably high with respect to the portion of an incentive adjustment charge in a target cost contract falling on the client.

Although, as noted above, penalties for late completion are already available in construction contracts in the form of liquidated damages, there is the possibility of introducing time-related bonuses as incentives to complete a project ahead of the specified completion date. To win such bonuses a contractor would need to invest extra resources in the work, and would only be willing to do so if the size of the potential bonus exceeded his greater investment in those resources. Quality of work, under existing admeasurement conditions of contract, is controlled according to the contract specification drawn up by or on behalf of the client. Under a target cost contract it can be achieved by the contractor being responsible for the design according to a detailed specification formulated by the Engineer on behalf of the client. The specification would be one of performance and fitness for duty (purpose), and the difficulty would arise in restricting the specification to one of performance without stepping into the area of design.

2.6 ICE CONDITIONS OF CONTRACT FOR MINOR WORKS 1988

Guidance notes state that this form of contract is for use where, among other points,

- the likely risks to both parties to the contract are judged to be small;
- the works are of a simple and straightforward nature;
- the contract value does not exceed £100 000;
- the period of completion of the construction does not exceed 6 months, except where the method of payment is on either a daywork or a cost-plus-fee basis;

Rules for conciliation procedure are given in the conditions of contract, and Clause 11 of the contract deals with all disputes, including arbitration. Further detailed study of this rather less publicised contract form is recommended.

2.7 INSTITUTION OF CHEMICAL ENGINEERS FORM OF CONTRACT

There has been a growing appreciation of the benefits of the Institution of Chemical Engineers' Model form of conditions of contract for process plant (1992) (the 'Green Book') as a practical means of adopting a reimbursable, often 'open book', target cost style of contract. (The IChemE 'Red Book' (1981) concerns lump sum contracts with a Contractor's design facility and the IChemE 'Yellow Book' concerns coordinated forms of 'back-to-back' subcontract.) The 'Green Book' was formulated for use on process plants with reimbursement of the cost of most of the goods and services provided, as would be the case in a civils contract. A major difference in principal between the IChemE contract and the ICE 5th and 6th Editions is that the Green Book reimburses the 'cost' actually incurred as opposed to the 'value' of the work executed under an ICE 5th or 6th Edition contract. 'Costs' would include those incurred as a result of errors, and would not therefore add value to the project. Under a Green Book contract (Clause 39.2) a contractor is reimbursed for estimated 'expenditure', not the cost, for the succeeding month, in contrast to the 'value' of the work completed in the month against which the application for payment is made.

'The *Contractor* shall submit a request for payment to the *Project Manager* towards the end of each month. The request for payment shall be in the following form:
(a) the *Contractor's* estimate of the total of expenditure that he will incur and the sums that will be due to him in the month succeeding the month in which the request is made, provided that in the first month the *Contractor* may submit his estimate for that month at its beginning;
(b) the total of the *Contractor's* actual expenditure and sums due to him for the month preceding the month in which the request is made; less
(c) any amount previously certified by the *Project Manager* in respect of the

Contractor's estimated expenditure and charges for the preceding month.

The *Contractor's* request for payment shall be supported by all relevant documentary evidence, including a statement showing the manner in which the total requested has been calculated.'

Payments are then more closely related to the contractor's cash flow than to the progress of the work. The client has a much greater influence over the progress of the work and its completion date by his control of costs in the form of increasing or decreasing expenditure (refer also to Guide Note Q on page 48 of the IChemE Conditions of Contract). He can instruct the contractor to spend more, or spend less, in a defined way, and so exert a control both on the completion date and the means whereby that completion is achieved. Thus, on the one hand the contractor is given incentives and reimbursement assurances and on the other the Project Manager is given a greater management role in the construction. The Project Manager acts on behalf of the Purchaser (Employer), and 'in all matters where [he] is required or authorised under the contract to exercise a discretion or make a judgement or form an opinion he shall do so to the best of his skill and judgement as a professional engineer' (Clause 10.1(e)). However, he is not, of course, required to act impartially in the manner of the Engineer in the ICE 5th and 6th Editions of the Conditions of Contract.

It is also claimed that particular advantages stem from faster procurement and economies of design cost, the greater likelihood of early completion, and better risk sharing between the contracting parties, but that for the contract to operate successfully the parties must be well experienced and operate in an atmosphere of mutual trust. One disadvantage of a Green Book contract seems to be that in its unamended form it automatically reimburses poor workmanship, but makes no provisions for retention monies. Advance payment of anticipated expenditure imposes a financial burden on the client and also leaves him exposed to problems resulting from contractor failure to complete the work.

One example of the adoption of an IChemE form of contract following problems while working on a standard ICE 5th (1973) Edition tender bill of quantities admeasurement contract was the Norwest Holst work on Stage 1 of the Mersey Estuary Pollution Alleviation Scheme (MEPAS), initiated in 1974. The original £17 million contract, for a 2.3km length of tunnel in the area of Liverpool docks, formed the first phase of North West Water's planned 26km long interceptor sewer driven along the banks of the river Mersey. The sewer was designed to intercept about 26 outfalls, which discharged into the river and were responsible for fouling the Crosby, New Brighton and Wirral beaches, and to transfer the effluent to a new £52 million primary treatment plant at Sandon dock. The short Norwest Holst 2.4m diameter tunnel (2.7m internal diameter with 1m wide tunnel rings), driven beneath Liverpool's North docks and close to the city's Pier Head at an average depth of about 15m, was designed to intercept six of the largest outfalls and to carry about 40% of the total dirty water flow.

The tunnel was driven through a glacial valley, the ground at the face comprising mixed sediments of silts, sands and clays. Underlying the soils is the Sherwood Sandstone which outcrops towards each end of the tunnel drive to provide good tunnelling conditions but which appeared intermittently at the face throughout the rest of the drive to create excavation difficulties. A high water table required a constant pressure on the face of up to 100kPa to maintain dry working conditions behind. The £1.8 million Fives-Cail Babcock 3.2m diameter, 55 tonne full-face pressurised bentonite slurry non-articulating shield, built under licence in France to a Kawasaki design for the work, was equipped with 24 roller cutters for dealing with rock plus a further 30 picks mounted behind the cutters. It operated on a 24 hour, two-shift basis for 5 days per week and achieved a maximum progress of 60m per week. However, it suffered major steering problems, and on a 450m radius curve drove 3m out of line and dived 300mm. Ultimately after only 700m of tunnel had been completed the machine had to be abandoned and removed from the ground via an enlarged nearby shaft. In a ICE 5th Edition Clause 12 claim on the region of £5 million, it was maintained that the steering problems were caused by the mixed face conditions, where firm boulder clay was in the invert and loose running sands in the crown, and that such conditions could not have been foreseen by the contractor. On the other hand, it was contended by the client that because the 5.5m long tunnelling machine was not articulated, the high 1.6 to 1 ratio of the machine's length to diameter created the steering problems.

The remainder of the tunnel in 'hard' rock (maximum unconfined compressive strength 52MPa with an average of 20MPa) was excavated by drill and blast techniques for a 420m section within the protection of a short open shield, and a mixed ground section was pipejacked by Belgian contractor Denys. The Denys TBM, of hybrid design, incorporated a 5m long articulate shield leading the pipejacked concrete lining. A roadheader type boom cutter protruded through a forward closed bulkhead of the Denys machine and allowed the face to be supported by a combination of bentonite, water, or compressed air. There was a boulder-crushing facility, comprising a combination of teeth and rotating screws, located forward of the bulkhead and immediately below the cutting boom. A single jacking point, providing 1260 tonnes thrust, was used to advance the shield and the 2.8m external diameter concrete rings.

As a result of this experience, the client adopted a target cost contract of IChemE form for the adjacent £25 million second phase (Stage 2) of the interceptor sewer (fee plus percentage of on-site costs plus a maximum bonus as a percentage of tender target cost), although most of the 4km of tunnelling was expected to be in firm rock which would not be prone to the same level of dispute and contractual claims. Procurement was by competitive tender and Lilley Construction was selected to proceed with the negotiation of a tender target cost. Under the contract, on encountering unforeseen ground conditions the contractor's cost would be automatically paid and, rather significantly, the client assisted the contractor (Lilley) with the choice of the most suitable tunnelling machine for the job. This was a 3.1m outside diameter Decon monolithic shield with an articulated tool section and a Dosco 105 roof-mounted roadheader. The tunnel 2.4m internal diameter, 15m deep, was driven through glacial till at the Sherwood Sandstone boundary.

It is useful also to note that Stage 2 saw the application of the well established 'just-in-time' manufacturing concepts to the supply of lining segments to a shaft bottom and spoil removal.

Another example of the use of this form of contract relates to the £16.5 million Thames Water's Camberley sewage treatment works. The new works, which is capable of dealing with the sewage from 96 000 people, is the largest of Thames' investment in upgrading plants that discharge into the River Blackwater. A feature of that particular form of contract, which was won by Gleeson on a competitive tender, was a formula for bonuses and penalties as a variant of the pure cost-plus form of the IChemE contract and which was introduced in order to speed up the work and more obviously share the risk with the contractor. In fact, the contract was completed on time and below target cost.

Thames Water's London Water Ring Main provides two interesting and contrasting examples of contractual conditions. Miller Construction's 3.8km drive, starting from a 55m deep access shaft at New River Head to Barrow Hill, was in mixed ground, notably comprising sand lenses filled with water, of the Thanet sands which are below the Woolwich and Reading beds. London Clay was encountered only near to Barrow Hill. Although the work was let under a conventional ICE 5th Edition form of contract, Thames Water was involved in the choice of the type of tunnelling machine - a full-face, 70m long, 3.37m diameter Canadian Lovat Machine but not operating in the earth pressure balanced mode. Under normal conditions the TBM removed the spoil from the tunnel face along a conveyor, but it could be converted to operate with an EPB screw if cohesionless soil, typically sand lenses containing high groundwater pressures, were to be encountered. A maximum jacking force of 11 000 tonnes was available, and there were two 185kW motors to drive six hydraulic motors at the cutting face. The actual construction sequence was to drive forward 1m (one ring), remove the spoil, erect a concrete ring within the shield, and then grout up the annulus between the ring and the excavated ground.   The section of the tunnel passed under four tube lines and King's Cross station, and at its closest was only 7km below an underground tunnel.

In the Tooting section of the Streatham to Brixton Water Ring Main the contractor encountered Thanet sand of the Woolwich and Reading beds. The sand was subjected to high water heads that prevented the use of compressed air for temporary support. Using an IChemE form of contract with a target cost formula Thames Water, with the contractor, adopted a combination of freezing and a Lovat full-face tunnelling machine operating in earth pressure balanced mode.

Yet a further example of a 'Green Book' tunnelling contract is North West Water's Oldham interceptor sewer.   The Oldham contract was part of a well-established £50 million scheme to replace a partly derelict and overloaded late Victorian interceptor sewer. Client and contractor, Kilroe, jointly decided on a pipejack method of construction for the 1.8m internal diameter, 1.8km long tunnel, which required good ventilation and fireproof equipment because it passed through old mineworkings. A Herrenknecht SM2 road-header, jacking through a maximum distance of 600m with intermediate jacking stations, was used to tunnel through silt, clay, mudstone, siltstone and strong sandstone. Excavation was difficult due to the need to avoid old mine shafts and a requirement to traverse through minimum 200m radius curves. The roadheader design was sufficiently flexible to be adapted to segmental ring construction in the event of the pipejack getting stuck.

North West Water's Southport sewage interceptor tunnel is yet another example of a cost reimbursable IChemE form of contract. Pollutants were getting on to the beach even at low rainfalls, but since part of the area through which the sewer was to pass was a site of special scientific interest it was decided to construct in tunnel rather than by pipe in trench.   The cost of the 2.82m internal diameter tunnel was £26 million, and the cost of the associated treatment works (the first of which was constructed in 1907) was £49

million. Although the site investigation along the tunnel route comprised 83 shell and auger holes, and included 43 cone penetrometer tests, the contractor (Costain) requested further site investigation. Tunnelling conditions consisted of alluvial sands, usually with 7% silt but which could increase to 15% maximum, and mixed glacial till with mudstone rising in the face along the drive. A slurry machine was ruled out due to fears about possible bentonite contamination, and a Lovat earth pressure balance auger machine was chosen for the work.

Southern Water's 'Operation Seaclean' tunnelling scheme between Brighton and Hove in Sussex (Brighton and Hove Stormwater Project) to ensure compliance with the EC Bathing Waters Directive 76/160/EEC is another example of this style of contract. The contract was let on an IChemE Green Book design and build basis, allowing the contractor to assess a range of storage options. A storage tunnel concept, involving a 6m diameter, 5.2km long tunnel under the beach between Hove and Black Rock, Brighton, was chosen to provide a stormwater capacity of 132 000m^3.   Tunnel depth ranges from 26m in the west to 40m in the east, and at the Brighton end of the scheme where the tunnelling was through fractured Upper Chalk an earth pressure balance machine was required to resist 4 bars of water pressure.

It would seem that if the project definition is low at the outset a 'pure' cost reimbursement contract could leave the client financially exposed, with little possibility of controlling costs or estimating the final outcome. There is also the inescapable fact that the contractor does lack financial incentives to keep expenditure to the lowest practical level. As in the case of the Thames Water contract above, it would seem that the realistic solution is to adopt cost targets as specific amendments to Green Book documentation in order to create the requisite incentive for the contractor. Savings or over-expenditure on target cost would be shared between client and contractor according to a well-designed formula which embodies the effective risk to each party to the contract. The share formula can either be imposed by the client or left for contractor tendering. A simple formula would provide for a 50-50 sharing of the change from target. Another would provide for a 50-50 sharing of any savings achieved but allow a weighting for any over-spend, say a 60-40 share, with the contractor carrying the greater risk up to some maximum specified price. A further variant would be to reduce the client's contribution progressively as any over-spend increases. Useful Introductory Notes in the Green Book include a listing of the comparative advantages and disadvantages of lump sum and wholly reimbursable contracts (*see* Table 2.2).

Although lump sum contracts are a relative rarity in the UK construction industry the information is valuable in condensed form for reference purposes.

There is provision in the Green Book for disputes between the parties, such as a challenge by the contractor to a decision by the Project Manager, which cannot be resolved by discussion, to be referred to either 'The Expert' (Clause 44) or Arbitration (Clause 45). Under Clause 44.5:

> 'The *Purchaser* and the *Contractor* hereby agree to be bound by any decision of an Expert under this clause and shall comply with any direction given therein and shall not question the correctness of any such decision or direction in any proceedings.'

The value of this form of dispute resolution lies in its relative cheapness, the 'Expert' being 'one who may reasonably be expected to form a correct opinion on the matter in question by virtue of his own skill and experience.' This seems to be the preferred option, but if an expert is not used then the parties to the dispute should go to arbitration. Alternatively, there may be a desire to delete both of the above Clauses (44 and 45) and to insert special Special Conditions as per Guide Note JJ on page 60 of the Conditions of Contract.

Although it is often contended (but may not always in practice turn out to be so) that the adoption of either the expert or the arbitration means of resolution should be less expensive than resort to a court of law, there may be a desire to use the latter on the grounds that the normal processes of law are to be preferred for their well-established and unambiguous procedures in respect of the evidence that may be submitted and the usual provisions for appeal to a higher court. Right of appeal against an arbitrator's decision is in general limited to aspects of 'law' as distinct from 'fact'.

2.8 NEW ENGINEERING CONTRACT

The New Engineering Contract (NEC), first launched in draft form at the beginning of 1991 for industry discussion and the first edition being dated 1993, is a single package of conditions designed to be used by a wide range of engineering disciplines and for almost any type of contract. Subsequently, in 1994, two further contract forms, the Adjudicator's Contract and the Professional Services Contract, each with Guidance Notes, were added to the package (Institution of Civil Engineers, 1994). In essence the NEC comprises a variety of basic clauses to which a number of secondary clauses can be attached to build up into a range of different contract packages. An 'activity schedule' is introduced as an alternative to the bill of quantities and so makes possible lump sum type contracts. The NEC projects a project management approach to the running of a contract in a manner which differs fundamentally from that of existing contract forms. Following a close determination of the roles played by all parties to the contract, and an analysis of their interactions, clause function statements, defining what each party is expected to do and be responsible for in the contract, were formulated. Flow charts have also been included in the package to assist a user of the contract to progress logically through the procedures.

At the core of the package are 59 clauses setting out general conditions, such as times, costs, quality and risk, which are basic essential parameters in all contract forms. To these core clauses can be bolted on one of up to six sets of option clauses which serve to convert the base contract into a particular form: conventionally priced, target cost, design and construct or management. There are further secondary clauses to accommodate specific requirements, such as contract price adjustment, retention bonds, or the use of multiple currencies. Other features of the flexible contract package are noted below.

Table 2.2 Advantages and disadvantages of lump sum and wholly reimbursable contracts (after the Institution of Chemical Engineers, 1992)

Advantages	
Lump sum	**Wholly reimbursable**
1) Purchaser knows his expenditure commitment. 2) Responsibility for the project is vested in a single body (the Contractor). 3) Allows fullest competition between contractors. 4) Bid evaluation is straightforward.	1) Requires minimum enquiry definition. 2) Shortest possible bid time (few days). 3) Complete flexibility - Purchaser participation entirely practicable. 4) Purchaser/Contractor conflict of interest is minimised. 5) Purchaser has control over costs incurred. 6) Purchaser can assess Contractor's rates. 7) Purchaser can use Contractor to evaluate alternative schemes. 8) Purchaser can terminate at will without incurring substantial costs.

Disadvantages	
Lump Sum	**Wholly reimbursable**
1) Purchaser/Contractor interests are more divergent than in other forms of contract. 2) Long bidding time is required (2-4 months), also long enquiry preparation time. 3) Crucial process design phase compressed into a very short period. 4) Lack of flexibility - changes are difficult/expensive. 5) Purchaser participation in project is difficult. 6) Cost to Purchaser may be unnecessarily high due to contingencies for risk and escalation. 7) Emphasis on low bid price may result in an unsatisfactory plant.	1) Contractor has no monetary incentive to minimise cost to Purchaser. 2) Purchaser has no assurance of final cost. 3) Purchaser has to check and verify Contractor's manhour and expense records. 4) Bid evaluation may be difficult. 5) Contractor competition is only on a very small part of the total cost. 6) Contractor has no monetary incentive to achieve early completion unless covered by Special Conditions.

Comments	
Lump Sum	**Wholly reimbursable**
1) Complete project definition is essential. 2) Contractor's bidding costs are very high. Purchaser should minimise contracting industry bidding overhead (and thus plant costs) by: (a) not inviting bids until there is a high probability of the project proceeding, (b) limiting the number of bidders, (c) pre-qualifying each bidder so that he will not be rejected later on grounds known to the purchaser before the enquiry is issued, and (d) considering reimbursing pre-qualified unsuccessful bidders for their costs of bid preparation.	1) The most flexible type of contract, allowing a very rapid start. 2) Flexibility may encourage Purchaser to introduce design changes, resulting in increased cost and longer programme. 3) Contractor's profit/loss is limited. 4) Conversion of all or part of contract to (eg) lump sum basis during the project is possible when scope of work becomes fully defined.

Notes on Table 2.2:

(1) Guide Note G in the 1992 Edition of the Conditions of Contract explains why 'the Engineer' in the 1976 Edition has been replaced by 'the Project Manager'.

(2) The terms 'plant' and 'purchaser' may be substituted by their respective civil engineering equivalents, 'works' and 'employer'.

(3) Under 'Wholly reimbursable Disadvantage (6)' the words '.....unless covered by Special Conditions' is an addition to the 1976 Edition wording. The 1976 Edition carried a Guide Note F relating to an incentive bonus by the purchaser to the contractor 'as a reward for early completion'. This note is discarded in the 1992 Edition in favour of the reference to 'Special Conditions'.

(4) Under 'Wholly reimbursable Comments (2)' above, reference should also be made to Clause 17, page 15 on the matter of variations (Guide Note P, page 47). Under a wholly reimbursable contract the Contractor carries out the work and the Purchaser pays. The Purchaser, through the Project Manager, may tell the Contractor what to do, how to do it, and when. This right, in Clause 11 of the Green Book (Project Manager's Instructions) means that the Project Manager does not have to issue and negotiate variation orders. However, provision is made for such orders under Clause 17 (Variations).

(i) There is no presupposition of UK legislation, so the contract can be adopted internationally and specifically in the European Community.

(ii) Unlike the old ICE 5th Edition (but not the 6th; see above) the contract does allow for the fact that the contractor engages in design, does not engage in design, or engages partly in design.

(iii) It incorporates a wide spectrum of risk allocation between the employer and the contractor, and, by selection of the contract options, requires the client, before the contract is drawn up, to decide on the level of risk that he is prepared to accept.

(iv) Language is simplified; excluded are long words and words such as liability and indemnify, and phrases such as 'in the opinion of'. Strict legal terminology and cross referencing are also omitted.

(v) The several job descriptions undertaken by 'Engineer', or 'Engineer to the contract' in the ICE form of contract, and sometimes generating problems, have been expressly defined in the NEC. The new job title of Project Manager is equivalent to that of employer's representative, and it deals with concentration on management of the project on behalf of the employer. Appointment of the Project Manager may be under the Professional Services Contract (Institution of Civil Engineers, 1994) The designer is a separate consultant. The former position of resident engineer becomes the (construction) Supervisor (acting for the employer and maintaining quality control). This appointment may also be under the Professional Services form of contract (Institution of Civil Engineers, 1994). The Engineer's role as referee in any contractual dispute is taken over by an independent Adjudicator whose appointment at the outset of the contract requires approval of both contracting parties. The Adjudicator, who will be named in the Contract Data, will normally be appointed jointly under the NEC Adjudicator's Contract (Institution of Civil Engineers, 1994) and will not be liable to either of the contracting parties for any breach of duty. Adjudicator fees are shared equally between the parties to the dispute, irrespective of the Adjudicator's decision.

(vi) 'Claim' is now replaced by the expression 'compensation event'. Sixteen listed 'events' are at the employer's risk, with the contractor being automatically reimbursed. Typical of these are 'a change to the Works Information' and the contractor encounters physical conditions that 'an experienced contractor would have judged to have such a small chance of occurring that it would have been unreasonable for him to have allowed for them'.

Risks are categorised as those relating to loss, damage or injury and to finance, and are entered under the headings of 'Employer's and Contractor's risks' and 'Contractor's risks'. Four main categories of 'Employer's risks' are given. Risks not listed are assumed to be borne by the contractor. Since the division of risk is thus (theoretically) prescribed, it remains only to determine the level of payment. Responding to a 'compensation event' the contractor submits what is, in effect, a mini-tender, quoting a likely cost and suggesting alternatives. Agreement by the employer holds the contractor to the quoted price.

The physical conditions comprise:
• the Site Information
• publicly available information referred to in the Site Information
• information available from a visual inspection of the Site and
• other information which an experienced contractor could reasonably be expected to have or to obtain.'

It is important that site information be clearly marked in order to distinguish it from works information. The site information should include:

• subsoil investigation borehole records and test results;
• reports obtained by the employer concerning the physical conditions within the site or its surroundings;
• references to publicly available information relevant to the site and its surroundings such as published papers and interpretations of the geological survey, the purpose of listing these references being to help the contractor in preparing his tender and deciding his method of working and works programme;

• information about piped and other services below the surface of the site;
• information about adjacent buildings and structures and about buildings and structures on the site.

Only factual information about the physical conditions on the site and its surroundings is normally included under site information, interpretation being 'a matter for the Contractor'. It is stated, however, that 'some Employers may wish to include interpretive information, such as inferred geological sections'. Under the Core Clauses document 1 General, 11.2(7) the site is defined as 'the area within the boundaries of the site and the volumes above and below it which are affected by work included in this contract'. It still seems, however, that the terms 'site' and 'surroundings' remain somewhat unresolved, and therefore open to interpretation as in the case of the ICE Conditions of Contract. Interpretation of the word 'site' is considered to be particularly prone to dispute in the case of a tunnel.

Also covered in the NEC is the possibility of the site information (including the information referred to in it) incorporating inconsistencies. In such a case the contractor is assumed to have taken into account those physical conditions more favourable to doing the work (i.e. application of the *contra proferentem* rule in English law, which means that any ambiguity, or other doubt, in an exemption clause must be resolved against the person who is seeking to rely on it; that is, against the person who is proffering it).

There are several comparative comments on this wording with respect to the equivalent wording in the ICE 6th Edition.

(a) The NEC overhauls, but does not radically change, the interpretation of 'physical conditions' as in other (such as the ICE 6th Edition) conditions of contract. The test now is whether an experienced contractor would have judged adverse physical conditions on site to have had a small chance of occurring rather than assessing whether he could or could not reasonably have foreseen their occurrence. There would thus seem to be a greater risk placed on the employer under the NEC form of wording even though, as inferred by the NEC, disputes on interpretation should be reduced by the NEC wording.

The fact that unforeseen conditions is a Compensation Event in a core clause means that the assumption of the risk of unforeseen conditions by the employer is intended to be a characteristic of all contracts under the New Engineering Contract whatever option is used and including contracts which involve full design and build work by the contractor.

The NEC also states that one method of reducing contractual disputes arising from physical (ground) conditions 'is to define in the contract the boundary line between the risks carried by the *Employer* and *Contractor*, i.e. establish what the tenderers should allow for in their tenders'. It goes on to take the example of tunnelling or extensive foundation works by requiring a statement of 'limiting boundary conditions'. However, this expression does not seem to be defined, but it is assumed to take the meaning generally of 'reference ground conditions' as envisaged in the draft NEC Guidance Notes. These conditions, which can be covered under the 'Special conditions of contract' (Option U in the Core Clauses), can include such matters as

- soil characteristics
- levels of rock/soil interface
- groundwater levels
- permeability limits and
- overbreak in rock construction

Tenderers will then be able to tender on a common basis, knowing that they must allow in their pricing for the occurrence of the physical conditions up to the stated limits.

(b) There is an inferred responsibility on the employer to provide the site information (as is quite standard under the 6th Edition, and is reasonable) and also explicit references to other (further) information that the contractor should access.

(c) In the case of information inconsistencies, the reasonableness of specifying that the contractor should assume the most favourable physical conditions for doing the work is somewhat open to question. There should really be a statement somewhere in the contract documents to the effect that if any inconsistencies in information are identified at the pricing stage, then those should be drawn to the attention of the designer before the contract is let. Some inconsistencies may not be noticed at that early stage in the contract, and the requirement on the contractor to disclose any inconsistencies that have been noticed does not move the balance of risk away from the employer, but it does place at least some moral constraint on the contractor. When the contract is underway there is a requirement under Clause 17.1 (Ambiguities and inconsistencies) for either party (project manager or contractor) to notify the other once any ambiguity or inconsistency comes to light. The project manager then gives an instruction to resolve the ambiguity or inconsistency. As stated in the Guidance Notes (page 34, Clause 17.1 and page 64, Clause 63.7), assessment of the resolution of ambiguities and inconsistencies is based on the *contra proferentem* rule.

(vii) Compared with the provisions in both the ICE 5th and 6th Editions, there is a stiffening of the requirements with respect to programming of the works. If the contractor does not update his programme in the light of events the project manager will make his own assessment of the effects of a compensation event on timing. This is a strong incentive for the contractor to adhere to procedures laid down for programming the works. In general the NEC requires the parties to act and respond to communications promptly, and there are sanctions for failure to comply.

(viii) Some of the major civil engineering problems have arisen in the area of subcontracting. The standard NEC subcontract includes conditions that are almost identical to those between employer and contractor, and should prevent one of the parties (the contractor) trying to pass on as many risks as it sees possible to another (the subcontractor) that is usually less able to support them. It encompasses contracts with the proportion of the work subcontracted ranging from zero to 100%.

(ix) As noted above, subsequent to the publication of the NEC package of documents there have been two additions (Institution of Civil Engineers, 1994): the 'Adjudicator's Contract' (AC) and the 'Professional Services Contract' (PSC). The AC is for an adjudicator to be available for calling in to resolve any disputes between the client and his contracted professionals. The PSC, as with the NEC, contains basic core clauses, main option clauses and secondary option clauses. The core clauses set out the main contract responsibilities, a requirement for the consultant to produce a programme of work, payment for work done, compensation events and their assessment, risks, insurance, disputes and termination. Main option clauses

cover the method of payment: lump sum priced contract with an activity schedule, a time-based (this is effectively cost-reimbursement at previously agreed prices) contract, a target price contract (cost-reimbursement time-based with a bonus share payable when the work comes in below budget) or a term contract, which is particularly applicable to compulsory competitive tendering (CCT). Secondary option clauses cover items such as payment in multiple currencies, copyright and price adjustment for inflation. PSC is seen as being particularly useful for the appointment of a consultant by a contractor under design and build conditions. During the tendering phase the consultant might be retained on a lump sum or target cost basis (with a bonus if the work is won) and then retained for the main design on a lump sum basis. After engagement under a lump sum style contract, if there are changes in a consultant's remit he will be paid for these compensation events, but the client must be notified as soon as they occur. Any disagreements between the parties will be sorted out by the adjudicator.

In summary, it is claimed that the NEC is a much more wide-ranging, more flexible document, likely to be more suited to a rapidly changing construction industry. Of particular significance is the fact that the importance of risk is fully recognised. Unlike the Green Book form of contract outlined above, the contractual options in the NEC include specific provision for cost targeting, with or without the framework of full cost reimbursement. The NEC is written in a non-legalistic manner ('for engineers rather than for lawyers'), and this is claimed to be a major advantage. On the other hand, the lack of legalistic precision, as tested by precedent, could prove to be a disadvantage.

There has been growing use of and international interest in this form of contract. It is noted that South Africa's electricity utility Eskom was expecting to let major contracts in 1992 under the ICE NEC form. Prominent among these would be contracts on Majuba power station, under construction. Interest has been expressed in Germany and the United Arab Emirates, and apparently the Australian General Conditions of Contract which were scheduled for issue in August 1992 are being prepared along the general lines of the NEC. In the UK the NEC is being used by BAA for its £300 million London Heathrow Express project and has also been considered by National Power for use on construction contracts.

Of special interest is the proposal by the Latham Committee (the UK Government's funded review of the construction industry's procurement and contractual arrangements) that a Joint Liaison Committee, formed from the Conditions of Contract Standing Joint Committee (CCSJC) and the Joint Contracts Tribunal (JCT - the equivalent committee in the building industry) should re-structure existing standing contracts in order to develop a 'family of documents' around the New Engineering Contract. This 'family' would share a number of common features such as a compulsory adjudication system independent of contract administration, a general duty to conduct business fairly with specific requirements relating to payment and associated issues, clearly defined work stages including 'milestones' or other forms of activity schedule, and the pre-pricing of variations. It is envisaged that these two bodies from the civil engineering and building wings of the construction profession would merge to form a National Construction Contracts Council. It has also been proposed the title of the NEC should be changed to the New Construction Contract.

Probably the most important of the proposals is that relating to the appointment of an independent adjudicator who will be required to make a quick decision on each problem as it arises and about which he may well have no detailed knowledge. As noted by Totterdill (1994), a weakness in this procedure could arise from the lack of consultation time and the opportunity for the parties in conflict to resolve the problems between themselves. He suggests that a practical consequence of this is that there will be a short period of assisted negotiation, the adjudicator having developed into a conciliator as used in the ICE

6th Edition and the contract with contractor's design. A suggestion is that all contractual problems be referred to a disputes adviser, or a disputes review board for a large project, appointed at the beginning of a project. He/she/it would seek to gain the trust of the parties from the outset, would meet regularly with the people concerned, be informed about progress and any problems pending, and then advise on the most suitable procedure to avoid or resolve disputes. Totterdill accepts that the adviser might have to act in an adjudicator role, giving instant decisions to avoid delays to the works and cash flow to the contractor, or as a conciliator, perhaps during negotiation of the final account. He also sees the disputes adviser recommending the use of independent technical or legal specialists in order to prepare an opinion for the contracting parties.

The intention is to give the Latham Committee proposals statutory backing by a Construction Contracts Bill planned to be introduced in the 1995-96 Parliamentary session. The Bill would outlaw particular 'unfair' amendments to standard form contracts, such as: amendments or deletions to sections in a contract referring to 'times and conditions of payment, including the right to interest on late payment'; attempts to stop adjudication or not being bound by an adjudicator's decision; attempts to set-off or contra-charge without advance notification, providing a reason, and being prepared to put the matter to an adjudicator's decision; attempting to set-off in respect of any contract other than the contract in progress; and introducing pay-when-paid clauses even in bespoke forms. The Bill would also introduce mandatory trust funds to be used for paying off subcontractors should a main contractor fail or for paying the entire construction team should a client become insolvent.

In addition to the above measures the Bill also provides for latent defects insurance for a period of 10 years from completion of all new commercial, retail and industrial building work above a certain commercial value. The Committee also recommends that statutory backing be given to a DoE working party majority conclusions on liability such that liability be limited 'to a fair proportion of the plaintiff's loss, having regard to the relative degree of blame', and that liability be fixed at 10 years from practical completion.

2.9   GC/WORKS/1 (EDITION 3)

This form of contract, particularly related to public building and road works, is noted for completeness, but as far as can be ascertained has had little application in tunnelling works. It could be more flexible for general civil engineering works than, say, the ICE 6th Edition, in that the role of the Engineer (designated as project manager under the terms of the contract) is open to interpretation. The client can act also as Engineer or can appoint one, and can so vary the control he exerts over the construction work, according to the nature of the project. Reference may be made to Powell-Smith (1990).

2.10   ABRAHAMSON FORM OF CONTRACT

Construction lawyer Max Abrahamson launched his alternative form of contract in April 1992. The contract, which uses graphics to assist understanding, comprises 5000 words compared with 40 000 words used in more established standard forms. It is argued that this reduction enhances clarity, every word having only one meaning, and reduces disagreements. Lines and arrows link clauses in the contract within a clearly defined chronological order. Boxes are used by clients to fill in basic details such a key dates or monies due for payment on completion of work.

A feature of the new form of contract is that the Engineer's liability for work done by the contractor is reduced; he is only responsible for actively disapproving planned working methods and the contractor is given more responsibility. This contrasts with other contract forms where the Engineer is responsible for approving or disapproving design changes, and in which he is liable for mistakes.

Although several bodies have apparently used this new form of contract (Yorkshire Water, the Royal Hong Kong Jockey Club and the South African Electricity Board, according to the New Civil Engineer, 9 April 1992, p6), some reservations have been expressed about

its use on highly competitive projects and there seems to be some concern that it might appear to favour contractors putting in claims for variations on complex jobs.

2.11   FIDIC

It is relevant at this stage to refer, but only briefly, to the FIDIC Conditions of Contract for Works of Civil Engineering Construction (Fourth Edition), 1989. Reference may be made to Bunni (1991).

Clause 12.2 sets out procedures to be applied in the event that the contractor encounters 'changed conditions'. For a Clause 12 contractual claim to be successful, the contractor must demonstrate unforeseeability, consequential delay, quantifiable delay or additional cost; that is, he must demonstrate entitlement, causality and quantum (amount). The first step in the preparation of a claim is to establish the reasonableness of the unforeseen. There are two components to this: the reference conditions and the contractor's site inspections. The reference conditions are contained in the geological report specific to the site. Under Clause 11.1, the employer provides to the contractor 'such data on hydrological and sub-surface conditions as have been obtained by or on behalf of the employer, from investigations relevant to the works...', but the contractor must do his own interpretation. Clause 11.1 also requires the contractor to do his own site inspection. In general, fore-seeability tends to be quite deeply investigated for proof or disproof, but, as noted by Huse (1993), causality and quantum are relatively neglected.

Clause 44 deals with extension of time, Clause 47.1 provides for liquidated damages payment in the event of delay attributable to the contractor, and Clause 47.3 provides for the possibility of a bonus in the event of early completion. If delay can be demonstrated to have resulted solely from unforeseeable ground conditions, then an extension of time can be requested. If the contractor is not granted an extension, and he then accelerates his performance, he may, in some forms of contract, be entitled to seek the costs of such 'acceleration'. These cost may reflect losses from a necessarily revised programme.

Costs comprise those both directly incurred and those attributable to overheads and (potential) profit (loss). At the basis of a calculation for losses in tunnelling is a comparison of drive rates in difficult (unforeseeable) ground with those that the contractor believes would have been achieved had those adverse ground conditions not been present. In a total cost method of calculation the contractor takes his total cost of performance and subtracts from it the amount that he believes the project should have cost, perhaps claiming that the latter cost would have been less than the tender price. In respect of the overheads/office costs, Huse (1993) refers to the US Eichleay step-by-step formula that is given below. Excessive claims on the contract, even when supported by appropriate calculations, usually turn out to be counter-productive. The inevitable effect of this is to cause protracted disputes when in very many instances, given realistic expectations and goodwill on both sides, there could usually have been a negotiated settlement.

US Eichleay formula

(1) $\dfrac{\text{Contract billing}}{\text{Total billings for the contract period}}$ x   Total overhead for = Overhead allowable to contract

contract price   (Eichleay)

(2) $\dfrac{\text{Allowable overhead}}{\text{Days of performance}}$ =   Daily contract overhead (Eichleay)

(3)   Daily contract overhead   x   Number of days delay   =   Amount claimed (Eichleay)

3

INITIAL INVESTIGATION PHASE

3.1 CONDITIONS OF CONTRACT AND SPECIFICATION FOR SITE INVESTIGATION

The Institution of Civil Engineers published its Conditions of Contract for Ground Investigation in 1983 (Institution of Civil Engineers, 1983). These conditions, which were based on the ICE Conditions of Contract for Civil Engineering Works 5th Edition (Institution of Civil Engineers, 1973), since revised as the 6th Edition of the ICE Conditions of Contract (Institution of Civil Engineers, 1991a), have been discussed in various publications (Uff, 1983; Fullalove *et al*, 1983; Cottington and Akenhead, 1984). Two later specifications for ground investigation were published, one by the Department of Transport (Department of Transport, 1987) and one by the Institution of Civil Engineers (Institution of Civil Engineers, 1989). More recently (1 September 1992) the Institution of Civil Engineers has formulated its Specification for Ground Investigation, which points the way for future practice in the profession (Site Investigation Steering Group, 1993). This latter specification, discussed by Greenwood *et al*. (1995), formalises certain well-used terms, the definitions of which may not be strictly adhered to in the present book. For example, 'boring' included percussion and auger boring. 'Drilling' includes rotary drilling techniques. 'Trial pits', 'inspection pits', 'exploratory pits' are the terms used for unsupported excavations; 'observation pits' and trenches are supported for the purposes of personnel entry. Because no truly undisturbed sample can be retrieved from the ground, the term 'undisturbed sample' is replaced by the term 'open tube sample'.

The person charged with formulating the site investigation specification should be experienced in both tunnelling and site investigation works. He or she should attempt to foresee the technical and consequential contractual problems that could arise within the works contract, and that vision should condition the specification. Under

ICE Conditions of Contract that person would normally be the Engineer on behalf of the employer and he or she should adopt an active role during this stage of the project. However, the role of the Engineer is tending to come under deeper scrutiny on several grounds. One is the development of quality assurance (QA) systems by contractors, so reducing the function of the Engineer to that of supervisor rather than adjudicator (*see* the implications of the New Engineering Contract in Chapter 2). There is comment on QA in Supplementary Information 1 in Part Two of this book. Another is the use of quantity surveyors as project managers where the traditional front-end (pre-contract) design function is either low (in many instances, for example, tunnels) or where the form of contract is much more contractor participatory from the outset (*see* Chapter 2 above). A third relates to government constraints on cost overruns in public works contracts (particularly roadworks) which can be at odds with the impartial judgement of the Engineer in, for example, the granting of extra time for completion. (There are examples of the ICE 5th being modified to allow the Department of Transport maximum contract control by the Department assuming the role of both client and Engineer, or by limiting the independence of the Engineer through an obligation to seek client approval for some decisions.)

There does seem to be one area of potential conflict concerning responsibilities for the provision of ground information, specifically information on in-ground structures. As noted earlier in Chapter 2, under the 6th Edition of the ICE Conditions of Contract the employer is now responsible for providing information to the works contractor on the presence and location of pipes and cables. This requirement in the Conditions of Contract is designated as a corrigendum (correction) rather than an addendum (addition). According to the Specification for Ground Investigation (Site Investigation Steering Group, 1993), General

Requirements Clause 3.3 (which effectively replicates Clause 3.3 in the Specification for Ground Investigation published by the Institution of Civil Engineers in 1983), and the later note on this clause, the ground investigation contractor must acquire information on 'The positions of mains, services, drains, sewers and tunnels owned by statutory undertakers, public authorities and private individuals......' This requirement may be taken as being consistent with the requirement/recommendation (implied) on the ground investigation contractor in BS:5930, Appendix A, page 121 (British Standards Institution, 1981, under review) which states, under 'General information required for desk study, A.1 General land survey' the following: '(f) Indications of obstructions below ground'. (Responsibility is not clearly defined, nor is the meaning of 'obstructions', which could be natural or artificial, nor is it absolutely clear that the obstructions referred to in the Standard are those which could affect drilling progress rather than those or in addition to those which could affect the progress of the civil engineering works.) Thus in the case of the works contract the responsibility in this area is vested in the employer while in the ground investigation contract the responsibility for acquiring the same or at the very least similar information is placed on the (ground investigation) contractor. The position is also somewhat complicated by Clause 6 of the Conditions of Offer for Works of Ground Investigation issued by the Association of Ground Investigation Specialists (AGIS) which states that 'The client shall be responsible for producing a plan or giving in writing, the presence of all known underground services, structures etc. The specialist shall take all reasonable care to avoid damage, but cannot be responsible for any damage to unknown services below the depth indicated.' The ground investigation specialist (contractor) will often make it clear that his conditions of offer are those of the AGIS, but the full significance of these Conditions, and especially this clause, may escape the scrutiny of the client until something goes wrong, typically the disruption of an electricity supply caused by the striking of an unknown power cable on private land during the ground investigation.

The three main areas of concern in respect of liability for the ground investigation contractor would appear to be contractual claims, safety against injury during the work, and payment for repairs to any damaged installation caused by the contractor's operations. Safety and health matters are well covered by existing legislation (*see,* for example, Section 6.1), and the contractor must ensure that he is fully aware of his duty of care responsibilities to staff on the contract and others that could be affected by the work. A major difficulty stemming from the likely presence of services is the insurance cost of claims for repair if they are intercepted and the operational costs of acquiring information on and proving the location of those services. Although the requirement and financial responsibility for inspection pits to prove the absence of potentially interceptable services are somewhat unclear, their cost is small compared with the cost of repairs to services and the cost of injury to persons. Typically, the cost of a hand-dug pit to a depth of 1.2m would be £20 to £30 at the time of writing, and to this must be added the cost of any break-out of hard paving and reinstatement.

There does seem to be some basis for contending that responsibility for obtaining information on buried services for the use of the works contractor *and* the ground investigation contractor would best be assigned to the client (employer) as part of the Engineer's searches for his design of the civil engineering works.

Whereas the Engineer requires ground information for the design of the permanent works, the civil engineering contractor's expertise is related to the methods and equipment necessary to secure the successful completion of the works to price (or below price) and on time (in order to avoid the imposition of liquidated damages). For equipment intensive projects, especially, for example, microtunnelling where the actual operation is remote from the control function and anomalous ground - typically the presence of large boulders - could severely affect progress, the style of the site investigation for that contract must be matched to the specific requirements of the tunnelling work, so ruling out early site investigations in isolation from any planning and design of the works. The works contractor is also ultimately responsible for the safety of the workers under the construction contract. It may be argued that since the experienced contractor tendering for the tunnelling work should know what ground information he needs in order to price his bid economically, responsibility for the investigation should logically be vested in him. Early selection of potential tenderers would be required, but because each of the tendering contractors would no doubt have different perceptions as to requirements, it could be difficult in practice to implement such a transfer of responsibility. In any case a further and obvious counter-argument is that the works should/can neither be designed nor billed before the results of the site investigation are known. However, proponents of the view that the contractor should be responsible for specifying the site investigation would state that if such responsibility is to remain vested in the employer (via his representative) then equivalent logic dictates that the employer should also be responsible for the design of the temporary works and specification of the method of working (Ackers, 1989). They would also argue that the subsoil information needed for design (and billing?) of the works is rarely the same in detail and extent as that required for construction, and that in only very few cases would erroneous information on the ground conditions, had it been apparent at the outset, have altered the original overall design. Under standard procedures and forms of contract (for example, Clause 11(1) of the ICE 6th which requires the employer to make available such site information, including the information on the presence and location of cables and pipes discussed above) it is difficult to envisage how responsibility for the site investigation could realistically be transferred to the contractor. As a half-way measure, the bidding contractors could be required, as a pre-condition for the contract and somewhat in the style of a Clause 14 method statement (ICE Conditions of Contract, 1991), to employ site investigation specialists who would provide a contractor's interpretation of the ground to be tunnelled. Such an interpretation would be issued actively to the employer for his own use or it could be used passively, unseen by the employer until a contractual claim is lodged. The difficulty with this system is that it could be expensive and might produce a different site investigation opinion from each contractor. Also, no doubt, the contractors would argue that the site investigation specialists are insufficiently familiar with the requirements of a works contract to offer a full interpretation of the site in

the context of the method of working. For this reason, the responsible, experienced tunnelling contractor will prefer to operate on the basis of a quality factual report and forgo an interpretation of the ground conditions however expertly assembled. Another possibility is for a mechanism to be established whereby a site investigation design (incorporating safeguards to ensure a cost-effective operation) could be agreed by all those tunnelling contractors who would later be invited to tender for the construction, and then be implemented by a specialist site investigation contractor, the ultimate objective perhaps being the express removal of the relevance of Clause 12 of the ICE Conditions of Contract 6th Edition (1991) to those geological conditions that are the subject of ground investigations but not the statement in Clause 12 relating to physical conditions other than those 'physical conditions' in the ground which are specified in Clause 11(1), that is, cables and pipes, and which are not the responsibility of the contractor. The experience of the tunnelling contractor would thereby be utilised in framing an investigation before the inception of the main works contract, perhaps providing data in support of an unusual method of construction that could be proposed for the Engineer's consideration prior to the design of the main works, and with more of the responsibility and risk being tilted in the direction of the Engineer and the employer, respectively. Again, in practice, it is quite unlikely that any agreement for cooperation between contractors in this respect would be secured. Even with a system such as currently operates in the UK, which does not charge the works contractor with responsibility for the site investigation, the needs of the contractor should always be at the forefront of concern during this phase of the work. Many would feel this concern to be best addressed by recognising that Clause 12 is such a fundamental element in the ICE Conditions of Contract that its removal would seriously disturb the balance of risk between the contracting parties that the Conditions seek to establish.

Example: On one trenched and towed long sea sewage outfall the project Promoter devoted a considerable sum of money to the site investigation (around 13% of the civils works contract). This sum arose partly from the adverse sea conditions during the ground investigation work. The Promoter then required the contractor to accept 'Clause 12 risks' (the conditions of contract being unique to the contract and not containing the specific ICE Clause 12) and agreed to pay a further substantial sum to him (around 5% of the contract value) to do so. In fact, it was the opinion of the Engineer during the currency of the contract, and subsequently, that the ground conditions were not substantially different from those that could be inferred from the site investigation information and that any claim based on that information would have been unlikely to have succeeded. Thus, from one perspective, the employer might be said to have wasted his 5%, but, from another, he had quarantined himself from the risk of a claim which could have been very substantial due to the necessary use of expensive marine plant.

In English law there is the question of tort and the matter of owing a duty in law independent of (apart from) contract to ascertain, for example, whether ground can be safely tunnelled. Courts have recently allowed greater scope for actions to be pursued in tort, which has had the effect of benefiting plaintiffs by extending the period of liability, facilitating suits against third parties and, in some circumstances, widening the scope for damages (Uff *et al.*, 1989). Tort law does not allow recovery of a purely economic loss but provides compensation for physical harm to people or property together with financial loss suffered as a result. The consequence of the law, as it now stands, is that, except for exceptional circumstances, the employer will take an unacceptable risk if he fails to ensure that an adequate site investigation is carried out. The question of actionable negligence could arise if Engineers initiate or recommend inadequate site investigations. Consultants can be liable in tort if their expertise is relied on by someone even if they did not have a contract with the person. (This matter is addressed later in the book when the question of general notes to accompany the ground investigation report is discussed.) On the other hand, working relationships between parties, such as between engineers and their clients, will be deemed effectively to constitute a contract, whether by a written agreement or

merely by the existence of an offer of services, an acceptance and a consideration. There is therefore a liability to a client for damages arising out of any breach of contract. Reference may be made to Cottington and Akenhead (1984) for information on the legal aspects of site investigation.

Internationally, and certainly in the UK, there is an increasing awareness of the need for and the value of quality assurance (QA) with respect to manufacture and services - *see* British Standards Institution (1994a: BS 5750; 1992c: BS 7850). In England, CIRIA (Construction Industry Research and Information Association) has produced a report (Power, 1985) on quality assurance in civil engineering and guidance notes have been written by Fountain (1991) - *see* Supplementary Information 1 in Part Two of the book. Similarly, FCEC (Federation of Civil Engineering Contractors) has published guidance on the adoption of BS 5750 in formulating site quality management systems, and reference should also be made to the Quality System Supplements Numbers 5000 and 8372 produced by Lloyds Register Quality Assurance Ltd (1989a, b) for assistance in interpretation of the assessment standard. The quality assessment schedule to be used in conjunction with Part 1 of BS 5750 (British Standards Institution, 1994a) - *see* Supplementary Information 1 in Part Two of this book - is concerned with assessing companies involved in ground investigation and foundations, and includes a discussion of the qualifications, requirements and general duties of the ground investigation specialist and any subcontractor(s). BS 5750 is triple numbered as an international and European document (ISO 9000 and EN 29000). Its projection as a substantive document ensures that overall quality in the site investigation industry will be raised by the promotion of common standards. Indeed, there has been growing pressure in Britain for a single national specification on site investigation, to be based on the current ICE specification and augmented to bring in the philosophy of Eurocode 7, and which would define limits of accuracy, quality of materials, equipment to be used, and operational procedures. The specification would also expressly address the problem of contaminated land which is assuming increasing significance in Europe. Many if not most in the industry will feel that this need has been satisfied by the 4-volume publication on Site Investigation in Construction by the Site investigation Steering Group (1993) under the auspices of the Institution of Civil Engineers.

It is important to understand the difference between quality control, a process which should take place during construction, and quality assurance, which is built into both the design process and indeed the structure of an organisation. The responsibility for quality assurance, is best placed with one of the most senior executives of the company, his or her status as quality manager ensuring that decisions in this area carry full weight, and in practice its implementation should rest with a department in the company independent of the construction (investigation) process, much in the same way as a company or governing authority (local or national) internal audit should operate. So far as BS 5750 is concerned, it is the quality of the management and efficiency of the entire production/operation system that is being controlled, checked and audited. In the UK construction field it is argued that much more responsibility for quality assurance should be vested in the contractor (both for civils work and for site investigation) than is the case under the current contractual conditions. Only then could the requirements for stringent on-site quality control procedures be lowered without lowering the quality of the finished job. It is indicated in Chapter 2 above that there is an expressed requirement for contractor quality assurance under Clause 8(3) of the ICE Design and Construct Conditions of Contract.

There can be a problem when quality assured contractors are working with resident engineers employed by a non-quality assured consultant. Quality assurance systems can come into direct conflict with contractual requirements, especially when a quality assured contractor does not agree with instructions given by a non-quality assured supervising engineer of perhaps limited site experience. For their part, consultants are tending to be less enthusiastic about site supervision, questioning its profitability and preferring to withdraw into the design function. A reduced Engineer role would see a reduction in his legal responsibility for defects shown to have resulted from inadequate supervision.

It has been suggested that quality assurance systems cannot operate properly within

construction projects if the Engineer's supervisory role under the ICE Conditions of Contract, 6th (and 5th) Edition 1991, is maintained. With an increased use of QA by contractors it may be considered that the function of the Engineer should be limited to occasional site checks to monitor contractors' QA systems, evaluating completed work, and making decisions on the cost and timing of work. The traditional role of the Engineer as independent client's representative is now seen to be becoming outdated and the contractual environment offered by the New Engineering Contract (1993), discussed earlier in Chapter 2, is considered by many to be more appropriate to current needs.

Larger organisations concerned with letting ground engineering contracts, and larger ground investigation contractors, should consider making provision for distilling their experiences into in-house manuals which highlight technical and contractual problems that have arisen during and after the currency of the contracts and which would be required reading by its engineers responsible for the specification, letting, supervision and reporting of site investigations.

Example: On one tunnelling contract there was a substantial claim for extra payment based on rock at the tunnel face being much stronger than had been revealed by the site investigation. Samples taken from the face showed unconfined compressive strengths of one lithology ranging up to 180MPa whereas the unconfined compressive strengths of the stratigraphical sequence as a whole had not been expected much to exceed 50MPa. On re-examination of the site investigation core samples, very strong rock was indeed revealed at horizons appropriate to the tunnel face level but for some reason it had been missed at the testing stage. However, the significant point is that this very strong rock had also been encountered on an earlier contract and had given rise to a successful contractual claim. The circumstances of this earlier case had not been fed back through the organisation and so the resulting financial penalty was great.

Unfortunately, the rate of change of staff in organisations and the ruthless attack on 'overheads' make such a logical and desirable objective difficult to achieve in practice.

3.2   PROCUREMENT OF SITE INVESTIGATION

Site investigation contractors may tend to be selected by competitive tender on the basis of price alone, or sometimes simply for convenience if a locally based drilling crew happen to be available for hire. This method of engagement will in many cases not give the best value for money. There is little scope for reducing realistic prices without seriously impairing the quality of the work and without incurring consequential increases in the final cost of the tunnelling works. Only those site investigation contractors having experience of tunnel site investigations, an adequate work force capable of providing quality field work and supervision, and with geotechnical consulting capabilities, should be considered for inclusion on a select tender list. Alternatively, there is a case for negotiating site investigation contracts in order to obtain specialist experience for specific jobs, or a specialist geotechnical adviser could be employed to supervise and interpret the site investigation. This latter point has been considered by Uff and Clayton (1986) and expanded as a recommendation in volume 2 ('Planning, Procurement and Quality Management') of the Site Investigation Steering Group (1993).

Public bodies and other authorities having large and continuing programmes of site investigation work may decide to adopt annual term contracts for site investigation work. A term contract may take the form of a schedule of unit rates for the provision of plant and personnel, and for which selected experienced site investigation contractors are invited to tender. In contrast to a bill of quantities form of contract with a specified method of measurement, a schedule of rates contract is probably a little looser but offers more flexibility for variation. On the other hand, this form is not favoured by the Site Investigation Steering Group (1993) which stated in its draft document 'Specification for Ground Investigation with Bill of Quantities' that 'Where there is a reasonable probability that an item will be required a realistic assessment of that quantity should be made and entered in the bill of quantities'.

Normally in a schedule of rates contract a geographical area of operation will be specified, and the availability of manpower and drilling rigs at short notice must be assured. A basic outline of the expected quantity of work will normally be given but in many cases the winning contractor can expect to undertake a greater work load. A typical contract will provide for trial pit excavations, boring, sampling, *in situ* testing and laboratory tests, together with the provision of geotechnical consulting services. When specifying test programmes it should always be remembered that the results of a site investigation could be influenced by the method of testing that is adopted.

Just as in tunnel contracting, where specialist elements of the work and sometimes all the constructional work may be subcontracted, so in site investigation some of the work may be let to another firm under 'subcontract'. Typically, this could comprise the laboratory testing work. The individual nature of the letting will determine the legal status of the subcontract with respect to liability in contract and in tort.

In America there are so-called 'flow-down' clauses and 'pay-when-paid' clauses. The 'flow-down' clause provides that the subcontractor will assume towards the contractor all the duties, obligations and liabilities which the contractor has assumed towards the employer. Main contract clauses generally contain many terms and conditions, and it would be unreasonable to expect all of these to be imposed upon the subcontractor. What really matters, and what would be taken into consideration by a court, is the *extent* to which a flow-down clause imposes obligations from the main contract upon the subcontractor. It is usually the case that the obvious administrative and procedural aspects of a main contract would not be considered as flowing down to a subcontractor, but even then a 'work requirement' cannot always be distinguished from an administrative or procedural matter.

In Britain, an employer, contractor and subcontractor are in a position of contractual privity and can look only to those with whom they are in direct contract for any contractual remedy. However, an employer can establish a contractual relationship by direct warranty, between himself and a subcontractor, in the form of an express direct warranty or as a warranty of quality. In an express direct warranty the subcontractor, providing a written warranty for the main contractor, normally agrees to perform his work in such a manner that the main contract will not be breached, and if he is in breach of such an obligation then he may be sued by the employer as if he were the main contractor. Provision of such a warranty would be a requirement under the main contract. The employer needs to provide a consideration for the warranty. This may simply take the form of his approval, but there is more security for the employer if a deed is executed or if he also provides a nominal fee to the subcontractor to ensure that sufficient consideration has been provided. In a warranty of quality there is reasonable assumption that a subcontractor is giving an undertaking that will be acted upon by the employer and there is also sufficient consideration to support the existence of a contract.

Although not usually expressed in a site investigation subcontract, and often not applied to site investigation contracts generally, the status of a subcontractor vis-à-vis liquidated damages can arise. 'Liquidated damages' are sums, calculated at a pre-agreed rate, allowable or payable by a contractor to the employer if in the event the contractor fails to complete the work (works) by the original or any extended date for completion. These represent the maximum claimable damages, but they must not be set as to penalise a breach of the contract excessively otherwise they would be seen by a court as a penalty clause to be treated as unliquidated and to be assessed by the court. Experience in America indicates that the wording of subcontracts makes the subcontractor liable for delay in completion of the work (works) irrespective of the extent to which he contributed to the delay, and courts in that country have tended to enforce such clauses. Where they have not been enforced, the reasons have been different; for example, that the clause constitutes a penalty and as such is unenforceable. There are distinctions between 'damages' and a 'penalty'. 'Pay-when-paid' clauses provide for the subcontractor to be paid for his work only when the main contractor has been paid for that work by the employer. However, should the employer fail to pay the contractor, courts in America have accepted the argument that, in the absence of any contractual indication to the contrary, the risk for

any such failure to pay should not be borne by the subcontractor. The general rule is that non-payment by the employer does not excuse non-payment by the contractor to the subcontractor. On the other hand, if there is an express term in the subcontract that payment by the employer to the contractor is condition precedent to the contractor's obligation to pay the subcontractor, the latter must then bear the risk of such non-payment. Even without the condition precedent wording, the courts have on occasions ruled that the wording of the clause clearly reflected the intention of a contractor to move the risk of employer non-payment on to the subcontractor.

The sums of money involved in site investigation may not be large in the context of construction works generally, but they could be substantial with respect to the continued well-being of a site investigation firm. It is important to be aware of the responsibilities, liabilities and cash flow implications within the framework of a subcontractual role.

3.3 SCOPE OF INVESTIGATION

Design considerations to be taken into account in determining the scope of the investigation are discussed in Dumbleton and West (1974, 1976); British Standards Institution (1981); West *et al.* (1981); Weltman and Head (1983). Although procedures for carrying out investigation work tend to be formalised and codified by national standards, constituting safe practice and providing a firm defence in the event of disputes under contract law, they are prone to having the undesirable effect of stifling innovation.

There is need for a clear specification of a construction requirement before the relevant questions of the ground can be asked. Only then can the site investigation be formulated with a view to providing the information essential for the design of the works. This information, sometimes conducted in a step-by-step feedback manner, may lead to a re-appraisal and perhaps re-location of the project. Probabilistic methods of site investigation analysis offering an iterative approach to the definition of the ground have been discussed by Attewell and Farmer (1976), Norgrove *et al.* (1979) and Attewell (1988a). Some tunnelling techniques are more sensitive to the nature of the ground than others and will therefore require a higher level of site investigation information both for their specification and implementation. Two examples of this are the use of slurry and earth pressure balance shields (refer to Section 4.3.3 on boulders and 5.4 on compressed air) and the adoption of the New Austrian Tunnelling Method (*see* Brown, 1981a and Muller, 1990; and Section 6.8).

The following items constitute particularly important requirements for all tunnel investigations:

(a) The field work should be carried out under the full-time supervision of a ground engineering specialist - an engineering geologist or a geotechnical engineer - to allow an initial evaluation of the results and to suggest any modifications to the scope of the investigation. (The site investigation contract needs to be drawn up in such a way as to allow this flexibility.) Drillers employed on the contract should normally hold a certificate of competency for percussive boring or rotary drilling from the British Drilling Association or an equivalent body in the European Community, but there will often be provision for non-accredited drillers to be employed provided that the drilling rigs are supervised by an experienced geotechnical engineer. A tentative geological section or sections, based on the ground information as revealed, should be drawn up at this stage and updated during the course of the investigation.

(b) Under the ICE Site Investigation Conditions of Contract (Blue Book) Clause 8(3) (Institution of Civil Engineers, 1983), the Engineer's approval of testing laboratories is needed. Seemingly there is no mandatory specification under BS 5750, which states in Part 1 Section 4.6 (Operating Procedures) 'The use of a laboratory holding NAMAS accreditation for the testing required will provide assurance that those parameters related to testing are followed by the Laboratory.' Although site investigations for tunnelling works do not usually require the full panoply of modern techniques to be employed, the sampling and testing must be very well performed and should be specific to the problems of tunnelling (including the sinking of shafts). For example, samples of the strata should be taken at every shaft position and tests performed on the material from ground surface down to the

shaft bottom. Tests should be performed on samples taken from the tunnel horizon (the planned face section) and at least one diameter above and below the tunnel soffit and invert, respectively. This extended vertical 'snapshot' of the tunnel face geology not only provides a better understanding of potential problems but also allows for any later changes in the planned tunnel level. The importance of accurately setting out the position - and recording the ground surface level - of each borehole from a temporary bench mark must be stressed. It is important that the records of any work carried out to BS 5750 standards should be made available to the Engineer on his request.

(c) Emphasis should always be placed on obtaining information as accurately as possible on the presence of groundwater, piezometric pressures and their possible variations with time, from which inflows into the tunnel can be estimated. There should be an awareness of the type of ground, geological structure (for example fault lines) and construction practice that could promote gas generation, transmission and solution under pressure into groundwater, and particularly the effects of a drop in the prevailing air pressure causing methane gas release at explosive mixtures. Changes in the groundwater table, during or after construction, can also liberate gas inside a tunnel or shaft. (Prediction of the likely presence of methane gas would require the specification of flameproof and/or intrinsically safe equipment in the tunnel, with sensors being placed along the tunnel to give an evacuation alarm at a 1% concentration and to provide for automatic isolation of all machinery at a 2% concentration.) There should be an awareness of the possible presence of radon gas in certain parts of the country, although the short exposure times experienced by tunnel workers are unlikely to create a problem. In the context of this gas, reference may be made to Moses *et al.* (1960), Evans *et al.* (1981), Kathren (1984), Eisenbud (1987), Hughes *et al.* (1989) and Gates and Gundersen (1992). Fortunately, radon emissions are not significantly affected by atmospheric pressure changes (Baver, 1956). Less obvious, perhaps, is the fact that heavy rainfall when tunnelling in chalk can promote the production of CO_2 which may tend to make the operatives light-headed and impair judgement.

(There should be provision in the tunnel for alarms to sound when the O_2 content of the air falls below 22.5%.) The position of the water table (the likely water pressure at the tunnel face and whether the pressure will be artesian or sub-artesian), the possible presence of perched water tables, and the ground permeability are essential elements of information that assist the Engineer to formulate the contract specification and the contractors to price the work. Reference may be made to Supplementary Information 2 in Part Two of the book for some further comment on the gases that could be encountered when tunnelling in coal mining areas.

(d) In urban areas, before any drilling or trial pitting takes place, full account has needed to be taken of the requirements of PUSWA (Public Utilities and Street Works Act, 1950). This Act came into operation on 26 April 1951, but since then minor amendments have taken account of subsequent highways legislation and the Local Government Act of 1972. Professor Michael Horne was later appointed by the Department of Transport to review PUSWA with a view to producing a new model agreement and specification (the 1985 Horne Report). Although there was some evidence of political pressures militating against its implementation, and some of its recommendations were uncomfortably placed for being dropped (for example, utilities to incur possible fines up to £15 million/year as an incentive to complete work quickly), a new Act was scheduled for implementation on 4 January 1993. Under the new Roads and Street Works Act 1991, which has been incorporated into the ICE Conditions of Contract, 6th Edition as a revised Clause 27, utilities (statutory undertakers) should assume full responsibility for excavation and reinstatements of all their work to new national specifications and performance standards. For instance, reinstatements will need to be guaranteed for not less than two years and a specified material must be used to backfill the trench. Other expectations concern properly trained and certificated supervision on site, the application of a 'need to know' principle so that the 39 notices prescribed under PUSWA can be removed, a central data store of information to improve efficiency, better mapping and recording of underground plant, advance coordination of major projects, and generally better planning and

consultation on diversionary works and cost-sharing arrangements for diversionary works. The Bill, outlined in the Queen's speech in November 1990 and having had its first and second readings in the House of Lords, received its Royal Assent in May 1991 and had been expected to be implemented in July 1992 if the controversial private road proposals in the Bill could have been removed. It is expected that the requirements imposed on the utilities with respect to excavation and reinstatement will apply equally to ground investigation. Where buildings and/or buried services could be at risk during the tunnelling works, consideration should also be given to estimating the temporary and permanent components of unavoidable (in the sense of Clause 22(2)(d) of the ICE Conditions of Contract, 1991) surface settlements caused by tunnelling in soft ground (Attewell *et al.*, 1986).

It is important to stress that the person responsible for implementing the ground investigation should obtain written instructions as to where to drill. These instructions must contain a map, of sufficient detail, specifying borehole locations. Verbal on-site instructions from the Engineer/Project Supervisor for, say, additional holes are insufficient because if anything goes wrong and an underground service is disrupted during drilling operations the responsibility for the problem may well be assigned to (the insurers of) the ground investigation company even though logic would indicate that the latter would not drill without an instruction, in whatever form, from the former. The ground investigation contractor must proceed with the utmost caution both in practice and in respect of instructions when drilling on private land where accurate records, and indeed records at all, of underground services may not exist. Reference should be made to the discussion on this subject earlier in this chapter of the book.

(e) If drilling or trial pitting is to take place in land that could be contaminated then reference should be made on health and safety matters (the Health and Safety at Work etc Act 1974; Health and Safety Commission, 1992 which came into force on 1 January 1993) and to the procedures given in the draft British Standard for Development DD 175 'Code of practice for the identification of potentially contaminated land and its investigation' (British Standards

Institution, 1988). Reference should also be made to the British Drilling Association Guidance Notes for the safe drilling of landfills and contaminated land (1992) and note should be made of the requirement for COSHH (Control of Substances Hazardous to Health Regulations, 1988) statements to be prepared to cover all investigation activities. In addition to site personnel being equipped with suitable clothing, and perhaps safety lines and retrieval harnesses if working in trial/inspection pits/trenches or near suspected old shafts, there are several other factors that need to be taken into account if there is a suspicion of contamination:

- Materials in the ground could be toxic, corrosive, or inflammable, or a combination of these. The presence of toxic metals should always be suspected on sites known to have a previous history of use as scrapyards, smelters, foundries, chemical works, and on many other old industrial sites. One site may contain toxic residues from several different industries.
- Dusts and carcinogenic fibres may be present. Asbestos is particularly hazardous, and it is less appropriate to use trial pits in the presence of such a contaminant. Face masks may need to be worn in the area.
- Precautions should be taken if there is any possibility of lead compounds being inhaled or ingested.
- Tarry materials may contain carcinogenic compounds, and so skin contact must be avoided with both liquid tars and more viscous tars present in powder form.
- There is always the potential for infection by pathogenic micro-organisms. Personnel might need to be given anti-tetanus injections. A combination of sewage and rats could lead to a risk of Weil's disease, and so there should be no skin contact with water from which there is any possibility of infection. Rubber gloves should be worn.
- Land that might previously have been used for tanneries or for the burial of diseased animal carcasses should carry a suspicion of anthrax, and so enquiries should be made of the local public health officer.
- Any suspicions of radioactivity must precipitate a close-down of any further site investigation until clearance has been given on the basis of specialist advice and confirmed by the local environmental health officer.
- The presence of any carbonaceous material, assisted perhaps by methane gas, can give rise to combustion which in turn could lead to the formation of cavities below ground level.

There is a minimum amount of site investigation required on health and safety grounds. Without such input a project cannot reasonably proceed. As a guide, the cost of the investigation for tunnel projects would typically range from 0.5 to 2.5% of the total contract value (Attewell and Norgrove, 1984b), possibly increasing to 5% for small value projects or where ground conditions for tunnelling are expected to be particularly difficult or for specialised investigations such as for offshore schemes. Uff and Clayton (1986) have suggested 1% for site investigations generally. It would be expected that the cost of a site investigation should be at least recouped by an equivalent reduction in total project costs, either as a result of improved design or through a reduction in contractual claims arising from unforeseen ground conditions (Institution of Civil Engineers, 1991c).

3.4   PRELIMINARY SITE APPRAISAL

It is important that an adequate appraisal of available information should be undertaken in order to assess the geological and general ground conditions and to establish any earlier uses of the site together with the locations of any old features which could influence construction (Dumbleton and West, 1976; British Standards Institution, 1981; West *et al.*, 1981). This preliminary appraisal is important for proper planning of the detailed site investigation. The preliminary appraisal, involving both a study of available documentation and an inspection of the site, could involve all or some of the following items:

- Examination of all editions of topographic (Ordnance Survey) maps, such as 1:2500 or 25 inches to 1 mile, relevant to the line of the tunnel. These maps may show important past features that could affect construction progress and long-term stability of the tunnel and its surroundings - evidence of earlier human activity, infilled ponds, clay/sand/gravel pits, quarries, mine shafts, original topography and drainage beneath tips, made ground and embankments, changes in stream and river courses, coastal erosion and deposition, the liability to flooding from river, lake and sea, changes in landslip areas, and so on. Place and street names could provide an indication of ground conditions; for example, 'Springwell Road', 'Quarry Heads Lane', 'Gravel Pit Terrace'. The number of the map or maps must be recorded for reference in the site investigation report(s).

• Examination of geological maps (solid and drift) and memoirs, where available, to provide as clear a picture as possible of the local geology. Full references to the maps and memoirs will be needed in the site investigation report(s). Care must be taken with terminology, which may only express a dominant lithology of a stratified sequence comprising several lithologies. For example, London Clay consists not only of the characteristic overconsolidated clay, but there are also nodular layers and hard limestone beds in places. Again, the Lower Lias clay shale of the Jurassic in Cleveland includes not only stiff clay but also intercalated beds of strong limestone seen as scars on the foreshore at Redcar. The contractual implications of misunderstanding such terminologies could be severe.

Special attention must be paid to the presence and attitudes of faults, to buried valleys, igneous intrusions, aquifers, rock jointing and cambering which could affect both the excavation process and the support of the tunnel. Faults are of major concern in tunnelling, especially if associated with zones of fractured material, with open (perhaps water-filled) fissures, and with fissures filled with soft gouge material under water pressure. Investigation for the presence of faults must not merely be limited to the ground that is considered likely to affect the construction but must also be extended to adjacent sites, otherwise negligence could be attributed to the person or persons that carried out the investigation (*Batty* v *Metropolitan Property Realisations Ltd* [1978] QB 554).

Certain features may not be detectable from maps, surface observation or by a later ground investigation drilling programme, but their presence may, to some extent, be inferred from the nature of the general geology. A typical example is the occurrence of isolated voids which, in certain parts of the world such as Greece and the former Yugoslavia, would be expected in karstic terrains. The importance of examining all available exposures at this early stage cannot be over-emphasised, but such exposures will not necessarily accurately reflect what is below ground level. For example, as noted by Wood and Kirkland (1987), in arid environments the behaviour of saturated material may be totally unlike its desiccated crust. Old infilled quarries may also be shown on geological maps. In Britain, discussions with the British Geological Survey located at Keyworth near Nottingham, and with other geologists, always accessible in appropriate institutions of higher education, may be useful, and there may also be helpful information in university theses and dissertations.

• Examination of aerial photographs which may show important geological features such as linear traces, typically bedding, seam outcrops, and faults, which could represent difficult-to-support and excavate, heavily fractured shear zones needing to be drilled to prove their existence, together with useful information on past developments such as old shafts. Oblique air photographs, making use of long shadows in suitable sunlight, can reveal subtle landscape details in areas of subdued topography. Light aircraft, helicopters, or even radio-controlled model aircraft can be used for this purpose. Air photographs are generally available in Britain from a central registry (Department of the Environment and the Welsh Office) at a scale of 1:25 000 (30km^2) but conventional black and white vertical air photographs taken for topographic survey and mapping purposes will usually employ a scale of 1:10 000 or 1:3000. Stereopairs can be particularly helpful in the hands of a trained observer. Infrared colour photographs have been used to detect infilled swallow holes in limestone areas and also springs and seepages on hillsides (West, 1986). Use of other types of air photography and imagery have been discussed by Amos *et al.* (1986). One point seems to be emphasised in the literature: it is preferable to undertake the desk study of an area and the study of any air photographs that may be available before performing the site reconnaissance (*see* (f) below), and then to repeat the review of the air photographs.

• Acquisition of information on old mine workings, old shafts and old opencast mineworkings from abandonment plans. Such obstructions could lead to ground instability and water hazards, and if not identified and suitably treated would generate contractual claims during the actual tunnelling works. All old shafts and adits may not be located on old plans. Many may have been abandoned and filled in, and/or covered up by needling in to the side walls just below ground surface an insubstantial wooden raft that may have rotted with age and thereby have allowed much easier surface water penetration. If infilled, the original fill will have compacted/consolidated to leave a void above it which would be particularly hazardous if intercepted by a tunnel face. It is also important to realise that earlier mineral extractions below tunnel level will have generated forward and side projections of tensile strain (Boden, 1969) that could severely affect the tunnelling works, particularly if under water, by opening up joint systems or reactivating movements on geological faults.

• In some instances, advertisements placed in local newspapers requesting information on such unrecorded features as old mine shafts and old well shafts. People whose forebears have been in the area for many decades can often provide valuable information.

• A site reconnaissance, which involves examination of relevant soil and rock exposures in the general area of the tunnelling works. In the vicinity of the tunnel line or alternative tunnel lines, special attention should be paid to the possible presence of faults (which may be seen or be inferred at outcrop), the presence of springs and poorly drained areas, swallow holes if the earlier map studies have suggested that the geological conditions could be conducive to such features, depressions generally in the ground, evidence of made ground, and changes in the appearance of crops and vegetation cover from which the subsoil conditions might be deduced through experienced interpretation. In some areas, geomorphological mapping may be of assistance. In the case of soils it is especially important to have some idea of the degree of cohesion or granularity, whether it is normally consolidated or overconsolidated, what are its

compressibility, excavatability, and self-support capacity, and if a till, the distribution of cobble/boulder material in it and, if possible, by examination of suitable exposures, the boulder sizes. Note must be made of any evidence of slope instabilities (hummocky ground, tilting trees, displaced fences, kerbs, drains, and so on. It might be possible to interpret any cracking, tilting and settlement in structures above the planned tunnel line in terms of the foundation geology, one simple example, after Attewell and Farmer (1976) being outlined in Supplementary Information 3 in Part Two of the book. Examination of such features is necessary anyway because it could otherwise be claimed that the tunnelling operations had initiated them, not merely exacerbated them. Careful examination must be made of any exposed interfaces between till and bedrock if tunnelling is to take place at this level. Movement of the ice sheet may have rotated bedrock blocks by exploiting the joint sets, so creating an irregular rockhead that could be misinterpreted by examination of the borehole evidence. Note must also be made of access routes for drilling rigs (need or otherwise for improving the ground; widths of fences), locations and heights of overhead cables, and the locations of buried services. Confirmation will also be required that the actual site area will remain free from flooding during the winter season.

• Enquiries to all organisations that might have carried out site investigations, tunnelling contracts or excavation works in the area. Of statutory importance are the enquiries to undertakers such as the water, gas, electricity, and telephone companies/authorities. Reference may be made to the British Geological Survey of borehole records, but, as noted in Section 7.2 below, there is as yet no national data base although in Britain the Association of Geotechnical Specialists is formulating a standard for data interchange. Notwithstanding the availability and provision of such information, there must be no absolute reliance on it. The person who accepts it must exercise reasonable skill and care in assessing its relevance and quality and will be deemed negligent if he fails to do so.

• Location of old landfill areas through which

or near which the tunnel will pass (*see* Sections 4.3.4 and 6.6). Tunnelling through filled ground, particularly ground comprising non-inert fill, will be a last resort, the aim of the tunnel designer being to avoid such hazards and to comply with the UK Code of Practice for Safety in Tunnelling in the Construction Industry (British Standards Institution, 1990a), the approved Code of Practice concerning Safety and Health which came into force on 1 January 1993, and European regulations concerning Health and Safety. It will be necessary at this early stage of the investigation to determine the nature of any toxicity such as might arise from factory chemical waste or even asbestos, and which would require the use of special ventilation, breathing equipment and special clothing in a tunnel passing through the fill and perhaps in a tunnel in the proximity of such fill. The Control of Substances Hazardous to Health Regulations (Health and Safety Executive, 1988a, b), which came into force on 1 October 1989, will be especially relevant in these circumstances. A substance hazardous to health is one that has been classified as very toxic, toxic, harmful, corrosive or irritant. Also included in the definition are micro-organisms, any kind of dust when present as a substantial concentration in air, and any other substance not covered by the above definitions that creates a comparable health hazard. There is a discussion on COSHH in Kirby and Morris (1990). Any old ash could absorb oxygen from the tunnel air, and from domestic waste there is a methane gas hazard that would require the specification of flameproof machinery and electrical panels. The age of the fill also needs to be known since this will determine its settlement characteristics should a tunnel be driven through it. Because of the low compacted density of much industrial and domestic waste, a careful check needs to be made on its capacity as a foundation for the lined-out tunnel.

Overall, there is a duty on a designer to investigate the history of a site (*Balcomb* v *Wards Construction (Medway) Ltd* (1980) 259 Estates Gazette 765), and failure to do so would be deemed negligent.

At this stage of the project, several routes may be feasible, and each must be assessed. Choice of a likely candidate for more detailed investigation will depend largely but not entirely on cost, and also upon the range of possible gradients. For road tunnels, acceptable gradients and curves must be related to design speeds and recurrent traffic costs. Whenever feasible, metro tunnels will be designed with the stations at a higher level than the running tunnels so as to promote deceleration on entering and acceleration on leaving the stations. Sewer tunnels must be designed with gradients that provide adequate dry weather flow velocities, and with a minimum number of pumping mains and siphons. Possible tunnelling methods - hand or machine (selective heading or full-face) excavation, slurry shield, microtunnelling, pipejacking, even open excavation where ground surface circumstances permit - need to be borne in mind.

Following completion of the preliminary appraisal, an overall impression of the geology and construction history of the area should have been obtained and a critical report written. An initial, tentative, geological profile along the tunnel alignment could be prepared to assist in the cost effective planning of the ground investigation. More detailed preliminary studies might be required in old mining areas (Section 4.7) and areas of carbonate rock, particularly if tunnels are to be driven for water transfer, sewage, or communication (particularly transportation) purposes.

Example: An old brick clay pit, adjacent to a tunnel route and shown on an old Ordnance Survey map, was not subjected to any form of investigation. The pit was later found to be infilled with waterlogged waste material, and this had caused a heavy discharge of water into the tunnel face. A Clause 12 claim under the ICE Conditions of Contract 5th Edition (1973) was substantiated although, had all relevant Ordnance Survey sheets been carefully examined, there is no doubt that measures could have been taken to avoid the submission of the claim.

Example: On a 1984 tunnelling contract, a pond located on an old Ordnance Survey sheet was interpreted as having been a shallow, near-surface feature. At the time of the site investigation, no evidence of the pond was visible on the site although several ponds were discernible in the area and these were known to be near-surface features. Because of temporary problems of access to the land no boreholes were put down within the old pond area, although boreholes were sunk to either side and confirmed firm to stiff glacial clays to the full depth investigated. Since the depth of cover to the 1.5m diameter tunnel crown was of the order of 10m, it was assumed that the old pond would not affect the tunnelling works. Subsequently the tunnel was driven and fill material comprising clay, timber, ash, and other waterlogged loose material was encountered. It transpired that the old pond was a backfilled clay pit, the clay, as in the above example, having been extracted for brick manufacture. The pit had been worked at some time between 1856 (when only fields were shown on the O.S. plan) and 1897 (when the pit was shown as a pond). The tunnel drive beneath the old pond became the subject of a Clause 12 claim. Had the pond been properly investigated, it is likely that the tunnel line would have been changed to avoid the made-ground deposits.

Staff in site investigation organisations have varying experience and abilities. In order to assist the relatively inexperienced members of staff it is always good practice for the organisation to develop and maintain an aide-mémoire (checklist) of all matters that need to be inspected and considered within the site investigation package. There would be provision on the list for the engineer to tick off each item when it had been dealt with, or to indicate that it was not applicable for the particular construction works in hand. Alternatively the list would be construction-specific - in the context of the present book relating the site investigation needs expressly to tunnelling works.

4

GROUND INVESTIGATION

4.1 PRELIMINARY GROUND INVESTIGATION

As complete an understanding of the ground/groundwater conditions on the site of the future tunnelling works as is reasonably possible within the constraints of cost and time is a fundamental requirement in the planning of the site investigation. In areas having uncomplicated geology, or for projects where the engineering problems are not expected to be difficult to solve, sufficient information may be obtained from the preliminary site appraisal. However, in other instances a preliminary ground investigation may be required or there may be no preliminary ground investigation phase at all. A flexible approach is needed in planning and carrying out the preliminary investigation.

Any preliminary ground investigation may typically consist of a limited number of boreholes (often at future shaft locations), perhaps accompanied by a few trial (exploratory or inspection) pits (see, for example, Cox et al. 1986) and/or trenches where access permits this, when the tunnel is at a quite shallow depth and when it is considered that such inspection pits/trenches will yield sufficient additional information to justify their use, put down to confirm the geological structure (strata dips, degree of weathering, frequency of joints and fissures) and groundwater conditions, and to permit the acquisition of large samples for test purposes. The Institution of Civil Engineers (1993) says that such pits should have a minimum base area of 1.5m^2, and notes that the pits must be kept free from surface water entry and from groundwater by sump pumping. Measurement of pits is by linear metre of depth and trenches by cubic metre. Alternatively, pitting and trenching may be billed on a daily basis. There may be restrictions on the use of pits in urban areas, particularly in terms of suitable location

and access for plant, and in many cases the cost of such pits could be the same as, or exceed that of boreholes. The pits will need to be sealed with concrete on completion of the investigation, and they should not be regarded as the normal form of investigation in built-up (urban) areas.

Drilling for soils and for weak rocks (see Supplementary Information 3 in Part Two of the book) and weathered rocks (see Section 4.4.2) can be accomplished by the *light cable percussion* method, with samples being obtained by means of a 100mm diameter sampler fitted, in the case of weak rock, with a strengthened cutting shoe. In some cases rock will be shattered by the percussion effect, rendering identification of the rock sample structure difficult.

Rotary core drilling, using diamond (surface set or impregnated) or tungsten carbide bits, is to be preferred for all rock strata. Surface set bits, in which the individual diamonds are set in a matrix, tend to be used for weaker, homogeneous rocks. The softer the rock, the larger are the diamonds that are needed, otherwise the bit clogs up. Impregnated bits, in which fine diamond dust is mixed into a matrix, are used for harder, heterogeneous and broken formations. Tungsten carbide tipped bits can be used for very weak uniform rock; they are less expensive than diamond bits but not as effective in operation.

In order to maximise core recovery a swivel arrangement is needed so that the core barrel remains stationary while the drill continues to turn. A double-tube swivel-type system in which the outer tube rotates to perform the cutting action is a minimum requirement. In this system the clean drill water passes between the inner and outer barrels, and the return water moves upwards between the outer barrel and the drill-hole walls. The particle-charged return flush water is less likely to erode fine particles and affect the quality of the core. In the case of friable formations which can be damaged by water, face-discharge bits, which receive water in the drilling annulus through ports in the face of the bit, can be used. The types of double-tube system are WX: internal discharge; WM: annular discharge, and modified WM; and F: face discharge, which limits washing of the core with the flushing medium.

However carefully the systems are chosen and the drilling proceeds, in the case of weak rock, and with rock where discontinuities may be filled with gouge material, there is always a possibility that water flush pressures will erode away the evidence. In such cases air flush drilling is to be preferred. Reverse circulation of compressed air via dual-walled pipe allows cuttings to be transmitted to the ground surface through the inner pipe rather than up the annulus between the hole wall and outer pipe. This prevents any tendency towards a cuttings build-up and potential contamination of the sample from higher up the borehole. If a coring string is used inside the dual-walled pipe, core can then be taken ahead of the main drill string at intervals (10 metres, for example) before the main drill string reams the hole down. Uncontaminated core is then retrieved by wireline.

Rotary holes drilled using air flush also seem to register more accurately the presence of groundwater and the rate of seepage towards a newly created free surface; that is, how much water comes up the hole and how quickly. This can be especially important in the case of drilling into old mine workings when it is required to know if they are flooded. Air flush can, however, influence the moisture contents of some rock cores and may be difficult to use under certain groundwater conditions, in which case a foam flush might be considered. Another possibility is to use about 6-10% bentonite clay by weight in fresh water as the drill medium. It supports the hole directly and by depositing a filter cake on the sidewalls, and it also facilitates transport of the cuttings to the surface. Different additives are available for use in the flushing medium; they are generally high molecular weight long-chain hydrophilic polymers, such as natural gums or synthetic alternatives. But whatever flushing medium is used, the choice must always be compatible with the need for good core recovery.

Core diameter depends on both the bit size and the type of core barrel. Samples taken with an N-size core barrel are 54mm diameter for a standard double tube (NX) core barrel and 48mm for a modern triple tube (NQ) barrel. A reaming shell serves to maintain a constant diameter hole even when the core bit wears during service. The reamer provides adequate clearance for the flushing medium and cuttings, and ensures that a new core bit will not jam in the hole. Reference may be made to BS 4019 Core drilling equipment, Part 1 Basic equipment (British Standards Institution, 1974).

Rate of penetration can be an important indicator for establishing the presence of voids in the ground, since a drill string will tend to drop under its own weight when little or no resistance is offered at the base of the hole. The supervising geotechnical engineer or engineering geologist can request that the drill string be lifted and then observe its free fall in order to derive a better judgement as to whether an open hole or one filled with loose material has been encountered.

Up to 150mm rotary drilling by wireline using a fully lined borehole and triple tube barrels can be used successfully for coring cohesionless soils, very weakly cemented cohesionless soils, and weak rocks, especially if combined with a polymer fluid to provide additional support for the borehole walls at the bit. The wireline system, which is particularly suited to deep holes, allows the inner tube to be withdrawn up to ground surface through the hollow drill rods after each run, and is needed for deep boreholes. Retractable triple tubes are especially useful for sampling rocks comprising alternating hard and soft bands which may occur, for example, in lava flows where significant weathering has occurred at the top of each flow before being covered by subsequent flows. In a triple tube core barrel the extra inner tube for retaining the core projects beyond the face of the bit and has a sharper edge. It can then be used to recover the cores of soil or very soft rock, and in some cases it retracts a little when hard rock is encountered. The so-called pitcher sampler has a spring loaded inner barrel which is pushed into soils or very strong rocks in advance of the outer cutting barrel. Other non-standard designs exist, some of them having been specially developed for air flush.

Core barrels are usually 1.5m long. Cores are retained in removable, semi-rigid PVC inner liners within the inner steel barrel. The core and liner are slid out of the barrel and the liner taped up to seal the core. Two disadvantages of the system are that a thicker kerf is needed (or a smaller diameter core is cut) and the plastic can crumple up inside the inner tube. The core may also swell as a result of stress relief. A hydraulic core pusher is needed to remove the core, or a split tube inner barrel can be used. There are also variations in the design of the core spring or core lifter; basket-type core springs are available with finger-like springs rather than a taper system.

A polymer fluid may form a thin skin on the surface of the cores and along open discontinuities, and serves to preserve moisture content and the integrity of the core. Some care might be needed during the logging operation as a result of the presence of this skin (Manby and Wakeling, 1990). The use of reverse circulation rotary drilling generally has been discussed by Manby and Wakeling (1990).

Holes in soils and weak rock need to be cased. N-size casing fits inside an H-size hole (100mm diameter). In turn the N-size casing admits an N-size drill bit which gives a hole of about 75mm diameter. B-size casing fits in an N-size hole and a B-bit, 60mm outer diameter, passes through the B-casing. Deep holes need such staged casings.

Groundwater levels, the rates of inflow, and the times of groundwater encounters should be recorded during drilling, which should be stopped after every water strike (even a seepage) and sufficient time allowed for the determination of the rate of inflow to the hole and a final standing level. During water observations the contractor would be paid at standing time rates. It is recommended that recordings be taken at 5 minute intervals over a halt period of 20 minutes. If the water level is still rising after that time, this fact should be recorded and, unless instructed otherwise by the Engineer, the exploratory hole continued. Special attention needs to be drawn to any

artesian conditions exposed by the drilling process. As a general procedure, water levels should be recorded before and after breaks in drilling, and particularly at the beginning and end of each shift. A careful record should be made of casing depth, and if the casing forms an adequate seal against groundwater entry then the site investigation contractor should record the depth at which no further entry, or only insignificant infiltration, of water occurred. If an adequate seal is not achieved between the borehole casing and the ground, then it is possible for inflow rates to be overestimated. Piezometers may be installed (*see* Section 4.5.2) and pumping-in or pumping-out tests, using a single or double packer system, may be specified in order to estimate permeability (*see* Section 4.5.3).

For the actual drilling process in soils, water should not be added to the hole in order to assist penetration except in the case of dry, coarse sands. However, if the hole is below the water table and disturbance from inward groundwater movement is likely, then water should be added to maintain a higher head in the hole than in the surrounding ground.

Monochrome or colour photography, with overlapping prints, and closed circuit television (CCTV) can be used to study ground stratigraphy and structure in boreholes that do not require casing support. More simply, Hencher (1986) has described the successful use of a borehole periscope to depths of 30m in vertical and inclined water-filled holes. But in all these techniques any water in the borehole will need to be relatively clean in order to achieve the requisite vision or, in the case of a pumped hole which does not recharge quickly, it may be possible to pump it out and replace it with clean water. Of special interest will be the closeness of discontinuities in rock, and it may also be possible to gain information on discontinuity trends. Of particular interest in this latter context is the use of down-hole impression packers (Hinds, 1974; Barr and Hocking, 1976; Harper and Hinds, 1978). Orientation data for discontinuities may be needed in rock tunnels where it is felt that roof and sidewall blocks could be unstable upon excavation and analysis is needed in order to decide upon the best method of temporary support - bolted anchorages, for example.

Reference on this subject may be made to Priest (1985, 1993), Goodman and Shi (1985) and Goodman (1989).

These in-hole visual techniques are also particularly useful where tunnelling is to take place in areas of old mine workings where decisions have to be made on the nature of any ground treatment (for example, bulk infilling and post-infill grouting) that is needed and the volumes of materials that need to be entered into the bill of quantities. There is a laser-based total station system (Analysis Geotech Ltd) for use in active or abandoned mine workings, ideally through a borehole of greater than 120mm diameter. It is claimed that the system can survey areas to 10mm accuracy and can then generate volumes, sections and three-dimensional views by means of computer software.

Under normal drilling conditions it is not possible to determine the directions of potentially significant structural features, such as persistent discontinuities in rocks, from an examination of cores. There are, techniques, however, of providing an orientated core. For example, in the case of shallow vertical holes the north direction can be periodically marked on top of the core by dropping an indenter through a hollow guide rod held against the north wall of the drill hole. There is a Craelius core orienter described by Hoek and Bray (1977) and Gamon (1986) - a mechanical device that operates only in inclined holes and a Christensen orienter which scratches the core along reference directions and takes orientated photographs of the scribing knives in the hole. This technique can be combined with calliper logging and geophysical techniques such as electrical logging and sonic logging.

A preliminary investigation will often provide sufficient information for an outline design of the works, the likely plant to be used, and an approximate assessment of the project cost. It should also provide reasonable confirmation of the proposed line and level and the shaft/portal locations. The results of the preliminary investigation would then be used in planning the main investigation.

4.2   MAIN GROUND INVESTIGATION

Most site-specific ground information will usually be obtained from adjacent exposures (nature of the soil and/or rock) and boreholes, although exploratory (trial) pits, on occasions trial headings (to assess the feasibility of, for example, rock bolting, grouting, shotcreting, and so on), and geophysical surveys may provide useful supplementary information in some cases. (For the Kielder project rock tunnels in north-eastern England an experimental tunnel described by Ward *et al.* (1976) was driven before the main tunnel construction so that the contracting parties would have early knowledge of the ground conditions and the stand-up capability of the unsupported tunnel - a factor of major importance with the full-face tunnelling machines that were used for the Tyne-Tees water transfer link.) The locations and spacings of the boreholes will be determined partly from geological considerations, partly by engineering requirements and partly by problems of access. Boreholes would normally be put down at each shaft location, with intermediate boreholes along the tunnel alignment.

The procedures outlined in Section 4.1 above also apply to the main investigation. Borehole spacings will generally be not greater than 100 metre intervals in soil along a tunnel route. Closer spacing may be required in areas where it is strongly expected that adverse ground conditions could lead to difficult tunnelling and ground support operations. The requirement is to achieve identification of the lithologies and structure of the ground to be tunnelled and of groundwater regimes within the boreholes, and to permit reasonable interpolations between boreholes. Any site investigation funding for deep tunnels in rock will probably accommodate only a few boreholes along the tunnel line but with a concentration of probing at the tunnel portals. There may be a desire to obtain information on the magnitudes and orientations of the *in situ* principal stresses and the stiffness of the rock so that estimates can be made of the possible relaxation deformations for the purposes of lining design. At such depths, however, the interpretation and reliability of any results from hydraulic fracturing or overcoring tests may be questionable. However, Ball *et al.* (1993) have noted the use of a wireline coring system with a double tube core barrel in order to install a re-usable Borre probe (Christiansson *et al.*, 1993) for measuring the complete state of stress in the Calder and St. Bees Sandstone at Sellafield in West Cumbria at depths up to about 1000m. These measurements were for the purposes of nuclear waste repository design. The Borre probe, which measures the complete state of stress using electrical resistance strain gauges bonded to the walls of the pilot borehole, can withstand water pressures equivalent to a 1100m head. This is in contrast to the CSIRO hollow inclusion stress cell which is not re-usable and is susceptible to water. An alternative to the Borre probe for wet conditions would be the USBM borehole deformation gauge which can be installed under water.

Without a reasonable density of direct ground information along the tunnel route from exploratory boreholes, laboratory and *in situ* tests the style of the works contract must then reflect such uncertainties.

In general, boreholes for tunnels in soil and rock should usually be taken about two diameters below the tunnel invert and be located at least one tunnel diameter from the tunnel perimeter. Careful consideration must be given to the expected geological structure when planning the investigation. For example, in dipping rock strata or for boreholes remote from a tunnel in congested areas, boreholes should be drilled to examine particular strata that may intersect the tunnel horizon. In steeply dipping beds, heavily faulted ground, or in strata which lack distinctive marker horizons, it may be necessary to locate boreholes at quite close centres and to drill to depths much greater than two tunnel diameters below invert level (Deaves and Cripps, 1990). Alternatively, the use of inclined boreholes may be considered for examining strata dips.

Boreholes at shaft positions should normally be located on the shaft centre line unless there are reasons to suppose that any water at borehole depth could be at artesian or at adversely sub-artesian pressure, in which case the hole should be sunk outside the shaft perimeter. All site investigation boreholes, with or without standpipes/piezometers installed, should be carefully backfilled with

cement grout, tremied in, so that water/air movement up and down the hole, which could affect the tunnelling work, is prevented. Inclined boreholes in rock are sometimes advocated for better resolution of joint sets where one of the dominant sets is vertical (Coats *et al.*, 1977) or where access for a vertical borehole is not possible because of buildings. Only for a few deep holes and high cost projects could such drilling be justified economically. Large diameter boreholes for man-entry and inspection of the ground directly have been suggested (Knights, 1974). This form of inspection and sample retrieval could be important when machine tunnelling is being considered. However, a preferred alternative, when practicable, is to sink the access shafts first, the cost of putting these down early for inspection purposes being part of the general construction costs. Alternatively, shaft construction could be undertaken as a separate part of the construction programme or perhaps as a separate contract. If explosives are used for shaft sinking in rock or frozen ground in an urban area then maximum permissible vibration levels that take account of building type and human perception need to be written into the contract documentation. On-site environmental problems from vibration (and noise) can best be overcome by way of consent to work agreements with local authorities based on current legislation (the Control of Pollution Act 1974) and for vibration limits before a contract is put out to tender (*see* Section 6.10 and Supplementary Information 20 in Part Two of the book).

As in the case of boreholes, any trial pits (*see* comment in Section 4.1 above) need to be properly backfilled and the fill compacted. Lean mix concrete may be specified where there are potential ground settlement and water problems. Their locations, areas, depths, and the type of reinstatement used need to be accurately recorded. The techniques for site investigation using trial pits have been discussed by Cox *et al.* (1986). Because of the relatively low hire costs of excavation equipment such as back-hoes, it is often considered that this method of investigation is less expensive. However, Cox *et al.* (1986) have suggested that trial pit costs are similar to those of boreholes. In fact, the cost of trial pits can be considerably higher.

Example: Nine ground investigation boreholes at relatively close spacings on the tunnel line were put down for one Tyneside contract. In the course of a claim for extra payment the contractor argued that trial pits should have been used, citing recommendations in BS 5930:1981. No such objections to the style of the ground investigation were raised before the contract was let. The actual cost of the site investigation was calculated to have been between 1% and 3% of the tender price. Because of the highly congested nature of the tunnelling site and the depth of the tunnel (6 to 7 metres) it was calculated that to have used trial pits instead of the boreholes would have produced a site investigation cost of between 30% and 40% of the tender price. This latter figure is obviously unacceptable. Even with a single trial pit in a built-up area there needs to be assurance that the additional value of the information retrieved is commensurate with the additional cost incurred compared with the use of a borehole.

Example: In another tunnelling contract on Tyneside, the lack of a borehole at a shaft position brought about unexpected extra costs when old mine workings were encountered during shaft sinking.

Example: In a further rock tunnelling contract an error in the site investigation led to the omission of a borehole at one shaft location. In its absence the prediction from a relatively remote borehole was that the tunnel drives should set off in a mixed face of sandstone overlying mudstone, with the possibility of a thin layer of coal between the beds. The sandstone in the crown would be of poor quality (RQD = 38%). There was no indication of groundwater. It was the opinion of the Engineer that, based on this evidence, no significant overbreak would occur, although some 'peeling off' along the bedding planes could be expected. The face conditions that were actually encountered as soon as the tunnel was advanced were quite different from those predicted by extrapolation. The sandstone occurred in the top 900mm of the crown, but was highly disturbed, having irregular bedding and joint planes. Both open and clay-filled fissures exhibited shearing with respect to an upper coal seam. Ground below this seam was undisturbed and there was a second 300mm thick coal seam 600mm below the upper one. Resulting overbreak in the sandstone provided the basis for a contractual claim. Remedial measures comprised close boarding using steel lagging (Tylag), packing with stone to shoulder level, and filling the cavities with pozzament or bagged stone over the crown, allowing for grouting points in the sheeting.

Choice of the frequency, depth and spatial distribution of the investigation boreholes also depends very much on the nature and projected capital cost of the construction. At the project design stage there may be several tunnel depth and alignment options that cannot be suitably resolved without adequate in-ground information. Sewer tunnel depths are controlled by the gradients necessary to maintain gravity flow at a suitable velocity under dry weather flow conditions, and by the need to minimise the number of pumping mains and siphons, and hence recurrent costs. For a sea outfall or a cable tunnel, a greater range of depth options should be incorporated in the investigation design.

There are European Community requirements, accepted by the UK Government, for increasing restraints by the year 1998 on the disposal of sewage at sea. The first long sea outfalls were introduced in the UK in the 1950s and more recent developments have seen the use of 1mm or lower mesh screens to reduce the marine disposal of non-biodegradable solids. However, around the coast of the British Isles there are still numerous short outfalls of Victorian age. Alternatives to these, and eventually to long sea outfalls, will involve major plant on land, perhaps on crowded coast locations, or underground, but in any case at considerable expense. More compact methods of treatment include wet oxidation, fluidised beds, and membrane processes. But before these can be introduced to any great extent there will have to be improved methods of sedimentation, perhaps flotation, secondary and tertiary treatments for discharge into the more sensitive waters, perhaps disinfection, or incineration of sludge (perhaps co-incineration with domestic and trade waste), together with ozone injection into scrubbing towers in order to remove unpleasant odours from the operatives. However, notwithstanding such developments, it would seem that long sea outfalls will continue to form a major method of disposal for many years (Reynolds, 1990), but with an increasing requirement for sewage disposed of in this way to receive primary and secondary treatment, with nutrient removal from the resultant liquid effluent 'where appropriate'.

Investigations for sewage disposal by sea outfall are a special site investigation case. Difficult decisions often have to be made, not entirely on cost grounds, as to whether the outfall will be in tunnel or whether it will take the form of a pull pipe, the latter involving construction of the pipe on land for it to be hauled out to sea in strings about 200m long by means of an anchored lay barge. The pull-pipe option is often less expensive, but there are hazards in that work can be 'weathered off' due to storms and gale force winds during the pull, winter work is often not possible, and it can have an adverse environmental impact on sea-bed fishing, such as crabbing. For the tunnel option, a relatively small number of deep holes will be drilled at sea along the proposed tunnel line using a pontoon. Because of the cost of the operation, good core retrieval is of paramount importance. The drilling must cover a range of possible depths because it is usually the case that the more shallow the tunnel, the less costly it will be, but the chosen depth will often depend upon the inferred depth of sea bed gullies which, if intercepted by the tunnel face, would lead to an inundation and possible loss of life.

Southern Water's £450 million 'Operation Seaclean' includes two offshore storm outfall tunnels, 1.2m in diameter. Bored tunnels were chosen rather than pulled-pipes on both cost and environmental grounds. The 1.2m diameter tunnels at Deal and at Ramsgate in the south-east of England, of 520m and 360m offshore length respectively, were driven using a remotely operated Herrenknecht AVN 1200 machine. The Ramsgate tunnel, in chalk and flint material, was able to be driven from a 12m deep shaft on the shore at only 3m below the sea bed due to the sealed nature of the boring machine. The machine was retrieved by first dredging a pit to about 1m below invert level at the seaward end of the tunnel, filling the pit with sand, driving the machine into the sand, disconnecting the hoses and control lines, flooding the tunnel, sucking away the sand to expose the machine, and then lifting the machine for removal to land and maintenance. Divers were needed for positioning the steel diffuser.

It will often be the practice for a client, when faced with decisions on outfall construction options, to commission a sea bed survey by a team of experienced divers. The results of the survey will be important also for the design of a pull-pipe option. In addition to the drilling and diver survey, it will be the usual practice to institute a geophysical sparker survey to probe the ground and correlate the results with the borehole evidence. Tunnel construction would require the specification of a programme of forward probing from the face (perhaps 30m in advance), with a suitable means of resisting the water pressure at the drill should a water-filled gully be intercepted. Ultimately, a programme of advance grouting may need to be instituted in order to cut off the water inflow locally and stabilise the ground, and grouting will also be needed to facilitate the coupling up of the sewer pipe to the diffusers which will usually have been installed in advance. (The diffusers will often comprise tubular steel of about 750mm external diameter capped with a reinforced concrete head the underside of which, at the sea bed, will need to be plugged by means of grout bags. The gap between the tube and the drilled hole will need to be grouted up, and this can be achieved by welding grout tubes to the extrados of the diffuser pipe.) A large diameter tunnel housing a sewer pipe, although advancing more slowly than a pull-pipe, does offer the advantages of extra space for inspection and maintenance of a smaller diameter pipeline installed in it, and it also allows for the possible addition of another pipe at a later stage to accommodate expanding populations. All these factors need to be anticipated both at the site investigation stage, by conducting a wide range of geotechnical tests, including grain size analyses on the gully infill material (retrieved by the divers), and by suitable provisions written into the tunnel contract documentation.

If the results of the preliminary and main investigations alert the designer of the tunnelling works as to potential problems, both during the construction phase and after, stemming from inadequacies in his initial design, then he could be deemed negligent if he fails to take further precautionary measures in the form of extended site investigation procedures and/or changes in design (*refer to* the New Zealand case of *Bowen* v *Paramount Builders (Hamilton) Ltd* [1977] 1 NZLR 394).

4.3   SOIL SAMPLING

4.3.1 Sample protection

Most soil samples of a cohesive nature are retrieved and retained in U100 tubes. Any free space at the ends of these tubes must be packed with suitable material that will prevent movement of the soil inside but not absorb moisture, the ends sealed with wax, and the end caps screwed tightly on as soon as possible. Thereafter, the tubes must be handled carefully and stored under conditions that are neither extremely hot nor extremely cold. Whenever the sample is partially extruded from the tube for test samples to be taken, the remainder of the soil must be re-inserted in the tube, the ends re-packed and wax-sealed, and the end caps screwed tightly on again.

Small bulk disturbed samples of soil should not weigh less than 0.5kg and should be placed immediately in airtight containers. Such samples should be taken at each change in soil type and midway between successive open tube samples or standard penetration tests (*see the* special topics section). In the case of sand and gravel samples, it is possible to use thick-walled, lined, open-drive samplers, of internal diameters 104mm and 150mm respectively, in conjunction with a light cable percussion rig. When silts are encountered, U100 samples

should be alternated with standard penetration tests at 0.75m intervals.

Open tube samples should be taken only after the bottom of the hole has been cleaned out and below the bottom of any casing. The standard recommendation is that samples should be taken at 1m intervals down to 5m depth and thereafter at 1.5m intervals or at every change of stratum. For tunnel site investigations there may be merit in taking more frequent, even continuous, samples throughout the depth of the tunnel face and for one tunnel diameter above and below the tunnel face. ICE (1992) states that following a break from the work exceeding one hour the borehole must be advanced 250mm before tube sampling is resumed. One might consider using the valvate system described and patented by Vallally (1986), the main feature of the system being a plastic 'valvator' catcher disk which allows soil to pass in one direction only. During boring, segmented and ribbed 'petals' that form the disk compress, so that the soil encounters little resistance as it passes through, but equally the petals resist movement in the opposite direction. As a result, both intermixing and loss of sample are restricted. The disk system has been adapted for use with boring tools and samplers, and the incorporation of a liner into the design makes sample handling in the field easier. Such samples of granular material should be transported in plastic liners, frozen in the laboratory using liquid nitrogen, dry cut with a masonry saw for logging, photographed in colour, and then unfrozen for grading. In hard or stony clays a U100 tube may not achieve full penetration, in which case a standard penetration test should be performed.

Groundwater samples should be taken from each hole and from each horizon where groundwater is encountered. In the case of granular soils where water has been added it will be necessary to bail out the hole before sampling so that only groundwater will be taken into the sample. A sample of not less than 0.25 litre will be needed.

4.3.2 Description

Soils may be described as indicated in BS

5930:1981 in terms of the mnemonic 'MCCSSOW':

M: Moisture condition (A factual descriptor)
C: Colour   (A factual descriptor)
C: Consistency   (A factual descriptor)
S: Structure   (A factual descriptor)
S: Soil type   (A factual descriptor)
O: Origin   (Requires interpretation)
W: Groundwater conditions
               (Requires field observation)

The meanings of these descriptors are given below.

Moisture condition

This descriptor is not included in BS 5930 and is not widely used in the UK. Nonetheless it is important, especially in drier climatic regions. Some suitable descriptive terms are:

dry, slightly moist, moist, very moist, wet

There is no direct relation between the descriptive terms for *moisture condition* and the actual quantitative measurement of *moisture content,* since moisture content depends upon the soil type. A clay soil having a moisture content of 10%, for example, could

be described as *dry*, whereas a sand having the same moisture content could appear to be *wet*. Reference may be made to Ewan and West (1983).

Consistency

Descriptive terms for consistency depend on soil type. A single term, or a range, may be used. A single term might be *stiff* and a range *stiff to very stiff*. In terms of the general soil types:

very soft, soft, firm, stiff, very stiff (for cohesive soils)
very loose, loose, medium dense, dense, very dense (for granular soils)
firm, spongy, plastic (for peaty soils)

These descriptive terms acquire numerical significance in respect of the undrained shear strength of cohesive soils and the N-values derived from the Standard Penetration Test. Reference may be made to Tables 4.1 and 4.2 which include expressions which assist in recognition of the different shear strengths, strengths of cohesive soils and the penetration resistance of granular soils. There is also the

Table 4.1 Definition of terms for a cohesive soil

Term	Field recognition	Undrained shear strength (kPa)
Very soft	Extrudes between the fingers when squeezed	<20
Soft	Moulded by light finger pressure	20-40
Firm	Moulded by strong finger pressure	40-75
Stiff	Cannot be moulded by fingers but can be indented by thumb	75-150
Very stiff (Hard)	Can be indented by thumb nail	>150

Table 4.2 Definition of terms for a granular soil

Term	Field recognition	SPT N-values
Very loose	Very easy to excavate with spade	<4
Loose	Fairly easy to excavate with spade or penetrate with hand-bar	4-10
Medium dense	Difficult to excavate with spade or penetrate with hand-bar	10-30
Dense	Very difficult to penetrate with hand-bar; requires pick for excavation	30-50
Very dense	Difficult to excavate with pick	>50

Table 4.3 Definition of terms for peat

Term	Field recognition
Firm	Fibres already compressed together
Spongy	Very compressible and open structure
Plastic	Can be moulded in hand, and smears fingers

Table 4.4 Scale of bedding spacing

Descriptive term	Mean spacing (mm)
Very thickly bedded	>2000
Thickly bedded	2000 to 600
Medium bedded	600 to 200
Thinly bedded	200 to 60
Very thinly bedded	60 to 20
Thickly laminated	20 to 6
Thinly laminated	<6

matter of peaty soils which entrap considerable quantities of water and which pose special civil engineering problems, particularly those related to the construction of roads. Recognition of peat is expressed in Table 4.3 and any stratification of the deposit can be expressed in terms of the bedding thickness given in Table 4.4.

Colour

This is a visible feature that can be used to recognise strata in different locations on a site. It is also an indicator of chemical and mineralogical processes mainly associated with iron compounds and can also provide valuable information about weathering processes. Colour, which is a feature of the soil descriptions given in the British Standards Institution (1981), can change with moisture content. It should be assessed in the undisturbed state and in the wet remoulded state. An accurate description of colour can be obtained with a colour chart (a Munsell soil colour chart: New York: Macbeth, Division of Kollmorgen Instruments Corporation, 1992), but although it is helpful to attach a chart to a core box a high level of accuracy in colour description is rarely required. Reference may be made also to Joyce (1982) and the Building Research Establishment (1993).

The following basic hues are usually sufficient for descriptive purposes.

pink, red, purple, orange, yellow - bright colours
brown, olive, green, blue - dark colours
shades of grey

Secondary colours can also be used descriptively; for example, *reddish, brownish, bluish,* and so on, together with modifiers such as *pale, light, dark, mottled.*

Structure

Structure indicates the presence of bedding, discontinuities or shearing within a soil. It is identified by the description of the feature, the spacing, dip, dip direction and details of the surface finish. Terms used to define structure include the following:

bedding, lamination, fissure, joint, slip surface, shear zone, gouge, intact
Spacing (*see* Tables 4.4 and 4.5) and dip may be defined by descriptive terms or by quantitative observations. Some of the descriptive terms are:

Spacing: *very thick, thick, medium, thin, very thin* (for bedding and laminations)
Spacing: *very wide, wide, medium, close, very close, extremely close* (for discontinuities)
Dip: *vertical,   sub-vertical,   sub-horizontal, horizontal*
Surface finish: *polished, striated, slickensided, grooved, open, closed, tight*

Table 4.5 Scale of spacing of other discontinuities

Descriptive term	Mean spacing (mm)
Very widely spaced	>2000
Widely spaced	2000 to 600
Medium spaced	600 to 200
Closely spaced	200 to 60
Very closely spaced	60 to 20
Extremely closely spaced	<20

Table 4.6 Descriptive terms

Descriptive term	Component percentage
Slightly sandy GRAVEL	<5% sand
Sandy GRAVEL	5-20% sand
Very sandy GRAVEL	>20% sand
GRAVEL/SAND	about equal proportions
Very gravelly SAND	>20% gravel
Gravelly SAND	5-20% gravel
Slightly gravelly SAND	<5% gravel
Slightly silty SAND (or GRAVEL)	<5%silt
Silty SAND (or GRAVEL)	5-15% silt
Very silty SAND (or GRAVEL)	15-35% silt
Slightly clayey SAND (or GRAVEL)	<5% clay
Clayey SAND (or GRAVEL)	5-15% clay
Very clayey SAND (or GRAVEL)	15-35% clay
Sandy SILT (or CLAY)	35-65% sand
Gravelly SILT (or CLAY)	35-65% gravel

Soil type

This descriptive category relates to the range of particle sizes (*see* Table 4.6) of the relevant components of a soil. Reference should be made to the British Soil Classification System for Engineering Purposes (BSCSEP) which is given in Table 8, p107 of BS 5930:1981 (British Standards Institution, 1981) and to the Plasticity Chart for the classification of fine soils and the finer part of coarse soils (*note* Terzaghi and Peck, 1967, p41; *see* Fig. 31, p108 of BS 5930:1981).

Percentages are expressed in terms of component weight, and the main soil types are:

clay, silt, sand, gravel, cobbles, boulders, peat

- Sand, gravel, cobbles and boulders are visible to the naked eye.

- Silt is gritty to hand and teeth, exhibits dilatancy when squeezed in the hand, and disintegrates quickly in water.
- Clay has a soapy feel when rubbed with water in the hand, does not exhibit dilatancy, and sticks to the fingers with a slow drying action.
- Peat consists predominantly of plant remains.

Terms such as '*very silty*', '*slightly silty*', '*slightly clayey*', '*sandy*' can be used to describe lesser constituents of the soil. Other descriptive forms, not conforming with BS 5930, could be '*with a little.....*' or '*with some.....*'.

Additional information on grading, shape and texture can be incorporated in the soil description:

Grading: *Well graded, poorly graded, gap graded, uniform*

Shape: *Angular, sub-angular, sub-rounded, rounded, flat, elongate, irregular*

Texture: *Rough, smooth, polished*

Origin

An attempt should be made to assess the genesis of the soil; and, if possible, to identify its geological name or stratigraphic unit. If there is doubt as to the actual origin, the words 'thought to be' or 'possible' should prefix the name on the log.

It is particularly important to attempt to establish the structure and fabric of soils that might be encountered at a tunnel face or during the sinking of a shaft. The presence of water-bearing permeable lenses, such as interstratifications of sand, silt or gravel in laminated clays, as emphasised by Boden (1983), or clay/silt layers in permeable strata, can create a major problem in tunnelling and shaft sinking. On the other hand, a tightly laminated deposit of glacial clay might not be excessively troublesome at a tunnel face below the water table. Because, in part, due to the problems associated with sample disturbance during site investigation drilling (*see* for example, Clayton, 1986), a standard ground investigation is unlikely to reveal details of the soil structure, the volume of any permeable lenses and their interconnections, so rendering difficult any attempts to predict possible groundwater inflows - the rate of water 'make' and whether there will be a reduction in that rate with time.

In potentially problematic strata, the vertical distribution of the different soil structures can often be estimated relatively quickly and inexpensively by using a static cone penetrometer (*see* Supplementary Information 5 in Part Two of the book). High quality undisturbed sampling, such as the use of piston sampling in soft alluvial deposits, could be carried out in selected boreholes. Undisturbed samples which will not be used for testing in the laboratory should be extruded from the sampling tube, split and carefully examined. (For splitting, a light incision should be made axially down the core from its perimeter and then the core should be torn open by hand to its centre in order to create an open wedge.) It is good practice to photograph the internal

structure of the core in colour. The vertical extent and vertical distribution of sand lenses can also be determined approximately by the use of static cone penetrometers which offer a relatively quick and inexpensive method of *in situ* ground investigation.

Example: A tunnel was located below the water table in alluvial deposits, described in the site investigation report as 'loose silty fine to medium sands'. Conventional dewatering failed adequately to drain these deposits. Subsequent investigation showed that the deposits consisted of clean fine to medium sands with silt layers. The silt layers acted as aquicludes, retaining perched water tables above the well-point tips. The solution was to install additional well-points with filters to provide vertical permeability and to intercept all the water-bearing sand zones.

For the actual description on the borehole logs BS 5930:1981 suggests that the following approximate order of descriptors be used, although other schemes can be equally effective:

(a) *Mass characteristics*
   (1) Field strength or compactness and an indication of moisture condition.
   (2) Bedding.
   (3) Discontinuities.
   (4) State of weathering.
(b) *Material characteristics*
   (1) Colour.
   (2) Particle shape and composition.
   (3) Soil name (in capitals, e.g. SAND), grading and plasticity.
(c) *Geological formation, age and type of deposit*
(d) *Classification* (optional)
   Soil group symbol.

Two examples are suggested, first of a thickly-bedded deposit:

'Firm closely-fissured yellowish-brown CLAY of high plasticity. London Clay.'

and, second, of a stratified deposit

'Dense yellow fine SAND with thin lenses of soft grey silty CLAY. Recent Alluvium.'

4.3.3 Boulders

The primary purposes of ground investigation boreholes are to probe stratified soil and rock deposits in order to determine lithological sequences in boreholes, to provide samples for testing in soil and rock mechanics equipment so that parameters can be quantified for the design and execution of the works, and to provide access both for piezometers and for *in situ* tests. No physical probing of the ground will permit a satisfactory quantification of cobble/boulder (and other random rock) spatial density distribution nor a firm indication of the mineralogical composition and the strengths of the cobbles and boulders *in situ*. (The comment at the end of Section 3.8 concerning the statistical weakness of borehole evidence is particularly relevant to this problem.)

Glacial (undifferentiated) tills may be extremely variable in composition, containing not only cobbles and boulders but also sandy lenses the engineering behaviour and permeability characteristics of which may be unpredictable. These deposits may in addition conceal important underlying bedrock features such as a buried valley which, if not delineated by the site investigation, could pose contractual problems during any actual tunnelling works. Cobbles and boulders form obstructions to boring and will often require the use of a cable tool to penetrate them. A large casing size should be used to start the cable tool, but it may prove necessary to use a rock roller cutter. Experienced interpretation of chiselling times noted in the driller's log is necessary. Boreholes should be taken deep enough to differentiate between boulders and bedrock, a depth of 3 to 5 metres into the rock usually being sufficient except in special circumstances such as pertain, for example, in Hong Kong where the penetration may need to be substantially increased.

Chiselling times will not usually give a true indication of cobble or boulder size, and indeed the Specification for Ground Investigation (Institution of Civil Engineers, 1992) recommends that the technique be employed only for a period of up to one hour without obtaining further instructions from the Engineer to the contract. If the object is not fully penetrated by that time then the site investigation contractor will be instructed either to continue or to bring a rotary drill onto site or to abandon the hole. Unless defined as such in a preamble note to the borehole logs, terms such as 'occasional' should not be used on the logs or in any text to describe the presence of cobbles and or boulders in a borehole, otherwise a contractor may be moved to infer that the term also relates to a boulder clay *en masse*. The expression 'cobbles/boulders noted' or 'cobbles and/or boulders noted' is preferred. Even if the indications are that the obstruction is a cobble rather than a boulder, it would be unwise to designate it as such, and especially by using capital letters (COBBLE), because this might convey the message that the particular obstruction is a dominant feature of the ground.

Nearby exposures may be available to give an indication of boulder size and distribution. Such information, of course, becomes increasingly relevant as the tunnel size decreases. BS 5930 (British Standards Institution, 1981) states that 'within limits of cost, the best method of investigating clay containing gravel, cobbles or boulders is by dry excavation', but such a method will usually be impractical, not only on the basis of cost but also of depth, accessibility and traffic disruption in an urban environment. As pointed out by Weltman and Head (1983), it is also difficult to prevent settlement taking place after the pits are backfilled since even well-compacted backfill in pits leaves paths for transmission of water and may cause instability or seepage problems. They suggest that if pits are close to existing structures they should be backfilled with lean-mix concrete to within 300mm of the surface. Early, pre-contract sinking of manhole shafts is another option, but this would prolong the time for urban social and traffic disruption.

Identification of rock cuttings may also allow differentiation between boulders and *in situ* rock; for example, in a glacial till boulders often consist of different sedimentary, metamorphic and igneous rocks derived from remote areas and transported by the ice sheet. It is important to record the geological and mineralogical nature of any boulder rock in order to establish whether it could easily be broken by pneumatic/hydraulic equipment in a tunnel face or by full face machine. However, as noted above, the composition and strength, as well as size, may vary considerably from boulder to boulder, and so any ground investigation information on these parameters must be treated with caution and the tendering contractors should be notified accordingly, notwithstanding the fact that these are geological conditions which, from their experience, they should be fully aware. It is well known, for example, that tills in the north-east of England contain cobbles and boulders of limestone, sandstone and 'whinstone' (quartz dolerite), and so a statement on any one of these rock types and strengths in a report (such as SANDSTONE cobble and/or boulder noted) without accompanying qualifications might be seized upon by a contractor as grounds for a contractual claim if he encounters boulders having different compositions and strengths. In any case, borehole boulder rock is not usually in a suitable condition for the derivation of unconfined compressive strength in the laboratory.

The possible presence of boulders within glacial tills can carry further important contractual implications, and the implications of a preamble to any bill of quantities need to be given careful consideration. Because cobbles and boulders are rock there is the *implied* contention in a tunnelling contract that their excavation should be paid for at the billed rock rate even though the quantities in the bill upon which the tendering contractors determine their rates will have been calculated on the basis of *stratified* rock as determined from the site investigation. A cobble (<200mm size) or boulder (>200mm size) may be geologically a rock but technically under the contract a soil. This matter is further discussed in Section 5.4.

'Nests' of boulders in a boulder clay can create both a practical and a contractual problem, and so special efforts should be made to detect them at the site investigation stage. There is often substantial evidence at exposures of such accumulations, sometimes in the form of boulder beds, and these should be identified by the contractor during his pre-

tender survey. Furthermore, if the tunnelling is in boulder clay close to bedrock the miners might also have to contend with blocks of the bedrock that have been rotated by the ice sheet, but minimally translated, into the clay. Individually the blocks may be below the specified critical size but together they could form a formidable obstacle to progress. In those circumstances the obstruction would normally be paid at a billed rock rate, but it would certainly be prudent in the contract documentation to make express provision for this.

Example: The mid-1970s saw the construction in the north-east of England of a 100 million gallon per day river intake and pumping station for supplying water to an industrial conurbation. The intake and pumping station were to be constructed in separate sheet-piled coffer dams and to be linked with four mini-tunnels. Information from the site investigation confirmed the reasonableness of the scheme, and also indicated the presence of boulder clay with a boulder being identified at the level of a tunnel horizon. Construction of the coffer dams proved to be difficult with the presence of a number of boulders at tunnel level. So the tunnelling scheme was abandoned and a linking coffer dam was constructed to reveal an almost complete bed of boulders which would have rendered tunnelling extremely difficult, or impracticable. Thus, a seemingly quite reasonable programme of site investigation had failed to reveal the true extent of the boulder problem.

The presence of large boulders could severely affect the progress of a slurry shield even though some shields can negotiate cobbles and boulders up to about 250mm size. For example, the Iseki 'crunchingmole' can handle boulders up to 20% of the tunnel diameter; the 'unclemole' operating in mixed ground of clay, silts, sands, gravels under a water head of up to 30m incorporates a cone crusher head to break boulders of size up to 30% of the installed tunnel diameter. A Herrenknecht rock head (disks) is able to tunnel through very strong rock without losing line and level irrespective of shield diameter, but the larger the size of shield the less is the need for a rock head. A contractor using a Herrenknecht machine may tend to use a rock head most of the time, notwithstanding the fact that it is more expensive to refurbish disks if they are not actually needed. However, if a slurry shield (or earth pressure balance shield) is contemplated for tunnelling in glacial deposits, the risk of large boulders being encountered must be acknowledged by the designer of the works and a very thorough ground investigation implemented. This is especially important in the case of microtunnelling where the tunnelling operation is remote and where the tunnel diameter (<900mm) prevents access to the face for any necessary remedial works. Even so, as noted earlier in this chapter, it would be most unwise to infer absolutely the actual nature of such ground from the results of any prior investigation.

> **Example:** The £5.5m Aller Valley Sewage Scheme passes 7.5km along the Aller Valley from Barton, north of Torquay in Devon, through Kingskerswell to Newton Abbot. There is also a 2km spur from Abbotskerswell to allow a sewage treatment plant there to be abandoned. The pipeline for the scheme varied in diameter from 450mm to 1500mm. Hand excavation in compressed air was chosen as the preferred tunnelling method after the contractor, Delta Civil Engineering, had examined the borehole logs. The firm had considered using a slurry shield machine, but because of the likelihood of encountering large boulders decided in favour of hand excavation. In the event, although most of the tunnel drive was through silty sands and gravels, a large number of boulders that a slurry machine could not have coped with were encountered. The largest boulder removed was reported to be 900mm in diameter.

Flints in chalk can pose similar problems, both mechanically in respect of excavation and contractually in the sense of their randomness and (lack of) anticipation, to those presented by boulders in till.

4.3.4 Filled ground

When it is planned to tunnel through filled ground, special methods of investigation may be required, particularly in industrial areas where waste material of quite variable composition, sometimes including special waste, has been dumped, often at random, in natural valleys. It is important to investigate the following:

- The physical, geotechnical and chemical character of the fill.
- The distribution of water in the fill and its chemical composition.
- The profile and engineering properties of the natural ground beneath the fill, in particular the area close to the interface with the fill.

Although boreholes will be sunk through the material and samples taken, the nature of the waste is much more readily discerned by open pitting. Access to such ground for pitting operations is likely to be relatively unrestricted. Special equipment may be needed for the operatives - resistant clothing in the case of chemical wastes; breathing apparatus and flameproof equipment on the rigs in the case of domestic wastes where there may be a substantial output of methane gas. (Methane output from such tips is a function of internal temperature and water content; production of methane and other gases is an anaerobic process.)

> **Example:** On Tyneside, several of the denes, which are generally at right angles to the River Tyne and carry stream water down to the river, have been filled over the years with industrial and domestic waste. The main north bank interceptor sewer of the updated Tyneside sewerage scheme has passed through these denes which contain waste material of different ages. The fill on one Newcastle upon Tyne city centre street is of 19th century vintage, well compacted and not incurring excessive settlement when tunnelled. (There was precautionary re-laying of old high pressure gas mains with high density polyethylene (HDPE) pipe

through the fill before tunnelling, the cost of this work being shared by the Northumbrian Water Authority (as it was then) and British Gas taking account of the betterment element via the Bacon-Woodrow formula.) On the other hand, the fill in one of the burns was of 1930s age, readable copies of *The Times* newspaper and some interesting artifacts being recovered from the tunnel face. This fill also contained chemical waste and old paint from an adjacent paint factory - materials which could have posed severe problems for the tunnelling contractor and the miners. Quite detailed information of the ground - its chemistry, groundwater and gas composition - was therefore required in order to avoid difficulties during construction and possible Clause 12 claims under the contract.

4.3.5 Rate of soil deformation

Rate of soil deformation is an important parameter in settlement analysis (Boden, 1969; Attewell and Boden, 1971; Attewell and Yeates, 1984; Attewell *et al.*, 1986) and can be useful in assessing shield progress, yet few attempts have been made to measure this parameter under the sequence of stress change conditions that apply in tunnelling (*see* Section 6.3). The paper by Attewell and Boden (1971) outlines the laboratory technique, which has been used on several occasions subsequently. An assessment of this nature only becomes important when tunnelling in urban areas where calculation of ground movements then becomes essential.

Standard laboratory site investigation tests for tunnels in soil will include assessment of c_u (undrained shear strength), Atterberg liquid (L_w) and plastic (P_w) limits, and natural moisture content (w), together with the grading of granular soils. Attewell and Boden (1971) have stressed the importance of checking the relation between L_w and w, and assessing the liquidity index (I_l). Liquidity index is equal to

$$(w - P_w)/(L_w - P_w) = (w - P_w)/I_w,$$

where I_w, the difference between the liquid and plastic limits, is the plasticity index.

As the moisture content, w, of a cohesive soil approaches the lower limit, P_w, of the plastic range, the stiffness and degree of compaction of the soil increase. If the moisture content of a natural soil exceeds the liquid limit ($I_l > 1$), remoulding transforms the soil into a thick viscous slurry which would pose problems at an unsupported tunnel face. If the natural moisture content is less than the plastic limit ($I_l < 1$), the soil cannot then be remoulded. If the liquidity index is near to zero, the compressive strength of the soil will generally lie between about 100kPa and 500kPa. Silt contents, derived from the grading curves, need to be assessed in the context of the Atterberg limits and I_l for their likely effects on ground loss and face stability.

4.3.6 Soil strengths

It is important to establish, to a reasonable degree of accuracy, the soil strengths at the tunnel face by detailed site logging, *in situ* testing where appropriate, and laboratory testing according to BS 1377:1990 (British Standards Institution, 1990b). Unfortunately, there is considerable scope for sample disturbance during the drilling operation (Clayton, 1986) and so there is the possibility that the laboratory tests may produce somewhat pessimistic results with respect to soil strength. The standard laboratory test, notwithstanding its technical disadvantages, is the quick undrained triaxial test (UU) on U100 core samples from which a c_u (shear strength) value is derived. Alternatively, the laboratory vane may be used on the undisturbed U100 sample while still in the sampling tube. The obvious disadvantage of the test is, of course, that it takes no direct account of the soil moisture content nor of any larger structural features in the soil mass. Undrained shear strength is expressed descriptively in Section 4.3.2. Soil unconfined compressive strength is twice the undrained shear strength c_u.

Example: 'Boulder clay' was frequently identified on the logs of the site investigation boreholes sunk in connection with the Tyneside sewerage scheme. However, some of the boulder clay encountered in the tunnel faces proved to be extremely 'hard' and, in terms of its cuttability, could have been regarded as a weak rock. Whenever 'hard' boulder clay is logged at tunnel face horizon, the site investigation report should provide information on its shear strength.

Soil strength is assessed for its ease (or otherwise) of excavation and for its self-support capacity at the tunnel face via an overload factor (*see* Broms and Bennermark, 1967; Attewell *et al.*, 1986, p88). Both of these requirements for design and billing of the works make use of the soil undrained shear strength, and so it would not usually be considered necessary to engage in effective stress testing. The stability ratio, or simple overload factor (OFS), for a soil is expressed as $[\gamma z_0 + q - \sigma_i]/c_u$, where γ is the soil unit weight, z_0 is the depth from ground surface to the tunnel axis, q is any surface surcharge pressure, and σ_i is any tunnel internal support pressure, usually compressed air as a temporary works measure. With a stability ratio of between 1 and 2, the ground loss (and associated settlement) will normally be small and the ground will be stable. In fact, a cohesive soil can normally be tunnelled without undue problems if the stability ratio is less than about 4.

Much credence tends to be placed on the results of standard penetration test (SPT) results (*refer to* Supplementary Information 5 in Part Two of the book). The test results may be used qualitatively for a general assessment of layering and the types of subsoil and/or quantitatively to estimate the engineering parameters, typically the relative density, shear strength and compressibility, of non-cohesive (and cohesive) soils. Although such *in situ* tests should strictly be reserved for granular deposits, it has been shown that the test can be used to estimate the *in situ* strengths of a range of stiff clays (Stroud, 1974; Stroud and Butler, 1975) and weak rocks (Stroud, 1974). It is necessary to check the SPT values carefully to ensure that they are consistent with written descriptions of the material, both on the borehole logs and in the text of the site investigation report, and also with the contract billing in the case where weak rocks may be

weathered and where the billing defines coal measures rocks as being weathered. Care should also be taken to ensure that the test is properly conducted (Thorburn, 1986). For granular materials, the water level in the borehole should be maintained at or above groundwater level, otherwise sand could flow into the casing and erroneous values would be recorded due to the lower density of the intrusive sand. It must also be noted that the dynamic compressibility of granular materials is highly stress path dependent, that is, the stress path that operates during the test and not the *in situ* stress path that applies before the test disturbance. The current SPT hammer weight, after BS 1377 (British Standards Institution, 1990b), is 65kg but an international standard of 63.5kg is proposed. It may be necessary in the future to define the hammer weight in all site investigation reports, and it is not yet certain whether the small difference in weight will have practical significance. The item on SPT in Part Two Supplementary Information 5 expands these comments and relates SPT N-values to granular soil density.

The static cone penetration test (CPT) is also a very useful borehole profiling tool. It can be used to obtain data on one or more of the following: layer stratigraphy and homogeneity over a site; depth of the firm layers, location of cavities, voids, and other structural discontinuities; soil identification; soil mechanical properties; 'driveability' and the bearing capacity of piles. Correlations have been developed between the cone end resistance q_c and soil properties. Direct relations between q_c and soil strength are not as well developed as is the case with SPT. In order to make use of the substantial experience with SPT testing, empirical correlations between q_c and SPT N-values are often used (*see*, for example, that given by Burland and Burbridge, 1985). Reference may again be made to Supplementary Information 5 for further brief discussion on this subject.

Other down-the-hole tests include use of the pressuremeter, typically the Ménard pressuremeter (Baguelin *et al.*, 1978) and the self-boring pressuremeter such as the 'Camkometer' (Windle and Wroth, 1977), and the seismic cone (both providing shear modulus values for use in numerical modelling), screw plate (undrained shear strength), and the piezo-cone (static porewater pressures, permeability and consolidation properties, shear strength, and also overconsolidation ratio of cohesive soils) although these would tend to be used only for quite high-value and non-standard tunnel construction.

It is useful to understand the stress changes and porewater pressure changes that a soil in proximity to the cut surface of a tunnel might undergo as the tunnel advances. This subject has been analysed by Atkinson and Mair (1981), and reference to their paper is recommended.

Swelling clay soils and swelling clays within rock strata (Selmer-Olsen and Palmström, 1989) can create problems for tunnel linings. To these materials should be added clay shales. Swelling may need to be accommodated by crushable inserts behind the permanent lining. In smectites such as montmorillonites and vermiculites the type of cation affects the degree of swelling. Na^+, for example, has a high swelling potential whereas Ca^+ swells to a less extent. These cations can be detected by X-ray diffraction methods. Vermiculite and swelling chlorites do not show volume changes to the same extent as do the montmorillonites. Illite is a larger mineral than smectite, and kaolinite is larger still. Both of these minerals, and halloysite, show less susceptibility to swelling. In a mudrock perhaps 40% or so of these minerals may be of silt size. The larger the clay mineral the less is its swelling capacity. Swelling in the presence of water occurs not only in the pure phase of the mineral but also when it is interlayered with relatively inert mineral.

Other non-clay minerals such as anhydrite ($CaSO_4$) is also prone to swelling, as are some pyrrhotites (sulphides/pyrites, FeS_2) in calcareous shales. Anhydrite conversion to gypsum ($CaSO_4.2H_2O$) involves expansions of just over 100% and pressures up to about 20MPa (theoretical), with actual pressures of 14MPa having been measured. Anhydrite marl, responsible earlier for serious damage to the linings of the Belchen tunnel and the Hauenstein rail tunnel, has affected construction of the Mont Terri N16 state highway tunnel in the Swiss Jura between Ajoie

and the St. Ursane link-up. This 10.2m internal diameter tunnel for two lane, bi-directional flow is the fifth-longest motorway tunnel in Switzerland. Swelling rock was encountered over a length of 1700m (40% of the tunnel) and anhydrite marl over a length of 865m. These problems were compounded by the fact that the rock contains a large percentage of sulphate that is aggressive to concrete (*see* Building Research Establishment, 1991), and so sulphate-resisting cement was required. It was also difficult to obtain suitable samples for laboratory testing, the results from which could be used to estimate swelling pressures for design (40 bars pressure was adopted in the case of the anhydrite).

Gypsum as a bedrock occurs in England as a 100km long narrow tract of shallow bands extending from Darlington in the north to Nottingham. It is prone to the development of solution cavities which have created major foundation problems in some areas such as Ripon in North Yorkshire where groundwater is most mobile. It is also commonly found on bedding planes of weathered fissile shales, sulphate deriving from pyrite oxidation. Pyrite oxidises, with the generation of much heat, to form sulphuric acid which can react with other ions. In contrast to the disseminated pyrite in a soil, expansion pressures will be exerted anisotropically when pyritiferous material is distributed in a stratified manner in a host rock such as a mudrock where bladed crystals can easily grow along the laminations. Typical principal secondary product minerals resulting from pyrite oxidation are jarosite ($KFe_3(OH)_6(SO_4)_2$), giving 115% expansion, melanterite ($FeSO_4.7H_2O$), giving 536% expansion, and anhydrous ferrous sulphate ($FeSO_4$). Older, fissile marine shales are especially prone to expansion as a result of the presence of such minerals, but compared to these degrees of expansion the expansions from the relatively inert original mineral illite ($KAl_3Si_3O_8(OH)_2$) are only up to about 10%.

Acid produced as a result of pyrite oxidation can also react with contact grout around a tunnel lining to produce thaumosite and ettringite. Expansion of these minerals tends to cause cracking of the lining, creating cracks through which acid can penetrate. The Woolwich and Reading beds contain a relatively large percentage of pyrite, and these mineral products are corroding the cast iron lining rings of the London Underground Northern Line in the vicinity of Old Street station. Affected rings are replaced by a stainless steel lining and a more resistant grout.

At the site investigation stage, therefore, when it is planned to tunnel in clays, clay shales or mudrocks, there would be merit in seeking further advice, first on whether groundwater, ground support or excavation problems have been experienced in the past when tunnelling in the material, and then on the clay mineral content from X-ray analysis. It is relevant to note that mudrocks create a special problem for microtunnelling, since they do not break down completely at the cutting head in the presence of water but rather tend to 'ball-up' and clog the mechanism. Production rates tend therefore to be lower in that material than in boulders and very strong rock. There is information on mudrocks in Supplementary Information 6 of Part Two of the book and some points relevant to the X-ray technique are given in Supplementary Information 7 of Part Two of the Book.

4.3.7 Loss of fines

Percussion boring can result in loss of fines, particularly in glacial (fluvio-glacial) soils containing gravel and cobbles within a sand/silt matrix. Realistic information on particle size distribution, particularly in the silt size range, is necessary for decisions to be made on the use of ground improvement techniques such as compressed air, dewatering, grouting or freezing. (In a site investigation report the percentage of silt size material will be assessed in the context of the prevailing water head.) Because such samples may not reflect the true proportion of fines present *in situ*, they may give a greatly misleading impression of gradation and permeability. Any evaluation of *in situ* permeability should therefore be based primarily on the results of permeability tests (*see* Section 4.5.3). Some general comments on grout injection, with particular reference to the influence of grain size, are included in Supplementary Information 8 in Part Two of the book.

Example: Cement grout was used in an attempt to stabilise water-bearing gravel materials for a shaft and short tunnel drive. The site investigation data indicated that the materials comprised clean sands and gravels. *In situ* inspection demonstrated that the materials contained a significant proportion of sandy silty matrix. Because of this the *in situ* permeability was too low to allow effective grout penetration. For the use of an opc grout, typical permeabilities should exceed 10^{-5} m/s.

4.3.8 Chemical tests

Cement is prone to attack from CO_2, Cl, Mg, SO_4, and NH_4. Soil and water samples must always be tested to determine their water soluble sulphate content, pH value and chloride content. High sulphate levels (*see* Building Research Establishment, 1981, 1991; Harrison, 1992) and low pH values (Building Research Station, 1975) could lead to a rapid breakdown of an unsuitable concrete lining and high repair costs.

Building Research Establishment Digest 363 (1991) lists the nature
  * sulphates: calcium, magnesium and sodium;
  * sulphides, particularly pyrites (iron sulphide) oxidising to sulphuric acid and sulphate;
  * humic and carbonic acids;
and genesis
  * humic and carbonic acids from peaty areas;
  * sulphates from colliery spoil tips, older power stations (pulverised fuel ash), brick rubble, blast furnace slag, soil shale residues;
of these concrete-attacking materials.

Table 1 of Digest 363 specifies the concrete requirements depending upon the percentage sulphate in soil or fill and in groundwater. These tests for sulphate will be a standard part of the site investigation. Since certain concrete elements such as basements, culverts, retaining walls and ground floor concrete slabs are more vulnerable to concrete attack than are foundations and piles, Table 1a of the

Digest recommends increased forms of protection. However, the recommendation also is that this requirement for increased protection can be waived for basements, culverts and retaining walls if, after completion of normal curing, the concrete face that is to be exposed to air is protected from rain for several weeks before any initial contact with the soil is allowed. Table 1b of the Digest provides recommendations in respect of the sulphate problem for other types of concrete that may be used for civil engineering purposes, and Table 1c gives codings for the cement and cement additives that are used in civil engineering works. Table 2 provides BRE recommendations in respect of attack from acids having pH values greater than 2.5, as would be determined at the site investigation stage of a civil engineering operation, while Table 3 defines the aggressiveness of carbon dioxide as given in Table 2.

It is noted in the Digest that, although limestone in aggregates can increase the overall vulnerability of concrete to acid attack, under circumstances of very small quantities of acid being generated (for example, on the walls of sewage systems above effluent level) large amounts of neutralising hard limestone aggregate can prolong the life of the construction. There is special reference in Digest 363 to the methods of determining sulphate content and pH levels in soils and groundwater. The tests are described in BS 1377:Part 3:1990 'Methods of test for soils for civil engineering purposes'. There is also reference to other British Standards which bear on this problem.

These aggressive conditions are likely to be especially severe in ground underlying old industrial areas (chemical plants, gas works, coal stocking yards). For areas close to river banks, it will be necessary to check on the tidal range of the river - the movement of sea water up-river - because over a period of time this movement will bring with it the possibility of aggressive chlorides being introduced into the ground. In the case of sulphate and pH problems it will not usually be feasible to flush these contaminants out of the ground either by natural precipitation processes or artificially by flooding with water.

Example: In one low-lying flood plain area on Tyneside high water-soluble sulphate contents of up to 21.8g/litre (in a 2:1 water/soil extract) and pH values as low as 2.3 were found in certain soil samples taken from alluvial clay deposits through which a tunnel had to be driven. The source of the sulphates and acids appeared to be overlying fill deposits which contained fly ash, colliery discard and coal. In the past the area had been a coal stocking yard. The waste materials contained quantities of pyrites (FeS_2) which breaks down by oxidation to ferric oxide, thence in combination with water to form ferric hydroxide ($Fe(OH)_3$) with the liberation of acidic hydrogen ions and sulphate ions. Temperature is an important factor in the breakdown, and oxidation is more rapid for fine-grained, amorphous, and impure forms of pyrite. (Note that in addition to pyrite (FeS_2) there may be marcasite (also FeS_2) and pyrrhotite (FeS) present.) Any presence of organic debris can also be correlated with bacteria. The catalytic or enzymic action of the acidophilic *Thiobacillus-Ferrobacillus* group of bacteria can accelerate the chemical oxidation several hundred times.

A major interceptor sewer tunnel was to be driven through the area. From the ground investigation, there was a randomness of the sulphate concentrations and zones of acidity in the area of the proposed tunnel and it would not have been feasible using a borehole array fully to define the concentration distributions. It was accepted, therefore, that the concrete lining of the tunnel would be exposed to an aggressive soil and soil water environment. Atmospheric oxygen, available *in situ* at the time of construction, would not be needed to promote continuous long-term reaction between the acid soil and the concrete. The neutralised products of the reaction would have little inhibiting effect on the progress of the reaction and it was not reasonable to rely on any natural groundwater flow to secure adequate dilution. The adopted solution consisted of a conventional concrete pipe cast within a 6mm thick GRP protective outer skin.

In the same area, major building construction required the raising of the site level by means of an inert material, together with the use of cast-*in-situ* concrete piles, 14.0m to 17.0m long, taken down to an acceptable bearing horizon. Sulphate resisting cement, having a Class 3 resistance to sulphates (BRE, 1981; now replaced by BRE Digest 363, July 1991), was used for the pile concrete.

Reference should also be made to BS 8004:1986 (British Standards Institution, 1986) for information on foundation concrete when exposed to sulphate attack.

The problems of pyrite in soil have been addressed by Hawkins and Pinches (1986) quoting Temple and Delchamps (1953). A pyrite-oxygen-water mix may be expressed (Penner *et al.*, 1973; *see* also Russell, 1992) as

$$FeS_2 + H_2O + 3.5O_2 \rightarrow FeSO_4 + H_2SO_4.$$

As increasing amounts of sulphuric acid are produced, the consequences of the lower pH are that chemical oxidation of the iron cannot occur so readily. The presence of acidophilic bacteria (most probably *Thiobacillus ferrooxidans* and *T. thiooxidans*) means that the oxidation can continue to take place biologically. Both of these bacteria are autotrophic (i.e. they are able to utilise carbon dioxide in assimilation) and they obtain carbon for cell growth from carbon dioxide, a deficiency of which stops the growth. They are aerobic and able to derive energy from the oxidation of reduced sulphur compounds. Further, *T. ferrooxidans* is able to oxidise ferrous iron [Fe(II):Fe^{2+}]

$$2FeSO_4 + 0.5O_2 + H_2SO_4 \rightarrow$$
$$\rightarrow Fe_2(SO_4)_3 + H_2O.$$

Because ferric sulphate has a strong oxidising effect on sulphides:

$$Fe_2(SO_4) + FeS_2 \rightarrow 3FeSO_4 + 2S,$$

this sulphur usually being oxidised by *T. thiooxidans* to give sulphuric acid

$$S + 1.5O_2 + H_2O \rightarrow H_2SO_4.$$

or, the sulphur may react with ferric sulphate:

$$2S + 6Fe_2(SO_4)_3 + 8H_2O \rightarrow$$
$$\rightarrow 12FeSO_4 + 8H_2SO_4,$$

with the ferric sulphate also being hydrolysed:

$$Fe_2(SO_4)_3 + 2H_2O \rightarrow$$
$$\rightarrow 2Fe(OH)SO_4 + H_2SO_4.$$

Sulphuric acid output increases with increasing bacterial activity as the pH drops, *T. thiooxidans* flourishing down to a pH of 0.8 while the optimum for *T. ferrooxidans* is a pH of between 2 and 4. These organisms develop and act very speedily; numerous bacteria can develop in otherwise neutral water within only a few days. A test for bacteria involves inoculating a ferrous sulphate medium with a suspect soil. Because chemical oxidation should not take place in a low pH environment a change in colour from pale blue/green through yellow to red-brown may be taken as indicating the presence and activity of *T. ferrooxidans*.

There must be an awareness of the effects in tunnelling of the breakdown of pyrite in mudrocks and shales and the production of sulphuric acid. If limestone is present in any form, then the liberation of carbon dioxide from the reaction could be dangerous to miners working in the confined space of a tunnel if the ventilation is inadequate. This is thought to have been the cause of the deaths in a manhole related to the Carsington reservoir works in Derbyshire.

The importance in engineering, particularly to concrete structures, of the self-perpetuating cycle - pyrite breakdown to thiobacteria resulting in sulphuric acid production and with the lowering of the pH increasing the activity of the bacteria - should not be overlooked. The activity of the bacterial organisms is a function of the depth to which oxidising conditions exist. If there is a groundwater flow, then the sulphuric acid and other sulphate products may be removed and a neutral pH of 7 sensibly maintained. Unfortunately, sulphate may also be taken to ground surface by capillarity, perhaps promoted by underground activity, and thereby, with time, affect building foundations. Hawkins and Pinches (1986) have pointed out that anoxic conditions or treatment of the ground with an alkali, say flooding with potassium hydroxide solution to increase the pH and reduce its atmospheric content by raising the water level (Penner *et al.* 1973), may reduce or halt bacterial activity. However, as noted above, such a procedure may not be either economical or practical.

It follows from the above discussion that simple SO_3 determinations (Building Research Establishment, 1981, 1991) must always be accompanied by an appreciation of the total soil and groundwater chemistry and a recognition that continuing changes may take place through weathering agencies. In particular, it will be necessary to determine whether any growth of gypsum/jarosite $[KFe_3(SO_4)_2OH_6]$ could lead to excessive pressures on structures, whether there is a directional sense to the pressure, and whether the effect will be heave of building foundations. Treatments under adverse conditions may include the adoption of sulphate resistant cement, flooding (if feasible - see above) to inhibit oxidation and capillarity, and perhaps the local injection of an alkaline solution into the soil if pockets of suspect ground can indeed be identified. Resistance to sulphate attack can be achieved by increasing the density of the concrete. There is also the possibility of incorporating metakaolin, an alumino silicate produced by English China Clay International by firing china clay (kaolin) at about 800°C, as a cement replacement material. Metakaolin reacts more rapidly with calcium hydroxide than do other pozzolans such as pulverised fuel ash and silica fume; the porosity of the concrete is reduced, it is less prone to chemical attack, and less permeable to ingress of destructive elements such as chloride.

The problems of long-term deterioration of reinforced concrete lined tunnels due to chloride attack have not been so seriously addressed. There is growing evidence that pre-cast concrete lined tunnels in the UK, Japan, Hong Kong, Italy, West Germany and Dubai are suffering reinforcement corrosion problems through chloride ingress. Because sea water has chloride concentrations that are 10 times less than the salt used on roads in winter, the concrete linings of sub-sea tunnels have not been considered at risk. It is now thought, for example, that the linings used for the Channel tunnel on the French side may not be able to resist chloride corrosion over their 120-year design life.

With sub-sea tunnels, the problems arise with watertight design and pressures in excess of eight atmospheres (bars), the level at which it is claimed that reinforcement corrosion is inevitable. If the tunnel is leaky by design, as is the case with the British side of the Channel tunnel, then the lower differential pressures - inside compared with outside the lining - serve to reduce the problem, since less chloride is forced through the interstices of the concrete lining. (The 'trade-off' is likely to be some corrosion of the rolling stock - there is evidence of sea water corrosion of construction locomotives' pick-up systems, leading to failures and a few fires.) Additional problems with subsoil rail tunnels arise from high-speed trains pulling slugs of air along with them, causing reductions in internal pressure and thereby increasing the differential pressures. Further, in the case of a leaky-design tunnel, the slip streams of passing trains will tend to draw up ponded sea water into a spray (mist) which, as well as causing arcing between the overhead power supply and the tunnel crown, will wet the concrete lining surfaces. Repeated drying and re-wetting will then serve to increase the concentrations of damaging chloride. Reinforcement corrosion is further advanced by traction current leakage - return currents from electric trains leaking into the reinforcement cages.

Several solutions to these difficulties have been proposed. The first is to use unprotected lining segments manufactured with high quality pre-cast concrete. Eurotunnel, for the Channel Tunnel construction, have opted for high strength $60N/mm^2$ concrete with 35mm reinforcement cover, but it has been claimed by other outside experts that such a concrete lining could still be inadequate to prevent chlorides reaching corrosion initiation levels near to the reinforcement well within the specified service life of the lining. An additional or alternative measure provides an epoxy coating for the reinforcement, perhaps also with cathodic protection of the steel.

Chemical problems to both personnel (short-term) and concrete (long-term) must also be addressed when tunnels are driven through, or in some cases as a result of downward and lateral migration of contaminants close to, contaminated land. At the desk study stage special attention needs to be paid to the former uses of the land, since many old industries have implanted highly carcinogenic and corrosive chemicals in the soil. Reference may be made to Attewell (1993a) and to Supplementary Information 9 in Part Two of the book for additional comment on this subject. ICE (1992) lists the primary and secondary contaminants for which testing may need to be carried out in soil, the possible contaminants in groundwater, and the gases which may need to be analysed from samples. Attention should also be directed to the Occupational Exposure limits for materials and chemicals as listed by the Health and Safety Executive (1988a, b - *see* also the Health and Safety Commission, 1989). (The Occupational Exposure Limits document also carries a listing of other earlier COSHH documents, and of Guidance Notes in the Environmental Hygiene and Medical Series, with ISBN numbers for easy retrieval.) Very careful attention to the possible risks from contaminants needs to be given from the very early stages of the desk study right through to the design and implementation of the ground investigation.

4.4 ROCK SAMPLING

4.4.1 Core recovery/orientation

There are several textbooks describing site investigation drilling operations (*see*, for example, Joyce, 1982). Although information can be retrieved from relatively inexpensive open hole drilling, errors can arise when attempting to assign cuttings to particular horizons in the ground. It is usual to specify cored drilling and to require the best possible recovery of rock cores in order to define in a proper manner the rock structure, its strength, discontinuity spacing and discontinuity dip with respect to the core axis - 100% recovery should be the aim. Core recoveries of less than 90% would not normally be regarded as acceptable. It may not always be necessary to core

throughout the full length of an investigation hole if a particular section or sections of strata extraneous to the design analysis can be specified with confidence. Either double or retractable-type triple tube core barrels could be used to enhance recovery in difficult cases, possibly in conjunction with mylar linings or various flushing media such as a foam flush. Rock cores should be of adequate diameter, ideally not less than 100mm hole diameter to produce a 75mm diameter core, but in weak rocks a core diameter of not less than 100mm diameter should be specified. In small diameter cores there will tend to be a greater preponderance of fractures induced by unsteady drilling and by retrieval, especially in the weaker sedimentary rocks. In weak rocks it may not always be apparent that certain fractures have been induced, in which case reported fracture spacing values will tend to be low with respect to their *in situ* state. Since fracture spacing strongly affects both excavatability and the (temporary) self-support capability of the rock, the reporting will then be optimistic with respect to the former and conservative with respect to the latter.

The geological structure, defined by such features as bedding and joint dip and direction together with fault locations, can often be inferred after preparation of the geological sections from the borehole log information. However, in some high cost projects it may be deemed necessary to measure discontinuity orientations in specific boreholes. This can be achieved either by the use of core orientation devices in angled boreholes or by impression devices such as the Treifus impression device (Treifus Industries (Australia) Pty Ltd) for use in vertical or slightly inclined boreholes (*see*, for example, Gamon, 1986), or by down-the-hole cameras, although visual inspection (by large diameter boreholes if feasible) is to be preferred.

Dips of persistent discontinuities and bedding planes in rock, and the spacings of the intersections with those discontinuities relative to the tunnel face dimension(s), can affect the stand-up capacity of a tunnel face. Dips into the tunnel face and against the direction of advance can induce shear instability problems of a temporary nature as the face progresses through them. The problem is probably less severe when the dips of low cohesion/friction joints are in the direction of advance or when the tunnel drives along the strike of such major discontinuities.

4.4.2 Rock description

It is important that cores should be logged (described) and reported in a careful manner by personnel trained in the work immediately following extrusion (BS 5930: British Standards Institution, 1981). Some other long-standing references in this respect are the Geological Society Engineering Group Working Party Reports of 1970 ('The logging of rock cores for engineering purposes', QJEG, **3**, 1-24) and 1977 ('The description of rock masses for engineering purposes', QJEG, **10**, 355-388). The descriptions will normally be expressed according to some established classification scheme (*see*, for example, Terzaghi, 1946; Deere *et al.*, 1969; Bieniawski, 1976, 1989).

As in the case of soils, rocks may be described in terms of a mnemonic: CGTSWROS:

C   Colour
G   Grain size
T   Texture and fabric
S   Structure (of the rock mass)
W   Weathering state
R   Rock type
O   Other
S   Strength

The meanings of these descriptors are as follows.

Colour

This is the same as for soils (*see* Section 4.3.2 above).

Grain size

A description of rock in terms of grain size is given in Table 4.7.

Texture and fabric

Porphyritic, crystalline, cryptocrystalline, granular, amorphous, glassy, homogeneous

Structure

Sedimentary - bedding
Metamorphic - foliation, cleavage, lineations
Igneous - flow-banding, tuffs

Reference may be made to Tables 4.8, 4.9 and 4.10 for further definition and quantification of rock mass structures in terms of discontinuity spacing, block size, aperture width and aperture roughness.

Table 4.7 Description of rock in terms of grain size

Rudaceous			Breccia, conglomerate
	2mm		
Arenaceous		Sedimentary	Sandstone
	0.06mm		
Argillaceous			Sandstone, mudstone, chalk
Coarse			Gneiss
	2mm		
Medium		Metamorphic	Schist
	0.2mm		
Fine			Slate
Coarse			Granite
	2mm		
Medium		Igneous	Diorite
	0.6mm		
Fine			Basalt

Table 4.8 Discontinuity spacing terminology

Discontinuities	Planar structures	Spacing
Very widely spaced	Very thickly bedded	>2m
Widely spaced	Thickly bedded	600mm - 2m
Moderately widely spaced	Medium bedded	200mm - 600mm
Closely spaced	Thinly bedded	60mm - 200mm
Very closely spaced	Very thinly bedded	20mm - 60mm
	Thickly laminated (sedimentary)	6mm - 20mm
	Narrow (metamorphic & igneous)	6mm - 20mm
	Foliated, cleaved, flow-banded etc (metamorphic)	6mm - 20mm
Extremely closely spaced	Thinly laminated (sedimentary)	<20mm
	Very closely foliated, cleaved, flow-banded etc (metamorphic & igneous)	<6mm
		<6mm

Table 4.9 Block size

First descriptive term	Maximum dimension
Very large	>2m
Large	600mm - 2m
Medium	200mm - 600mm
Small	60mm - 200mm
Very small	<60mm

Second descriptive term	Nature of the block
Blocky	Equidimensional
Tabular	Thickness much less than length or width
Columnar	Height much greater than cross-section

Table 4.10 Discontinuity aperture width and wall rock roughness

Aperture size	Aperture width (mm)	Wall rock roughness	
		Category	Degree
Wide	>200	1	Polished
Moderately wide	60 - 200	2	Slickensided
Moderately narrow	20 - 60	3	Smooth
Narrow	6 - 20	4	Rough
Very narrow	2 - 6	5	Defined ridges
Extremely narrow	0 - 2	6	Small steps
Tight	0	7	Very rough

Table 4.11 Secondary permeability estimated from mass discontinuity frequency (Anon, 1981)

Rock mass discontinuity spacing	Term	Permeability k (m/s)
Very closely to extremely closely spaced	High permeability	$10^{-2} - 10^{0}$
Closely to moderately widely spaced	Moderate permeability	$10^{-5} - 10^{-2}$
Widely to very widely spaced	Slightly permeable	$10^{-9} - 10^{-5}$
No discontinuities	Effectively impermeable	$<10^{-9}$

Discontinuities may also be described in such terms as are given below.

- Tight, infilled, open
- Planar, curved, irregular
- Rough, smooth

Table 4.11 gives an indication of the level of permeability to be associated with discontinuous rock. Dip magnitude and dip direction will also be required for describing the nature of the discontinuities. The methods of measuring these are given in Hoek and Bray (1981) and Priest (1985, 1993).

Weathering

Weathering grades and descriptions are given in Table 4.12. In some descriptive tables, zones 1A and 1B are not distinguished from one another.

Other

- Stratigraphic location
- Fossils
- Coal seams

Intrinsic strength

Unconfined compressive strength and point load strength for rock material, the latter for correlation against the former, are fundamental mechanical parameters, particularly for excavation works. As a first approximation, following Broch and Franklin (1972), the unconfined compressive strength (UCS) may be taken to be 24 times the point load strength (I_s) for a 54mm diameter rock specimen, or, more generally, UCS = kI_s (Bieniawski, 1975), where the index to strength conversion factor $k = 14 + 0.17d$, where d is the distance between the compression platens, equivalent to the specimen diameter. In any site investigation report, the terminology used for descriptive purposes, particularly on the borehole logs, must accurately match the quantitative value for unconfined compressive strength or point load strength as the case may be. Terminology and strengths are shown in Table 4.13.

Cargill and Shakoor (1990) have produced correlations between UCS and other rock strength parameters. For example:

$$UCS = 23I_{s \text{ (54mm diameter)}} + 13$$

$$ln \ UCS = 4.3 \times 10^{-2} (R \times \rho_d) + 1.2$$
for sandstones
$$ln \ UCS = 1.8 \times 10^{-2} (R \times \rho_d) + 2.9$$
for carbonates

where R is the Schmidt hammer rebound number, and ρ_d is the dry density in Mg/m^3.

$$UCS = 1450 \ (\% \ loss/\rho_d)^{-0.91}$$
from the Los Angeles abrasion test.

UCS is also correlated against slake durability of rock.

Cores often do not receive the detailed logging that the particular circumstances of the tunnel contract are later shown to merit. Such words as 'occasional', 'abundant', 'thin', and 'rare', or 'hard' in the context of discontinuity presence and rock strength, are imprecise and should be avoided unless quantified in a site investigation report. Discontinuity spacings, for example, are quantified in Table 4.8.

Without adequate quantification of the parameters the information thus cannot satisfactorily be used in rock engineering design. Suitable definitions of all terms that are used on the written log for presentation to contractors bidding for the tunnelling work should be included as a standard table at the front of a site investigation report.

Example: Argillaceous 'rock' horizons may contain weak bands, stronger fossiliferous layers, calcareous clayey horizons or limestone bands. On one contract in Cleveland involving hand excavation within the protection of a shield this type of material was logged as a 'mudstone with occasional thin bands of limestone'. Because neither the frequency of the stronger limestone bands, nor their thicknesses, were quantified in a site investigation report, a contractual claim for extra payment was successfully prosecuted. Difficulty of tunnelling in such rock and the

Table 4.12 Engineering classification of weathered rock

Zone	Description	Definition	Likely engineering characteristics
1A	Fresh	No visible sign of rock material weathering; no internal discoloration or disintegration	Normally requires blasting or cutting for excavation; may be self-supporting in excavations, but may sometimes require support in tunnels if closely jointed
1B	Faintly weathered	Slight discoloration of major discontinuity surfaces	
II	Slightly weathered	Some discoloration on and adjacent to discontinuity surfaces; discoloured rock is not significantly weaker than undiscoloured fresh rock; weak (soft) parent rock may show penetration of discoloration	Normally requires blasting or cutting for excavation; suitable as a foundation rock but with open jointing will tend to be very permeable; not suitable as an aggregate rock
III	Moderately weathered	Rock is significantly discoloured; discontinuities will tend to have been opened by the weathering process and the discoloration penetrated inwards from the discontinuity surfaces; less than half the rock material is decomposed or disintegrated to a soil; samples of rock containing discolouration are noticeably weaker than the fresh undiscolored rock; an original weak rock will comprise relic blocks of mainly weathered material	Can sometimes be excavated without blasting or cutting (i.e. by block leverage at the discontinuities); will be relatively easily crushed by construction plant moving over it *in situ*, but with care is suitable as a foundation rock; suitable as a low permeability fill; joints may have acquired lower friction characteristics (check whether infilling is degraded host rock or imported frictional/cohesive soil) so rendering side slopes and tunnel roofs unstable
IV	Highly weathered	Rock is substantially discoloured and more than half the material is in a degraded soil condition; the original fabric near to the discontinuity surfaces will have been altered to a greater depth; a deeply weathered, originally strong, rock may show evidence of fresh rock as a discontinuous framework or as core-stones; an originally weak rock will have been substantially altered, with perhaps small relic blocks but little evidence of the original structure	Can be excavated by hand or ripped relatively easily; the rock is not suitable as a foundation for substantial structures, but may be appropriate for lightly loaded structures; weathered rock may be used as a fill; new discontinuities may have formed in the fraction as-yet unweathered; may be unstable in steep cuttings and will require continuous support in tunnels; exposed surfaces will require erosion protection

Table 4.12 (Continued) Engineering classification of weathered rock

| V | Completely weathered | Rock is substantially discoloured and has broken down to a soil, but with the original fabric still largely intact; the soil properties are a function of the composition of the parent rock, e.g. a sandstone with a substantial quartz content goes to, say, an SM-SC soil; in the case of an originally weak rock, the soil product will be discoloured and altered, with little or no trace of the original rock structure remaining | As in IV above |
| VI | Residual soil | Rock is discoloured and completely degraded to a soil in which none of the original fabric remains; there is a resultant large volume change; the soil has not been significantly transported | Unsuitable for most foundations; any stability on slopes relies upon vegetation rooting and there will be substantial erosion without the presence of any hardcap or preventative measures; careful selection and purpose of use knowledge are needed. |

Table 4.13 Rock strengths and descriptions

Term	Field recognition	UCS (MPa)	I_s (MPa)
Rock			
Extremely strong	Rocks ring on hammer blows; sparks fly	>200	>12
Very strong	Lumps only chip by heavy hammer blows; dull ringing sound	100 - 200	6 - 12
Strong	Lumps or core broken by heavy hammer blows	50 - 100	3 - 6
Moderately strong	Lumps or core broken by light hammer blows	12.5 - 50	0.75 - 3
Moderately weak	Thin slabs broken by heavy hand pressure	5 - 12.5	0.3 - 0.75 (1)
Weak	Thin slabs break easily in hand	1.25 - 5	0.075 (0.1) - 0.3
Very weak	Crumbles in hand	<1.25	<0.075 (0.1)
Cohesive soil			
Very stiff	Can be indented by thumb nail	>0.3	

Notes on Table 4.13:
- UCS means unconfined compressive strength.
- I_s means point load strength. Reference may be made to Broch and Franklin (1972), Hassani *et al.* (1980), Norbury (1986), Panek and Fannon (1992), and Singh and Singh (1993).
- The brackets after some of the I_s values denote an alternative range that has been suggested.
- The undrained shear strength value for cohesive soil is half the respective unconfined compressive strength value quoted in the table. This 'very stiff' soil value is entered in this rock strength table because the material behaviour approximates to that of a rock.

rate of progress depends not only on these parameters but also upon such factors as the strength of a particular bed, the discontinuity density in it, its position in the tunnel face and whether it is juxtaposed by weaker bands of rock or clay. If a strong band is thin and/or replete with discontinuities and/or sandwiched by the weaker bands then it will be easier to excavate by a combination of direct pick action and leverage. It is sometimes claimed that a face could not be excavated by normal equipment (usually meaning FL22 jigger picks) and that heavy duty equipment (for example Haus Herr H11 'German jiggers') had to be brought on to the job. Such a potential claim should be assessed pre-emptively by formally setting up a ground reference mechanism (*see* Section 5.2.3) in order to determine the ease or difficulty of excavation tied in to payment for work done. On-site checks for placing the current face into a particular reference category would be conducted at specified intervals by a representative of the Engineer (often an ARE) and a representative of the contractor (usually the agent). These joint visits should take place at least once per shift.

A situation should not be allowed to arise whereby the Engineer's representative does not attend on one or more occasions, because in those circumstances the contractor will claim that his uncorroborated records fill in the lacunae (gaps) in the joint (common), or Engineer's own, records. Under the conditions of an arbitration, the arbitrator should normally accept the Engineer's records.

There will be a scheme of standard symbolism to be used to denote each rock type when reporting the borehole log information. Symbols to be used for the principal soil and rock types likely to be encountered in the UK are shown in British Standard 5930, Tables 11 and 12, pp117, 118 (British Standards Institution, 1981).

An appraisal of rock strength is clearly important and should take account of the groundwater environment from which the core has been taken, noting that even if the core is encapsulated on retrieval it will have suffered some degree of desiccation before being tested for strength. Unconfined compressive strength differences of 120MPa dry and 40MPa wet have been recorded for a sandstone.

Difficulties of excavation and of excavation support could be overestimated and underestimated, respectively, by the results of a laboratory test on such material.

Any terms quantifying rock strength on the borehole logs must be defined in tabular form according to, and referenced to, an established authority such as the International Society of Rock Mechanics (Brown, 1981b) or the Geological Society of London (Anon, 1970). There is more discussion on weak rock in Supplementary Information 4 in Part Two of the book.

As noted in Section 4.3.6, although standard penetration test (SPT) values are usually related to relative density in soils, such tests are also performed on rocks, particularly weak rocks. There should be a standard penetration test conducted when rock is first penetrated and thereafter at 1m intervals or at each change of strata. As noted in Supplementary Information 5 in Part Two of the book, if there is refusal (more than 50 blows), then the blow count and penetration should be recorded. U100 samples may be taken in soft rocks, such as chalk and marls.

There are various forms of pictorial log presentation (*see*, for example, Attewell and Farmer, 1976, pp469-470, 472-473; British Standard 5930: British Standards Institution, 1981, pp92-93, 96), each investigation organisation tending to develop its own house style of reporting. One possible suite of symbols to be used alongside the logs to define the sampling and test operations at the borehole are shown in Supplementary Information 11 in Part Two of the book.

Rock Quality Designation (RQD) (*see* Deere, 1964) is defined as the length of core recovered as solid cylinders, each greater than or equal to 100mm long, expressed as a percentage of the length of core drilled using an NX (57.2mm) or larger diameter core drill. Fracture frequency, the number of fractures encountered per metre length, may be related to RQD (*see* Supplementary Information 10 in Part Two of the book).

The degree of natural fracturing in a rock mass may also be assessed on a lithological basis. Hawkins (1986) has suggested that a lithological quality designation (LQD) could often be more helpful, and would be included on the borehole log sheet in relation to the material descriptive length rather than to the length of core run. LQD_{100} and RQD_{300} are defined as the percentage of solid core present of length greater than 100mm or 300mm within any lithological unit. These reference lengths are quite arbitrary and it has been recommended that beneath both the rock quality designation and the lithological quality designation the maximum intact core length (MICL) also be given. It has been argued, however, that LQD information is given anyway on the borehole logs by the fracture index, so making LQD somewhat redundant.

Logging should include total core recovery, solid core recovery, RQD and fracture frequency. Detailed logging of discontinuities - their dips with respect to the borehole axis, their surface characteristics, and so on - should be carried out for cores at tunnel horizon. It is important that the bases of the RQD values be defined. For example, when - as is often the case - the discontinuities are inclined to the borehole core axis then the distance measurements for solid stick lengths should be taken, in effect, at the core axis location. Also, some account should be taken of the lithological factors which affect the RQD value. Tightly banded rocks having differing strengths will tend to show low RQD values. Priest and Hudson (1976) have related the critical RQD solid stick length of 100mm to a broader scheme of fracture frequency which may form a better understanding of the influence of intrinsic fractures on both mass shear strength and the process of rock mass excavation.

In addition to descriptions of rock type (SANDSTONE, LIMESTONE, and so on), an indication should be given of the degree of weathering to which the rock has been subjected since this will affect its strength and perhaps its permeability. The standard terms are: 'fresh rock' (Grade I), 'slightly weathered' (Grade II), 'moderately weathered' (Grade III), 'highly weathered' (Grade IV), 'completely weathered' (Grade V), and 'residual soil' Grade VI). A fuller description is given in Table 4.12. In site investigation reports, terms such as 'slightly', 'partially', and 'wholly' are often used, but it is rarely clear whether these refer to the *degree* of weathering or to its *extent* (spread). The difference can be

important; the degree of weathering is more readily discerned in a core sample than is the extent to which the weathering has infiltrated the rock mass and which might affect both excavation and support. It is most important to get the weathering description correct on the borehole logs because a higher degree of weathering implies a greater ease of excavation.

When in the core box the cores should be photographed in colour and identified by means of a job board. A standard colour chart and graduated scale should also be included in the same photograph as an aid to identification at a later stage.

The description of a rock core, as it appears on the log, would typically include the following in sequence: weathered state; structure; colour; grain size; subordinate particle size; texture; alteration state; cementation state, if relevant; intrinsic strength; mineral type; rock geological NAME. (Even if strength data are available, the use of a descriptive term on the logs in lieu of tabular data in the text of a site investigation report avoids the implication of precision where, in fact, such precision rarely if ever exists.). Examples of such a description could be: 'slightly weathered, thickly bedded, yellowish, medium-grained, strong, dolomitic LIMESTONE', or, 'fresh, medium-to-thickly bedded thinly laminated closely-to-very-closely jointed, dark grey, fine-grained, strong, brittle, effectively impermeable SHALE. Reference may also be made to the British Standard order of description in BS 5930 (British Standards Institution, 1981).

A scheme of routine logging may not be sufficient for expensive and potentially hazardous projects such as tunnelled sea outfalls. In these situations the cores spanning the tunnel face area should be re-examined and carefully re-logged in much more detail.

Example: For a sewage sea outfall scheme in the north-east of England, the site investigation borehole rock cores were removed from their boxes and then re-assembled in plastic guttering. The guttering was placed on lengths of white paper with borehole depth and the proposed tunnel horizon clearly marked. The cores were laid out in the guttering in such a way that their positions were fixed with respect to ordnance datum and the several lithologies were correlated between cores by stringing coloured tape, matched to lithological boundaries, across the cores. Because of the depths to which each borehole had been sunk, this exercise involved the hiring of an old factory having sufficient floor length and width, but the operation proved to be worthwhile in that it did allow the cores to be examined in great detail and in circumstances which replicated as closely as possible their *in situ* state.

The important procedure for examining clay cores and the use of colour photography for maintaining records has been mentioned in Sections 4.3.1 and 4.3.2.

4.4.3 Rock core protection and handling

Cores are expensive to obtain. Deterioration of cores may occur as a result of exposure, weak mudrocks being particularly vulnerable. The cores may provide misleading information if they are not properly labelled, protected and handled. They should be packed carefully into custom-built boxes to prevent sliding during transport. The core boxes should be of solid construction and equipped with latched lids; the weight of box and cores together should not exceed 60kg. Boxes with cores should be handled carefully, stored horizontally, protected from the weather and not be allowed to experience temperatures below 5°C. In every case the contract title, exploratory hole reference number and the depth of coring contained in each box should be clearly marked in indelible ink inside the box, on top of the box, and on the right hand end of the box. The shallowest core should lie at the left hand side of the box near the hinge when viewing the open box from the catch (non-hinge) side, and with depth increasing from left to right and towards the viewer. The cores should be properly labelled, and blocks marked clearly with the depth should be inserted at the top and bottom of the total core inside the box and between core runs. It is also advisable to mark the tunnel horizons, top and bottom, in the core box. All the core should be sealed with

brushed wax or wrapped in plastic or foil to prevent moisture loss. In the case of weak cores a shrouding with polyurethane foam could be advantageous. After logging, some core samples will be removed for testing, and substitute labelled spacers (sawn timber, for example) having the same lengths as the removed core samples, inserted in the boxes. The remaining cores should be re-waxed or re-wrapped to await inspection by the contractors tendering for the tunnelling work. Following the inspections, which must be recorded since they will have relevance in the event of later claims for extra payment by the winning contractor, the cores must be wrapped yet again for long-term storage under controlled temperature conditions. Under no circumstances should they be allowed to dry out or freeze. Care should be taken when handling and stacking the core boxes. It is essential that the cores should be retained, re-packed, carefully protected and stored, for later inspection by the parties to the tunnelling contract in the event of a contractual claim.

4.4.4 Rock strengths

In a tunnel, rock strength is manifest in terms of its deformability and its excavatability. Deformability may be expressed, after Terzaghi (1946), as 'squeezing' (a physical process) and 'swelling' (a chemical process).

Squeezing rock moves into the tunnel slowly in a viscous manner without significant volume increase, a prerequisite for squeeze being a high percentage of microscopic and sub-microscopic particles of micaceous minerals or of clay minerals with a low swelling capacity. Less commonly, according to Terzaghi, the squeeze can be chiefly due to an increase in the water content of the material. Aydan *et al.* (1993) note that squeezing may involve three possible forms of failure of the surrounding medium:

• *Complete shear failure* with multiple shear lines around the excavation leading to splitting and detachment of the rock; this is widely observed in ductile rock masses or in masses having widely spaced discontinuities.
• *Buckling failure* which is generally observed in metamorphic rocks such as phyllite and

mica-schists or thinly bedded ductile sedimentary rocks such as mudstone, shale, siltstone, sandstone and evaporitic rocks.
• *Shearing and sliding failure* which is observed in relatively thickly bedded sedimentary rocks and involves sliding on bedding planes and shearing of intact rock.

Swelling rock advances into the tunnel mainly on account of intrinsic expansion, the capacity to swell being mainly limited to those rocks which contain clay minerals such as montmorillonite which swell substantially in the presence of water. Swelling clays, shales (Mesri *et al.*, 1994) and rocks are likely to exert much heavier pressure on tunnel supports than will ground materials without any marked swelling tendency. Swelling minerals are considered in Section 4.3.6 above.

Squeezing is also a stress relaxation phenomenon and is thus related to depth of burial and the general state of stress in the ground. Swelling is more protracted than squeezing, and they may take place together in particular types of rock and rock stress conditions.

It is important to ensure that sufficient tests are carried out on borehole core material to confirm the range of intrinsic strengths in the bedded rock present at and close to the tunnel face horizon since these strengths are frequently used to define important tunnelling criteria such as cuttability, and thereby the adoption of appropriate plant. Preconceived notions of rock strength are not always borne out by the test results. Although sandstone strengths are often below 50MPa and rarely above 100MPa, quartzitic sandstones may have unconfined compressive strengths of up to 250MPa. Unconfined (uniaxial) compressive strength tests (*see* Table 4.13 and its associated discussion together with Supplementary Information 12 in Part Two of the book on the determination of UCS) would often be supplemented by more rapid (less expensive, and thus more numerous) point load tests, but other options are available. If point load tests predominate, it will be necessary to establish a satisfactory correlation with unconfined compressive strength for individual projects, as noted above. Since a conversion factor from I_s to UCS is required for each rock type, the

point load test is not suitable for highly and closely variable strata.

Example: Maximum and minimum rock strengths were defined for a tunnelling contract on the basis of two to eight samples tested per borehole. Insufficient care had been taken to ensure that the full range of rock strengths had, in fact, been established. In the event, rock strengths greater than the maximum reported from the site investigation were encountered at the tunnel face. Re-assessment of the borehole cores, and particularly the strength profiles, following notification of a Clause 12 contractual claim demonstrated that the stronger rock could have been readily predicted. It is interesting to record that (fortuitously, albeit anomalously) the descriptions accompanying the borehole logs in the factual site investigation report did, in fact, accommodate the higher strengths, but because the strength values actually acquired had indeed been tabulated the Engineer decided that the claim should be paid following a re-calculation of productivity loss.

Unconfined compressive strength (UCS) is usually adopted as a convenient contractual criterion of *rock excavatability* by, for example, a roadheader or tunnel boring machine. Volume excavation rate by mechanical system tends to be inversely proportional to the square of the strength of the rock (Farmer, 1987). Thus, a doubling of strength will decrease the volume excavation rate by a factor of 4 approximately. As a very rough approximation, average tunnelling rate is normally found to be inversely proportional to the rock strength, usually defined as UCS.

In practice, the fragmentation process tends to be one of splitting by tensile force, shearing and abrasion, although crushing forces do operate, notably when discs are used in intrinsically stronger rock. A fundamental aim must be to minimise the specific energy needed to fragment the rock by keeping the resulting particle size high and being assured that brittle fracture occurs. When cutting with drag picks, an economic rate of advance depends, with other machine factors such as the number of tools, the tool lacing pattern, the tool radial/forward attack, tip geometry and grade of carbide in the tip, the cutter head speed and power, and the general rigidity of machine construction, upon the pick change frequency, and although rocks having unconfined compressive strengths in excess of 100MPa can be cut, a practical upper limit for an economic advance rate seems to be about 50MPa for a roadheader cutting quartz-bearing rocks. (Much stronger rocks can, of course, be cut using discs on a tunnel boring machine.) As the rock strength increases, a practical solution is to reduce the number of picks on the cutting head accordingly so as to increase the overall pick force. If a roadheader is to be used, a transverse head machine may be suitable for moderately strong ground but an axial head machine will be needed for moderately strong to strong rock.

An underlying, and incorrect, assumption here is that rock must necessarily be cut for it to be excavated. The closeness, and to some extent the orientation, of pre-existing discontinuities in the rock affect and assist the action of the cutter. At one extreme a rock of high intrinsic strength (say in excess of 200MPa) that is very closely fractured (say discontinuities at about every 100mm spacing) may be excavated by cutting machine in which the picks or discs serve merely to wedge, rotate and translationally displace the individual blocks of rock away from the mass. Work performed in the late 1970s by the British National Coal Board's (British Coal's) Mining Research and Development Establishment suggests that, for homogeneous and massive rocks having joint spacings greater than 150mm, cuttability depends substantially on rock material properties but that where the joint spacing is greater than 10mm but less than 150mm cuttability depends upon both intact rock properties and discontinuities. Evidence of rock tunnelling with discs suggests that the tunnelling rate increases out of proportion to the decrease in discontinuity spacing; that is, the rate of advance speeds up significantly as the joint spacing decreases. In the more general sense of excavatability (digging, ripping, blasting to loosen, blasting to fracture and displace), relations between unconfined compressive strength and fracture

frequency have been used as a basis of rock mass classification and reference ground conditions for contractor excavation pricing in which the client receives tendered rates from contractors for excavating a tunnel through specified ground conditions (for example, Carsington and Megget dams in Britain).

Singh (1989) has noted that machinability, expressed in terms of cutter wear, depends on the grain size of the rock, bonding strength of the matrix material, and the percentage of silica (actually, free quartz), and that the rate of penetration depends on the strength and hardness of intact rock, size, shape, number and geometry of the cutting tools, the thrust and torque available, the mode of excavation, geological features such as the state of stress, and 'environmental factors' which would include, as noted above, discontinuity density. In summary, he identifies four different but important parameters: the portion of free quartz, the point load strength index (MPa), the average grain size (mm) of quartz and hard minerals, and the Schmidt rebound hardness. The point load strength index represents the strength of the mineral components and the bond strength of the matrix material. Quartz grains of size less than 0.025mm have little or no influence on abrasivity, although the quartz fraction may be used as a general expression of hardness and abrasivity. (A coarse-grained sandstone causes up to 50 times greater pick wear than does a mineralogically similar, but finer-grained, rock.) West (1989), describing a laboratory test for abrasiveness, has confirmed the relation between that parameter for Coal Measures sediments and their quartz contents, noting that British rocks having abrasiveness values exceeding 4 or 5 may give rise to high rates of wear on tunnelling machine cutters. Braybrooke (1988) notes that quartz content is the dominant variable in pick wear but only when it exceeds about 50% by volume in the rock. If thought to be necessary, the effect of other minerals on wear may be expressed as an equivalent-to-quartz hardness (Rossival's scale), this hardness being related to Mohs' scale of hardness. Roxborough (1987) also notes that pick wear is a function of quartz grain size, quartz grain angularity, and the degree of cementing in the rock. Abrasive wear increases as the square of the quartz content of the rock and also increases linearly with one-dimensional grain size.

It is clear from laboratory work that a whole range of variables affects rock cuttability. To assist a practical assessment Roxborough (1987) has suggested that cuttability be based on a simple strength evaluation:

Specific energy = 0.25UCS + C

(for UCS not greater than about 30 to 40MPa)

where UCS is the unconfined compressive strength of the rock material,
    and
constant C relates to pick wear (abrasivity and shock spalling).

Laboratory-determined specific energy values have been tabulated by McFeat-Smith and Fowell (1977, 1979). In general terms, within a specific energy range up to about 20MJ/m³ the performance of a medium-weight roadheader will be 'moderate' to 'good' at specific energies up to 8MJ/m³. Within a specific energy range up to about 32MJ/m³ a heavyweight (heavy-duty) roadheader will perform 'moderately' well at specific energies up to about 17MJ/m³. These figures assume that the excavation takes place in massive, widely jointed bedded rock so that the intrinsic discontinuities can be assumed not to contribute to the cutting performance. Hawkesbury Sandstone (*see* Pells, 1985), which underlies and outcrops in the area of Sydney, Australia, is a practical example of a rock for which the parameters affecting cutting performance have been determined. In conjunction with the submerged tube tunnel underlying Sydney harbour, and which acts as a duplication for the heavily over-trafficked Sydney Harbour bridge, the sandstone has been tunnelled on the north bank of the harbour and on the south bank under the forecourt of the Opera House. The rock contains 68% quartz, and with specific energy values ranging from 8 to 14MJ/m³ cutter wear is in the range 2.2-3.3mg/m. As a target figure for excavation, with a specific energy value of 10MJ/m³ a cutting rate of about 40m³/hr was deemed achievable. Consideration of these parameters will again

come to the fore in connection with the rail link between the capital and the airport in Botany Bay.

Reference may also be made on the subject of excavatability to Section 5.4 and to Supplementary Information 14 in Part Two of the book. The paper by Tarkoy (1973), the theses by Glossop (1982) and Athorn (1982), and the reports by Morgan *et al.* (1979) and Ian Farmer Associates (1986), may also be consulted. 'Boreability' indices, comprising drilling rate, bit wear and cutter life, have been suggested by Chen and Vogler (1992). Tests formulated by the Norwegian Institute of Technology can also be used to evaluate 'boreability'. These include an impact test which assesses brittleness and measures the energy for rock crushing, a miniature drill test which gives an indirect measure of rock surface hardness, and a test which measures rock abrasivity using rock powder on a sample of cutter steel. For more general reading on excavatability and rippability, but in an open excavation setting, reference may be made to Weaver (1975), Megaw and Bartlett (1981), and particularly the paper by Robbin in volume 1 page 284 of that publication, MacGregor *et al.* (1994) and to Pettifer and Fookes (1994).

The strength of weak rock may also be assessed *in situ* by the use of down-the-hole standard penetration and cone penetration tests. Weak rock, and the subjects of SPT and CPT, are lightly covered in Supplementary Information 4 and 5 respectively in Part Two of the book.

4.4.5 Slaking tests

Slaking and slurrying can pose major problems in excavation and handling, particularly in mudrocks (refer to the paper by Varley, 1990). Hence, the potential for the material to break down in the presence of water could be a key factor in estimating realistic rates of progress. Slake durability tests should be regarded as essential for mudrocks, limestones, chalk and dolomites in order to quantify possible slaking problems in the tunnel. Such problems should be expected where slake durability values of less than 85% are identified. Dipping rock strata can cause problems with slaking where water runs into the tunnel face, and so such values should be appraised in the context of the local geology and hydrology. Slurrying could create an obvious problem when mechanised mucking is contemplated; tracked plant should be chosen in preference to tyred vehicles.

Slake durability tests (Franklin and Chandra, 1972; Brown, 1981b), involving abrasion (attrition) as well as exposure to water or other fluids if deemed necessary, can be supplemented by simple water tests, the base of a sample being placed in water and its rate of disintegration measured. In a slake durability test dry rock is rotated within a wire drum in water (or other fluid) and the dry weight retained after 10 minutes of revolution at 20rpm is recorded and expressed as a percentage of the original weight. Two-cycle tests have been suggested, and there are reasons for believing that a three-cycle test is required before a constant value of the slake durability index is achieved for strong mudrocks (siltstones). Such tests may provide a useful index of material degradability, reflecting the actual behaviour of a rock when wet and exposed to the atmosphere at a tunnel face.

> **Example:** A water-bearing mudrock, correctly identified at the site investigation, was encountered at the tunnel face which was excavated by a roadheader. The slurried rock proved difficult to remove and also caused stability problems to the primary support system. It would have been advisable to institute laboratory slaking tests at the site investigation stage and to note the results of the tests in the site investigation report, and also in an engineering report to have clearly identified the potential problems associated with slurrying. The tendering contractors could then have priced for more suitable plant and more realistically estimated the support erection problems.

4.5 GROUNDWATER AND PERMEABILITY

4.5.1 Groundwater levels

Because so many decisions involving the method of construction, such as the possible use of compressed air (*see* Sections 5.6 and 6.4), the alternative use of dewatering (and the likely effect of associated settlement on nearby property, especially in organic and other highly compressible soils) or a full-face machine, pricing of the work, and so on, depend on reliable information about groundwater, determinations of water level, artesian or sub-artesian pressures, the possible presence of perched water tables and ground permeability must be conducted carefully. The importance of such information, together with evaluations of the implications of groundwater pressures and inflows to the tunnel, cannot be over-emphasised. These factors will occupy an important place in the Engineer's report. The aim must be to establish a benchmark of mutual understanding, between Engineer and contractor, on the groundwater effects in a contractual setting. An alternative is for the contractors to be required to allow for water makes as a contract risk, perhaps up to a specified rate of flow, with any excess being paid extra.

Ground dewatering in advance of a tunnel face in order to assist construction draws down the water table to the sides of the tunnel and, in the case, say, of a tunnel centre line passing along a road in an urban area, will remove some foundation support from buildings either side the road and possibly lead to structural damage. In the case of *Langbrook Properties Ltd* v *Surrey County Council and Others* [1969] All ER 1424, involving a 13-storey office block which experienced differential settlements attributable to the dewatering of an adjacent excavation, Mr Justice Plowman held that there was an unfettered right to abstract water from one's land, and so an action in nuisance could not lie against any of the defendants who were each acting within their rights. This judgment is in contrast to the rights of mineral support (soil, rock, and so on) where the duty to support comes into being when the support has been provided for longer than 20 years.

Consideration must also be directed to the longer-term effects of possible rising groundwater tables upon the continuing integrity of tunnel linings. Many major cities obtain water by pumping from the ground. Changes to industrial practices and water supply systems over many years have led to reduced abstractions. Because of this, water levels which had been depressed have now risen beneath major cities such as New York, Tokyo, Paris, London, Birmingham and Liverpool. In 1965 the piezometric levels in west and central London were less than -60m O.D. compared with around +10m O.D. before the underlying groundwater was exploited. The plan area of that region having a groundwater level less than -30m O.D. has been approximately halved since 1965, the rates of rise having been greatest in west-central London. At Trafalgar Square a rise of 1m/year has been recorded over the last 15 years.

Design of new tunnels needs to take account of such factors as the influence of water changes upon particular types of ground (swelling and heave of clay soil being one potential problem area) and of increased water pressures on linings together with the greater potential for leakage through seals between segments, chemical attack, and also of accompanying buoyancy changes which would modify the design loadings. It has been estimated, for example, that about 30% of the existing 130 kilometres of tunnels under London, located near the clay base or in the sandy deposits, are at such risk from increased seepage, chemical attack, and increased lining loads. For these tunnels, the cumulative cost of remedial works, especially prevention of seepage, will certainly exceed £10 million if the groundwater level rises without control, and this figure takes no account of any disruption to public services. As an example of chemical problems, acid attack on the Northern Line of the London Underground, which lies between the Woolwich and Reading beds, will require lining replacement at an estimated cost of £1.3 million. It is thought that the acid is contained within the soil and its effect concentrated by the stream flows. There must clearly be substantial rises in the cost of new tunnels to take account of all these factors.

4.5.2 Piezometers

After retrieval of samples from cored boreholes, or after completion of open hole drilling, piezometers (hydraulic/pneumatic) and/or standpipes will often be installed by experienced personnel in ground investigation boreholes, generally located in the zone close to the tunnel face so that any changes in the groundwater flow regime can be monitored before, during and after construction. Smearing of clay soils and mudrocks at the sides of a borehole can render piezometer readings and permeability (*see* Section 4.5.3) calculations inaccurate.

Example: On a 1.5m diameter tunnel project in County Durham, piezometers were not installed within water-bearing sands and silts, since most of the tunnel was expected to be in clay strata. Boreholes did not indicate a high water table. When silts were encountered at the tunnel face, considerable quantities of silt carried by groundwater inundated the tunnel and led to large ground movements at the surface with some property damage. Piezometers, had they been installed, would have confirmed that a high water table existed and that a construction problem was likely to arise. Well-point dewatering was used to provide stable conditions for the tunnelling.

Where several water-bearing layers may be present in the area of the tunnel, piezometers should be installed to terminate at various levels in order to check piezometric pressures in the separate layers.

Example: Granular strata were encountered in a shallow tunnel driven beneath a railway embankment in County Durham. Shallow pumped wells were unable to cope with the large seepages. Boreholes had failed to identify the presence of two aquifers separated by a clay layer, since standpipes had only been installed in the upper aquifer. Deep pumped wells were sunk to dewater the lower aquifer.

As with simple standpipes, all piezometric water levels should be monitored on a daily basis until equilibrium is achieved. The levels should be recorded each week thereafter and be related to known precipitation conditions. Monitoring should ideally be extended over a winter period with levels recorded monthly or bi-monthly in order to evaluate seasonal fluctuations. In tidal areas, monitoring should be carried out at hourly intervals over a typical daily cycle in order to assess tidal effects, or alternatively an automatic recorder should be used.

4.5.3 Permeability

Permeability tests are important to establish likely inflows and to determine the suitability of possible treatment measures such as dewatering (*see* Section 4.5.1 above) or grouting (*see* Supplementary Information 8 in Part Two of the book). The actual test techniques and formulae to be used for interpretation of the results are given in numerous publications (*see*, for example, the British Standard 5930:1981, pp30-40) and will also be specified in detail within the in-house manuals of site investigation firms. Some of the formulae that can be used are shown in Supplementary Information 13 in Part Two of the book. Sufficient information can often be obtained from simple borehole permeability tests - falling head, constant head, or rising head tests. Rising head tests are generally to be preferred since they are self-cleansing, but some care is needed to prevent 'blowing' in, for example, loose granular soils. Gravel packs may therefore need to be specified. Permeability tests could be performed in piezometers within clay soils. When the efficiency of piezometers is suspect, they can sometimes be cleared using compressed air to remove debris clogging the tip. However, this operation must be done with care.

> **Example:** A 1m diameter tunnel on Tyneside was constructed through boulder clay containing a 0.5m thick sand layer which was present at tunnel level. Permeability values obtained from tests in boreholes were used to predict water inflows at the tunnel face. As a result, compressed air was specified for the tunnelling works.

Pumping tests should be conducted during the ground investigation if it is considered that water ingress could pose major problems during construction and that drawdown, as the tunnel acts as a sink, could affect superadjacent property. Definition of the drawn-down phreatic surface caused by pumping at one hole requires the sinking of additional observation holes. These tests are expensive to conduct and require careful interpretation, but they do serve to characterise an aquifer through the determination of the coefficients of permeability, transmissivity and storage, and often prove to be effective in predicting groundwater inflows.

> **Example:** A full-scale pumping test was carried out for a 1.5m diameter tunnel on Tyneside where ground investigation boreholes had indicated a 15m water head above the tunnel invert. The permeabilities derived from the test indicated the need for special dewatering measures, such as deep pumped wells, to be implemented for the tunnelling works.

In rocks, flush returns may provide a useful indication of *in situ* permeability, especially if correlated with a core fracture log. Packer permeability tests are also extremely useful for defining potential seepage zones and changes in rock mass properties.

4.5.4 Strata dips and groundwater

Care should be taken to establish the local dips and dip directions of sedimentary rock strata when tunnelling is to take place in that strata below the water table. Water can more easily be conducted down dip along bedding planes and can thereby more readily assist the breakdown of rock, particularly that which is more prone to slaking, in the tunnel invert.

> **Example:** Water ingress, ranging from a slight seepage to continuous flow, was encountered over long stretches of a tunnel in rock where the bedding dipped into the tunnel face. This water proved to be the major cause of machine handling problems, arising particularly through rapid deterioration of the tunnel floor. This problem was not identified at the site investigation stage, probably due to the use of water flush drilling in the ground investigation boreholes.

4.6   GEOPHYSICAL SURVEYS

Seismic refraction surveys may be useful for defining rockhead levels and would be particularly relevant for subaqueous tunnels where an otherwise adequate number of ground investigation boreholes would be too expensive. A special advantage of such a method is that the survey could cover a possible tunnel alignment corridor quite easily, accommodating design changes in tunnel line and level relatively easily. The survey information must be calibrated and confirmed, however, at regular intervals by the putting down of boreholes and careful assessment of the actual data.

Seismic reflection methods may not be generally useful for tunnelling investigations since the input wavelengths may be too long for near-surface resolution. Any decrease in wavelength leads to greater signal attenuation and more problems with the signal-to-noise ratio. However, it has been claimed that good results have been obtained from shallow seismic reflection surveys using standard portable seismic instruments and small software processing packages (Anon, 1990a). Steeples and Mileer (1990) regard the method as being particularly good for detecting bedrock at a depth greater than 20m, with seismic refraction usually being better for shallower depths.

Gravity methods are not usually adopted for this work, but may have application for determining the presence of sub-surface voids. For example, Darracott and McCann (1986) have discussed their use for actually locating a tunnel. It has been suggested that unrevealed old mine shafts, which would present a hazard (water, gas) if intercepted by a tunnel, may be detected by magnetic surveys. However, experience with proton magnetometry has not generally been very satisfactory.

The newer technique of ground probing radar (*see*, for example, Caldwell, 1986) can be successful under suitable ground conditions. A single antenna transmits a signal and receives reflections. Electromagnetic wave reflections occur where the water content of the ground changes abruptly. The time scale of the instrument can be calibrated in depth using control information from boreholes or from vertical wave velocity measurements. A maximum depth of search is about 30m for an overburden having a resistivity of $2000\Omega m$, and it becomes 15m at a resistivity value of $1000\Omega m$, 8m at $500\Omega m$, 4m at $250\Omega m$ and 2.5m at $125\Omega m$. Its usefulness is limited by the many reflections originating from the numerous subsurface reflectors and also by the fact that, in areas of water saturation and clay overburden (Cooke, 1975), absorption of the electromagnetic waves causes radar signal attenuation, with a corresponding poor depth (sometimes less than 1m) of penetration. The technique is suitable for sandy soils, and conductivities of less than 30 millisiemens/m are normally required for its successful implementation. One big advantage is that the technique is insensitive to a vibration background, so making it suitable for adoption in urban areas.

Other geophysical methods, particularly for rock investigations, as discussed by Halleux (1994), include the following.

• Borehole radar reflection, in which a dipole transmitter and receiver are lowered in the same borehole at the same interval stepwise with frequencies of 22MHz or 60MHz being used. Penetration, depending upon electrical resistivity, is only a few metres or less in clayey or silty ground (and is generally not applicable), about 40m in fractured rock, about 40 to 150m in massive rock, and, exceptionally, up to 300m in very intact rock. Current systems use directional receiver antennae and allow a full three-dimensional resolution (image) of a rock mass around the borehole within the investigation radius.

• Radar tomography (crosshole) in which the transmitter in one hole and the receiver in the other are lowered steadily and very short e.m. pulses again at 22MHz or 60MHz are emitted. Definition of radar velocity distribution between boreholes is achieved by tomographic inversion. Radar attenuation is similarly determined from the amplitude data. Grey scales or colour sections are used to show velocity or attenuation.

• Seismic tomography in which, as in the case of radar tomography, the seismic source is lowered in one hole and the seismic receiver is lowered in another. The propagation times are inverted to produce a section showing seismic velocity. Both P- and S-waves can be used, S-waves giving valuable information on the soil characteristics but they are more inconvenient and more expensive to use because the source and the ground have to be cemented together, as also does the casing. Below the water table, P-waves couple through the water. Typical sources of waves are explosives, electrical sparkers, and air guns.

• Time domain electromagnetic surveys are based on the decay of eddy currents induced in the ground by interrupting a current circulating in a loop at ground surface. Eddy currents produce a secondary magnetic field detected by a receiver coil, and curves similar to resistivity sounding curves are derived and inverted to give a vertical succession of resistivities. TDEM is much faster than conventional resistivity sounding and it requires a smaller size of surface array relative to the depth being probed. The result is better horizontal resolution but, as with all electromagnetic methods, it is sensitive to the presence of electrical interferences, cables, fences, and so on that are located both above and below ground.

• Nuclear magnetic resonance is a newer method for hydrogeology and engineering geology and is useful through being sensitive to groundwater.

Geophysical methods in general and ground probing radar in particular should be used only with caution.

> **Example:** In connection with the Tyneside sewerage scheme, the Northumbrian Water Authority used ground probing radar at the quayside in Newcastle upon Tyne to assist in determining the location of buried foundations of the medieval city wall. The method not only detected those foundations but also provided information on the extent of loose fill (mainly the tipping configuration of 'Thames ballast') and the locations of existing sewer pipes. (Thames ballast is the fill that was put into ships to provide the necessary stability for their return journeys to Newcastle upon Tyne having discharged their coal cargoes in the capital.)

4.7 ABANDONED MINING AREAS

In those mining areas where old abandoned workings may be present at or close to the tunnel alignment, it is important that the Engineer should obtain advice from a consulting mining engineer with experience of ground conditions in the area. As many sources of information as possible should be consulted at the preliminary appraisal stage (*see* British Standards Institution, 1981). These sources should include British Coal, the Opencast Executive, the British Geological Survey, local authority archives, and previous investigation records held by consultants and individuals, such as owners of large land holdings. There is also a *Catalogue of Abandoned Mines* (1931) and a volume of *Miscellaneous Mines in Great Britain* (1975). In many areas the actual extent of abandoned mine workings may be difficult to determine because of old unrecorded (and sometimes illegal) extractions.

If extensive abandoned mine workings are thought to be present along the tunnel line (noting that workings both above and below the tunnel level can create tunnelling problems of equal severity to those associated with abandoned workings at tunnel level) then a borehole investigation is unlikely to prove the full extent and layout of the voids even if borehole cameras are used and video records are taken. It may then be necessary to consider using a borehole sonar surveying system such as a Geosonde as used, for example, by Walsall Metropolitan Borough Council for surveying a disused limestone mine (Russell, 1994). There may, however, be circumstances when direct methods of investigation are the only realistic option. The system would have to be opened up and the mine workings surveyed in the region of the tunnel line. Great care would then have to be taken with respect not only to the dangers of roof collapse but also to the hazards of oxygen deficiency, carbon monoxide, and methane (refer to Supplementary Information 2 in Part Two of the book on this subject).

> **Example:** In the north-east of England suspected large voids in an area of old ironstone workings were encountered by a

tunnel driven for a coastal sewer. It was necessary to determine the full extent of the workings, which also underlay a major road and nearby buildings, by carrying out a quite extensive underground survey. As a result of this investigation a comprehensive grouting programme was subsequently carried out in order to stabilise the worked-out ground.

In the north-east of England problems have often been experienced in locating old mine shafts and bell pits situated on or very close to tunnel alignments. It is generally preferable to rely on a careful search of available information such as old mine plans, Ordnance Survey sheets, aerial photographs, and so on, together with site inspections and trenching to establish shaft locations. A borehole should be drilled at each likely shaft position to determine the nature and condition of the material used to fill the shaft (particularly the depth to which the fill has settled), the condition of the capping, which may be of wood supported by steel beams needled into the shaft walls, and the depth of the shaft. Such boring should be carried out in accordance with legislation and safety measures dealing with drilling through old shafts, and with the approval of British Coal.

Any residual uncertainty remaining from the investigation may be accommodated by specifying a regime, albeit expensive, of forward probing from the tunnel face.

4.8   LIMESTONE AND CHALK

Limestone rock and chalk may present particular problems for tunnelling engineers for several reasons including: the presence of solution cavities containing water under pressure; slurrying caused by excavation and the movement of plant; the presence of flints which could affect TBM progress; release of carbon dioxide caused by acid waters.

Solution cavities extending up to ground surface and sub-surface caverns in limestones and other calcitic rocks are unlikely always to be identified by conventional probings at the site investigation stage and so it would be natural to resort to geophysical techniques. Solution in limestone operates initially as a

result of rainwater mixing with carbon dioxide in the atmosphere to form carbonic acid, but the process is augmented by the addition of carbon dioxide from the air phase within the soil mantle and by humic and other acids derived from the percolation of water through the soil. In generally acidic water-bearing ground, solution will continue to the detriment of tunnel support. The rate of solution is modified by the prevailing temperature (more carbon dioxide being dissolved in water at lower temperatures), as well as the potential aggressiveness of the water, and the rock type. Generally, acids having the highest apparent solubilities possess lower reaction rates with limestone than do those acids having lower apparent solubilities.

Bacterial metabolisms may also produce potent acids, such as sulphuric acid, in bogs (Jennings, 1985). Sulphuric acid, produced by the weathering of sulphide minerals such as pyrite, may be developed in sufficient quantities by the weathering of interbedded shales as well as pyritic limestone. The finer-grained the pyrite, the more reactive it is. Other sources of sulphuric acid may involve hydrogen sulphide produced from hydrocarbon deposits in the vicinity, and the mixing with gypsum brine with fresh water. Sulphuric acid from these latter sources will be supplied in the saturated zone in contrast to the other supplies of acidified water such as may be supplied from surface inputs, typically from peat-covered ground. There is further comment on this subject in Attewell (1993b).

To assist the reporting of site investigations, Fookes and Hawkins (1988) have proposed a simple engineering classification of limestone solution features on the grounds that an understanding of the processes of solution weathering is an essential prerequisite to the planning, execution and interpretation of both desk study assessments and field investigations in limestone terrains.

An interpretive (or interpretative) report may need to include recommendations for ground improvement. It would be suggested that any voids on the tunnel line and at the tunnel level located by ground investigation boreholes or by geophysical probing need to be backfilled before tunnelling begins. Tunnelling progress will be slowed by the need for forward probing from the face (which will be included in the specification and which will require to be priced in the bill of quantities), with a suitable means of stopping-off high water pressures at the drill. Other voids will need to be suitably backfilled when intercepted, the form of backfilling depending upon void size and whether or not the void is water-free or, if not, can be drained. For low void densities, pumped concrete may be suitable, a high filler-to-cement ratio being used to enable the tunnel to proceed through the fill more easily. Larger cavities may need other bulk infill material such as sand, gravel or aggregates combined with low strength concrete, but the use of pulverised fuel ash should be avoided because it could pollute groundwater with heavy metals. Once the initial tunnel support is in place there will need to be a programme of substantial back grouting for the purposes of both water-stopping and lining support, the latter being especially important in the case of larger, cavernous voids.

Chalk is a soft, fine-grained limestone which may be shattered to depths of several metres as a result of frost action, and sometimes reduced to the consistency of a silt. The use of standard penetration test values is one simple, inexpensive way of classifying chalk for engineering purposes without actually examining it in borehole cores, but there have been instances where such an approach has led to contractual claims for extra payment. Chalk that was classified, for example, from SPT N-values greater than 30 as grade I or II, implying widely spaced fractures, has in some cases been so severely fractured that pile borings could not be advanced without the use of pile casing. A chalk having a SPT N-value above about 25 can be described as a stronger, unweathered chalk with tight, widely spaced joints (which means, in effect, a solid body). In a weak, weathered state the chalk will be fractured, and have open fissures. Its weakness, degree of weathering and fracturing, and width of fissures decrease with increasing N-value. Reference may be made to Ward *et al.* (1968) and British Standards Institution (1986) for chalk classification on the basis of SPT, stiffness (Young's modulus) and foundation bearing stress.

The problem of flints has been noted above. Their presence needs to be foreseen as possible random inclusions in the stratified rock and their influence on tunnel progress, especially when a cutting machine is to be used, assessed in the ground investigation report. Flints were encountered in the Chalk during Anglian Water's 2km long sea sewage outfall tunnelling for the North Norfolk Waste Water Management Project. Tunnelling was chosen in preference to the pull-pipe alternative on the grounds of environmental protection for fishing and crabbing in the area. A Dosco 2.64m outside diameter full-face earth pressure balance machine was chosen for the work. The EPB created a plug of remoulded soil in the excavation chamber to restrain the water pressure and allow a steady extrusion of material at atmospheric pressure through a 380mm diameter by 5m long screw conveyor. From a 350mm by 350mm guillotine gate in the screw conveyor casing the extruded soil was transferred by belt conveyor into trains of skips which were lifted singly out of the access shaft for emptying. Bullet-type picks and spade-type tools were used at the head which was fitted with hydraulically controlled flood doors for sealing the excavation chamber almost watertight when closed. The lining took the form of Charcon trapezoidal pre-cast concrete segments, six per ring, fitted with a hydrophilic water sealing gasket.

4.9 INVESTIGATION RECORDS AND REPORTS

4.9.1 Types of reports

It is paramount that good and accurate records of the investigation be maintained. These records must be attributable to the person who kept them, they must be legible, pay adequate attention to detail, and be kept in a secure place for several years after the investigation has been completed (a minimum of 6 years from the date of the origin of the records according to BS 5750 - *see* British Standards Institution, 1979, 1994a).

Example: Part of one tunnelling contract in the north-east of England gave rise to a

Clause 12 contractual claim under the 5th Edition of the ICE Conditions of Contract. The claim could be attributed solely to the erroneous labelling of a borehole (in fact, two boreholes had been given the same labelling). As a result of the error, the tunnel passed through 50m more of filled ground than were indicated on the contract drawings and which were entered in the bill of quantities for pricing by the contractor. At the time, the tunnelling contractor claimed an additional payment of some £29 000 for driving through the extra length of fill material, the claim being based on a variation of bill rate rather than Clause 12 claim. Ultimately the claim was granted on a revised extra over rate and it amounted to some £16 000. What is often feared actually happened; incorrect labelling of boreholes may prove to be expensive!

A common procedure is for two separate reports to be submitted to the project design engineer: a *factual site investigation report* and an *interpretive report*. This interpretive report considers the implications of the investigation evidence specifically with respect to the proposed tunnelling works. It is recommended that the person charged with the writing of an interpretive report be familiar with the work in hand at the outset, discussing with the client the way in which the factual and interpretive information can be used in the design and construction of the engineering works.

With or without the threat of liquidated damages, a report should be issued by the due date. Any failure to complete on time does demonstrate a lack of professionalism, notwithstanding the good reasons that can usually be offered for the delay. Furthermore, when a report is delayed the person responsible for writing the report may come under some pressure from the client to provide verbal information so that problems can be anticipated and detailed design pushed forward. Release of information in this way can be risky and should be resisted even though more goodwill, and perhaps future work, is lost.

All reports need to be independently checked before release. A robust checking procedure needs to be set up within the framework of the company quality system (*see* Supplementary Information 1 in Part Two of the book).

4.9.2 Factual report

The facts of the investigation are the material, statistics and properties which can be seen, measured or identified by means of generally accepted and preferably standardised criteria, classifications and tests. Facts should always be the same and be independent of the measurer, observer or tester. The factual report would contain background information on the area to be tunnelled, as gleaned from the desk study, the examination of the area, outcrop locations, samples and cores available for inspection by the contractors, the results of any literature searches and enquiries, and such photographic evidence as may be available. It would in addition contain the lithological names/descriptions of soils and rocks according to BS 5930 and BS 1377 and the logs of any exploratory holes (including the depths of any water strikes and rest water levels), together with laboratory and *in situ* test results (*see* Norbury *et al.* (1986) for a good review of soil and rock descriptions). Field descriptions, compiled under non-ideal environmental conditions, and laboratory descriptions, made, for example, when soil samples are extruded from U100 tubes, may not be the same, and so it is advisable for the person who logged the cores to check the material in the laboratory. A laboratory description should normally take precedence over a field description. Checking also provides an opportunity for the engineer in the laboratory to establish that the descriptive terminology employed under a standard classification scheme is in accordance with the actual test values derived for the soil or rock.

There would also be a longitudinal section showing the positions of the exploratory holes along the tunnel line together with the log of each hole. No lines should be drawn between adjacent boreholes linking what appear to be the same stratigraphical boundaries, and this restriction should be applied even to short lines projecting from the stratigraphical boundaries of the logs since any such markers might be taken to imply continuity of strata which may not exist between the boreholes. The test procedures should be quite specifically referenced, and any departures from standard procedures (for example, in the pre-treatment of samples) should be carefully noted in the report.

Facts need to be ordered and linked through recognition of similarities. Drawing inferences leads to correlations, interpolations and extrapolations, which constitute the geotechnical interpretation.

Interpretive data may be defined as information derived from competently made interpretation of facts using accepted and proven techniques, or reasonable judgement exercised in the knowledge of geological conditions or processes evident at the site (Anon, 1987). Engineering geologists and geotechnical specialists may well vary in their interpretations. Typical interpretations prone to variation are the borehole logs and the inferred stratigraphies between boreholes. Even though the samples themselves are facts, and even where there has been 100% recovery from a cored borehole, or the sides of a trial pit have been fully inspected, the logs remain an interpretation of the facts. It is re-emphasised that facts relate to sampling points. Between sampling points the performance of a drill, rates of penetration, requirements for casing, water returns, and so on are also facts when recorded by an experienced observer, and should be noted systematically in an approved manner on the borehole log. Also needed for the best possible interpretation, with the descriptions of the samples, is the other factual information, such as the laboratory and field test results and the information on time-related water levels.

4.9.3 Draft of possible notes to accompany a site investigation report

Whilst it is obviously unacceptable and contractually untenable to include in a site investigation report disclaimers which attempt to negate the responsibility of the site investigation contractor for the technical competence and quality of the work it is, nevertheless, both valid and sensible to point out any general and special constraints that could affect the application of the results of the investigation to the ensuing groundworks. In

particular it is necessary to point out that a ground investigation actually investigates very little of the ground directly, relying substantially on experienced inferences for its use in engineering design.

Two possible contractual situations are most likely to arise:

(1) A specialist site investigation company ('ground investigation contractor' or 'ground investigation specialist') is engaged by or on behalf of the client (employer) and produces a report or reports for the client to (normally) include in the contract documents for the Works.
(2) The client, consultant-designer or project management company for the client has his own in-house site investigation section or department, or even wholly owned company, that produced the site investigation report or reports.

In the first situation any general comments in the site investigation report will bear no reference to the Conditions of Contract for the engineering works. However, there are substantial grounds for supplementing the statements in the site investigation report with additional comments in the Specification for the Works. This may also apply to the second situation, but in this case it may be decided to incorporate within the general notes accompanying the site investigation report some statements which also relate to the works contract. In essence, the substance of the comments in the Specification are then interleaved with the comments on the site investigation.

It is considered that the use of general notes in the site investigation report together with separate statements in the Works Specification provide a more robust solution. Accordingly one example of some of the comments that might be included in the site investigation report and in the Works Specification are given below. These notes are obviously not intended to be, and cannot be, exhaustive, and their actual style depends to a great extent on geographical location, the type of the ground, and the nature of the construction works. The site investigation general notes should be applicable to all ground investigation work but the statements in the Works Specification will need to be adapted to suit requirements of the particular contract.

GENERAL NOTES TO ACCOMPANY THE GROUND INVESTIGATION REPORT

These notes, which accompany the ground investigation report, are intended to assist the user of the information contained in the report. They point out some inevitable shortcomings of any ground investigation and do not constitute a disclaimer of responsibility for the results obtained by the ground investigation specialist.

1. The information in this report is based on the ground conditions encountered during the ground investigation work and the results of any field and laboratory testing. The exploratory hole records describe the ground conditions at their specific locations and should not be regarded as representative of the ground as a whole.

2. Site investigations are performed by this Company in general accordance with the recommendations in BS 5930 (1981) 'Code of Practice for Site Investigations'. The testing of soils, rocks and aggregates generally follow the recommendations of BS 1377 (1990) 'Methods of test for soils for civil engineering purposes', the International Society of Rock Mechanics (Brown, 1981b) 'Rock characterisation, testing and monitoring, ISRM suggested methods', and BS 812 (1975) 'Methods of sampling and testing of mineral aggregates, sands and fillers', respectively.

3. The primary purpose of ground investigation boreholes and trial pits is to probe the stratified sequences of rock and/or soil. From the results of these probings no conclusions should be drawn concerning the presence, size, lithological nature and numbers per unit volume of ground of cobbles and boulders in soil types such as glacial till (boulder clay).

4. When cable percussion boring techniques are used in superficial and drift deposits some mixing of thin-layered soils inevitably occurs.

If strong randomly occurring pieces of rock are encountered in soil material then the rock may be either pushed aside or penetrated and broken up in which case the arisings that are recovered may not be indicative of the nature of the material *in situ*.

5. Rotary drilling techniques may sometimes be used for drilling through superficial deposits and rocks in order to provide a very general indication of the nature of the ground. Where open hole methods have been used for the ground investigation the description of the ground is based on the cuttings recovered from the flushing medium and the rate of progress in advancing the hole. Descriptions of strata and the depths of changes in strata may not be accurate under these conditions.

6. Groundwater conditions noted during boring may be subject to change through seasonal and/or other effects such as, for example, boring and constructional excavation. When a groundwater inflow is encountered during boring, work on the hole is suspended, typically for 20 minutes, and any change in water level is recorded. The groundwater level recorded on resumption of boring may not be the natural, pre-boring standing water level. When piezometers are installed in boreholes the reported groundwater levels may also be subject to variation due to seasonal and/or other effects.

NOTES BEARING ON THE GROUND INVESTIGATION FOR INCLUSION IN THE WORKS SPECIFICATION

• The Ground Investigation report accompanying the Contract documents does not purport to describe fully the nature of the ground which is the subject of the present contract.

• There is no expressed or implied guarantee that the ground excavated for the civil engineering works will necessarily be the same as that revealed in the boreholes and/or trial pits.

• It is the responsibility of the contractor, under Clause 11(1) of the ICE Conditions of

Contract (1973) or Clause 11(2) of the ICE Conditions of Contract (1991) or equivalent clauses in other forms of contract, to seek such further visual and documentary evidence as will assist in the realistic pricing of the bill of quantities. The employer does not guarantee that the ground excavated for the civil engineering works will necessarily be the same as the ground conditions revealed at any geological exposures and excavations.

• The site investigation report describes the ground material in geological and lithological terms. In the contract documents technical descriptions of the ground material may be used for the purpose and convenience of billing and pricing of the works.

• In order to minimise disruption to traffic and inconvenience to the general public it is the policy of the Employer not to sink trial pits in built-up urban areas. Where trial pits are sunk in open ground it is not the general policy of the Employer to sink such pits to depths in excess of 3 metres notwithstanding the fact that the excavation for the Works may take place at a depth in excess of this.

• It is the policy of the Employer to request that contractors tendering for the Works draw the attention of the Employer, well in advance of the due date for the lodging of tenders, to any perceived errors or deficiencies in both the ground investigation and the ground investigation report that could have influenced the nature of the billing for the Works and would, unless changed, affect the pricing of the Works.

An engineering geologist or geotechnical engineer will be potentially liable to a client if he produces a site investigation report without due skill and care. He will then be required to recompense a client for any losses that the client suffers as a direct result of that negligence.

There is also always the possibility that the engineering geologist or geotechnical engineer (the 'ground investigation specialist') could owe a duty of care to those people who read the report and rely on it for whatever purpose,

despite the lack of any contractual relationship between them. It would be prudent to add a disclaimer, which is reasonable in tone, in the report to avoid any such liability. O'Reilly (1993) has suggested such a disclaimer.

'This report has been produced by...[name of engineering geologist or geotechnical engineer]...for...[name of the client]...in accordance with that client's particular requirements. The contents of the report are confidential and other persons are not entitled to use the information contained within it. No liability whatsoever will be incurred to any persons using this report without express permission from...[name of engineering geologist or geotechnical engineer].'

This potential difficulty of third-party use of the factual ground investigation information, the use of the information on projects for which it was not intended at the time it was acquired, and the selective use of parts of the information can also be covered by the addition of two further notes to the General Notes listed above:

7. The factual information contained within the ground investigation report should not be used for any development project other than the one for which it was prepared unless a check has been carried out on its applicability. Where the ground investigation report contains an interpretation of the factual information that interpretation must be considered in the context of the stated development proposals and should not be used in any other context.

8. This report is valid only in its complete form for the use of the person or organisation that commissioned the work. [*Site Investigation Company name*] accepts no responsibility if the information in the report is used by any other party unless written approval has been obtained from [*Site Investigation Company name*]. The information is the property and copyright of the person or organisation that commissioned the investigation. It should not be reproduced or transmitted in any form without the owner's written permission.

4.9.4 Interpretive report

The interpretive report would include an evaluation of both the geological and hydrogeological conditions along and to either side of the tunnel line, together with an assessment of their effects both on tunnel design and construction (such as face stability, possible water inflows, choice of method of excavation, rock cuttability, tunnel support options, and so on) and on likely environmental effects such as the effects of water and soil pollutants on the permanent lining, of the tunnel atmosphere on the operatives, the effects of ground settlements on buildings and buried services above the tunnel and adjacent to the tunnel line, and of vibration and noise caused by construction upon nearby buildings and people.

Some of the facts, such as the groundwater levels and in some cases the laboratory test results, may need to be qualified in the factual report as being 'at the time of test'. It must be recognised that the factual site investigation report will usually be 'statistically weak' and that much poorer ground conditions than indicated by the investigation could occur locally even if a high borehole density has been used. A conclusion continuously reinforced by reviews of contract records is that reference benchmarks concerning ground conditions, as revealed and in addition as implied by interpolations/extrapolations from borehole evidence, must be established at the outset; that is, at the pre-tender stage. An interpretive report will assist the Engineer in his ground referencing (*see* Section 5.2.3 below). It will also form the basis of his design report (*see* Section 5.2.2 below) that will sometimes be released to the tendering contractors, usually but not always as part of the contract documentation.

A professional opinion, specifically in this case on or stemming from the results of an investigation into the ground conditions expected to pertain at a site of future tunnelling, depends on conclusions or recommendations, themselves based on considerations of the relevant available facts, interpretations, analyses and/or the exercise of professional judgement. Opinions may vary, but the greater the depth and breadth of

relevant knowledge held by the participants to the decision-making process the more likely is there to be substantial agreement. Examples of instances where opinion might be injected into contractual decision making include excavatability (Section 4.4.4 earlier) and the definition of suitable plant, likely water inflow rates (Supplementary Information 13 in Part Two of the book), stand-up time at the tunnel face and consequential (unavoidable) settlements (Supplementary Information 16 in Part Two of the book).

Although, on preliminary consideration, interpretation is a function only of the derived information, further reflection might suggest that it should be geared to a proposed or perceived method of working and also to the recommendation and/or choice of form of contract. In the former case, for example, special emphasis could be directed towards the ability of particular types of plant to excavate and support the ground, since there is reduced operational flexibility with equipment such as full-face tunnelling machines. In the latter case it would be preferable for the form of contract not to be chosen until the implications of the ground conditions and the relative risks to the employer are assessed and recommendations given. However, it is recognised that this sequence of events will often be the exception rather than the rule.

It is often considered that an interpretive report on the ground conditions when issued as part of the contract documentation does tend to increase the exposure of the client to the possibility of contractual claims for extra payment. In general, however, this should not act as a deterrent to the formal implementation of an interpretation and provision of that opinion to the contractors bidding for the work. Although a geotechnical engineer may be moved within an interpretive report to offer advice on temporary works this should be resisted because responsibility for such works usually rests entirely with the contractor. However, there seems to be no reason why any of the geotechnical aspects of temporary works should not be discussed in a covering letter or, preferably, orally with the Engineer.

As in the case of factual ground investigation reports the project-specific nature of an interpretive report needs also to be emphasised. This requirement is covered in General Note 7 above. Clients must be deterred from passing selected portions of a report on to a contractor because when used out of context by a contractor that information could be used in the formulation of a contractual claim. This eventuality is covered in General Note 8 above.

5

CONTRACT PREPARATION AND TENDER EVALUATION

5.1 PROCUREMENT IN THE EUROPEAN COMMUNITY

5.1.1 General

It is important to draw attention to the fact that through its Utilities Directive (90/531/EEC) the Council of the European Communities has introduced legislation to control the manner in which procurement of contracts is activated. This Directive has been incorporated into UK statute through the Utilities Supply and Works Contract Regulations 1992 (referred to below as 'the Regulations').

The main objective behind the EC Utilities Directive and the Regulations is the removal of nationalistic procurement practices which discriminate against other European suppliers or contractors. Achievement of this objective is through transparent and fair procurement practices which place all suppliers and contractors, whenever possible, on an equal footing when competing for all contracts having estimated values in excess of specified thresholds. Current (22 December 1993) thresholds are £3.75 million for works and £299k for supplies (the respective figures for 1 January 1993 were £3.5m and £283k). (The threshold sum for professional and consultancy services, maintenance and repair is ECU 400 000 (£280 000), but this limit was not operative until 1 July 1994.)

The effect of the legislation is to require all contracts having an estimated value above the stipulated threshold to be advertised in the Official Journal (OJ) of the EC. The may be through the mechanism of a Periodic Indicative Notice (PIN), Notice of the Existence of a Qualified List, or a Call for Competition. Tendering for work (or supplies) may be through the Open, Restricted or Negotiated procedures, and in the case of the latter two the tenderers must be selected on the basis of objective criteria. Post-tender negotiation should be prohibited. In open and restricted procedures there should be no negotiation with candidates or tenderers on fundamental aspects or variations and, in particular, on prices, but discussions would be allowed in order to clarify or supplement information. Actual specifications must relate to European standards, where they exist. Contracts must be awarded on the basis of either the lowest price or the most economically advantageous tender. If a contract is awarded on the basis of the 'most economically advantageous tender' then the criteria for the award must be stated in either the contract notice or the tender documents. These criteria may include price, technical merit, operating costs, response times, compliance with national standards and laws, the proposed programme for execution and completion and the technical support (back-up) offered. When an award is made above the threshold that applies at the time, it must be published in the Official Journal of the EC. Appropriate contracts publications and releases should be consulted for further information on this subject.

In the case of public sector authorities in England, Scotland and Wales, under the Local Government Act 1988 they have to publish, in the trade press, notices of any invitation to tender and are required to advertise when they are compiling a new list of approved contractors.

5.1.2 Procedures for letting contracts having estimated values above a current EC threshold

When estimated contract values fall *below* the relevant thresholds, standard in-house client procedures apply. A typical maximum number of tenderers invited to apply for the work would be eight.

For contracts above the threshold value the client may publish in the OJ a *Periodic Indicative Notice*. The PIN will indicate the approximate total contract value and the number of individual contracts expected to be within particular product groups. The aim will be to help potential contractors or suppliers to identify areas in which their services might be used.

An individual *Call for Competition* should be sufficiently detailed to allow contractors or suppliers to decide if they wish to submit a tender. There will usually be an invitation to apply for further information from the client organisation.

It is expected that most organisations will use a Restricted tendering procedure for the letting of work, but some may choose to use an Open procedure for particular supplies contracts where there are few risk implications. In general, an Open procedure would not be suitable for substantial civils works projects. A Negotiated procedure may be followed, but this method is not expected to be widespread. A Restricted tendering procedure with qualified lists of potential tenderers has advantages in that there is less risk of a poor contractor appearing and surprises are more easily avoided, time is saved for each contract when a client organisation is running several contracts, and it can be demonstrated to be fair. Once such a list is established, it must be used. It needs to be kept under constant review and the actual selection criteria need to be examined carefully. The question is also raised as to how the tenderers are selected for a limited entry list. This could perhaps be by rotation of candidates who are able to demonstrate their suitability by the provision of audited accounts and details of similar contracts successfully completed, evidence of a good health and safety record and the availability of skilled and experienced manpower, and by the absence of any reservations from the client organisation itself.

5.2 ENGINEER'S EVALUATION OF GROUND CONDITIONS

5.2.1 General

Experience on tunnelling contracts suggests that the preferred method of tender preparation is for the Engineer to carry out a careful evaluation of the ground conditions as outlined in the factual and interpretive reports. So the information provided to the contractors bidding for the work would be the factual site investigation report together with an Engineer's report should the Engineer think this to be desirable. This latter report would be developed from, and would not necessarily be the same as, the site investigation interpretive report. Further advice might be sought by the Engineer for the substance of his report. Typically, he might re-approach the site investigation contractor and/or consultant, and his own staff with experience of tunnelling in similar ground elsewhere. When writing his report the Engineer will be aware that he is not only providing a distillation of experience but is also highlighting issues that may subsequently prove to be contentious. The style of his report should be geared towards the form of contract that has been chosen for the work or, preferably, the form would not be chosen until the implications of the ground conditions and the relative contractual risks to the employer are assessed and the recommendations made accordingly.

So that the Engineer and the tendering contractors can more easily understand the geological and groundwater conditions, a three-dimensional physical model of the site can sometimes be helpful. The model would be available for viewing by the contractors when they examined the borehole cores. It would comprise quadrant-shaped timber dowel rods (representing the ground investigation boreholes), approximately 10mm in diameter and on the curved side of which would be marked the stratigraphical boundaries as recorded in the ground investigation. On one of the flat sides would be recorded the water strike levels and the final water level. The other flat side would be used for borehole identification. The individual dowel rods would be fixed at the borehole positions on a board, the surface of which would be covered with a 1:500 scale Ordnance Survey sheet. This board would be defined at a particular level above (usually) or below ordnance datum, perhaps +10m OD for a tunnel driven near to a river that is close to the sea, and so the heights of the dowel rods would vary according to both borehole depth and ground surface topographic height at the location of the borehole. A useful vertical scale for the rods would be 1:50, that is, a vertical to horizontal exaggeration of 10 times. Borehole data from earlier investigations might also be included in the form of additional dowel rods on the board. Further assistance in the interpretation is provided by the use of colour on the curved surfaces of the rods, coded to define the geological profiles, and by the adoption of coloured string between dowels to provide inferred linear interpolations of the geological boundaries between boreholes. It would be necessary to include a statement with the model pointing out that linkage of the stratigraphical horizons between boreholes in a linear manner should not be taken to imply that the geological boundaries necessarily follow this form *in situ*.

The release of an interpretive report (or an Engineer's report), as well as a factual report, to the tenderers is entirely in line with the views (p17) expressed by the Ground Board of the Institution of Civil Engineers in the report 'Inadequate site investigation' (ICE, 1991c). It also seems logical that the employer should be apprised by the Engineer of the financial risks, including those perhaps stemming directly from release of an interpretive report, at all stages of the contract from project conception to its completion, otherwise an uninformed employer could be unsympathetic to any speedy resolution of a realised risk for which he has to pay. The Engineer places the employer at greater contractual risk through the release of an engineering report as part of the contract documents, together with expressed or implied reference ground conditions (*see* Section 5.2.3), but by so doing he perhaps has a better chance of promoting a reduction in the overall level of the tender prices because the tendering contractors then have a broader perception of what risks they must undertake. In theory at least they should then be able to reduce the padding element in their pricings. For his part, the contractor should be required to demonstrate at the tender stage what interpretation he has placed on the factual evidence and what credence he has placed on the Engineer's understanding of the site investigation evidence and/or the reference ground conditions in order to build up his tender rates. This is the time for the contractor to query any matters about which he is uncertain in the site investigation report(s) and to express any dissatisfaction that he might feel about the conduct of the site investigation. Failure to do so could be used in evidence against him in the event of a dispute and claim for extra payment, but there must be suitable protection for the employer against frivolous demands from a contractor. One argument states that the successful contractor should be required to deposit, in a bank, details of his itemised pricing

beyond the simple numbers that he writes into the bill of quantities. This information, the extent and form of which would need to be agreed *a priori*, would be available for retrieval and perusal at a later stage in the event of a contractual claim and possible arbitration (*see* Section 7.8).

Example: A system of this nature operated on the Boston Harbor effluent tunnel contract in the USA, the contract being part of the £3390 million Boston Harbor clean-up scheme. A three-point plan was adopted. First, a high percentage (5%) of the tunnel's value was expended on site investigation with a view to reducing the incidence of dispute on the contract. The site investigation supported a geotechnical report in which the client took over the contractual risk by guaranteeing the ground conditions to the contractors. Second, a contractor's 'escrow' report, which broke down the components and rates making up the contractor's bid and which remained the contractor's property, was held in a bank vault. Third, in the event of a claim for extra payment a disputes board of three independent engineers monitoring construction from the outset decided if the claim was valid, with the value of any valid claim being calculated using the report. This latter was in contrast to the position of lawyers who come into the dispute only at the end. It has been considered that such a system could be simply bolted on to existing ICE contract documents at a cost of only about 0.5% of the contract price, and would thereby remove the need to change the form of document as is the case when, for example, the IChemE conditions are adopted. Apparently there is evidence of bid prices having fallen in the USA as a result of this system being adopted, presumably both because of the higher level of site investigation information made available to the contractor and the significant tilting of the balance of risk towards the client, and a greater degree of contractor confidence that any dispute will be resolved in a more equitable manner. Of the tunnelling contracts let under these arrangements in the USA, as far as the writer is aware at the time of writing none has led to litigation.

5.2.2 Engineer's report

As suggested in Section 5.2.1 above, the Engineer may decide to prepare an engineering report (equivalent to a 'design' report) for inclusion in the contract documents if he is reasonably sure that the geology and groundwater conditions along the tunnel route can be sensibly inferred from the available evidence. The report could refer to such matters as the ground conditions expected at the site of the tunnelling works and their probable effects on the tunnel construction, including likely face conditions, groundwater inflows, achievable rates of advance, and so on. As noted also in Section 5.2.1 above, the report would normally be prepared for two clear purposes: first, to assist in the design and construction of the works, and, second, to attempt to mitigate any contractual claims for extra payment, and/or an extension of time, based on (ground) conditions that could not have been reasonably foreseen. If during the tunnelling work the contractor encounters any adverse ground conditions that he had not foreseen from the information provided under the ICE Conditions of Contract 6th Edition Clause 11(1) (and also notwithstanding his responsibilities under Clause 11(2) of the ICE Conditions of Contract to examine the site, its surroundings, and relevant documents) he will then very closely re-examine the site investigation evidence with a view to submitting a claim for extra payment.

Using the engineering design report as a basis, the Engineer should be in a position to define his perception of the ground conditions, drawing attention of the tenderers to those areas where significant uncertainty or doubt exist and where his own interpretation may amplify some of the factual evidence, highlighting potential hazards, and possibly commenting on some of the deficiencies in the site investigation. He will wish to state the degree of reliance to be placed on the different categories of information given in the report(s). The Engineer should clearly refer to any factual information which has conditioned his interpretation. This (and indeed many of the observations in this document) is very much a counsel of perfection, unlikely to be entirely achieved in reality, but setting a framework for good contractual practice.

The report should contain a statement on the engineering practice which the Engineer feels could reasonably relate to the ground conditions, perhaps acknowledging that the contractors will often have more experience in a particular area than have his own staff. If the Engineer perceives any special health and safety hazards arising from the geological environment, including, for example, the possible presence of toxic/flammable/asphyxiating gas (*see* also Sections 6.6 and 6.9, and also Supplementary Information 2 in Part Two of the book), then he must identify them in his report, include measures for minimising their effect in the work specification, and require the contractor to address them, with other safety matters, at the time of his tender and in his method statement. In more general terms related to health and safety, the ground should be regarded as a substance that is provided by the employer to the contractor. In the case, for example, of ground that may have been contaminated by or with the knowledge of the employer (perhaps a chemical company or a toxic waste disposal company) such an expert knowledge of the contaminants imposes a particular responsibility on the employer to fully investigate and disclose (*refer to* Supplementary Information 9 in Part Two of the book). If the Engineer (and the employer via the Engineer) withholds information which could lead to injury and loss of life, he may be liable to common law penalties. The Engineer's report should clearly differentiate between that which is fact and that which is inference and interpretation. It is likely that the Engineer will require assistance from specialists, such as a geotechnical engineer/engineering geologist, a tunnelling engineer, a mining engineer, and so on, in formulating the report. As noted above, the Engineer should only prepare such a report for issue as part of the contract documents if he feels that the geological conditions can reasonably be inferred from the site investigation data. Nothing that is said in the design report nor the factual site investigation report should be assumed to absolve the contractor from fulfilling his Clause 11(2) responsibilities under the ICE Conditions of Contract 6th Edition (1991) nor his general responsibilities for acquiring site information that could affect the make-up of his tender, his performance on site, and his responsibilities for the health and safety of his workforce.

5.2.3 Reference ground conditions

A site investigation report may indeed be factually correct but it may not be interpreted in the same way by both Engineer and contractor in relation to the best working practices for overcoming problem ground conditions. It must be recognised that ground conditions poorer than indicated by the investigation could occur locally, even if a high exploratory hole density has been used. It must be clearly understood that the borehole information is representative only of itself, usually being only a fraction of 1% of the ground being investigated. References to that information should not purport to suggest that it 'represents' the ground beyond the boundaries of the borehole nor should it ever be claimed by a contractor that the employer guarantees that the conditions at the tunnel face will be the same as those exposed in the ground by any (especially the nearest) borehole or trial pit. The task of the contractor and the Engineer is to use experience and geological knowledge to infer such ground conditions by interpolation and extrapolation from very limited evidence, since the volume of ground exposed by boreholes/trial pits is usually only a very small percentage of the ground actually being investigated. Experience suggests that reference ground conditions should be established at the tender stage, provided that the Engineer is reasonably convinced that the detailed nature of the geological conditions can be so inferred from the available evidence. This evidence includes, in addition to that revealed in the site investigation report(s), the visible and other (documentary) evidence deemed to have been inspected by the contractor as part of his Clause 11(2) responsibilities under the ICE Conditions of Contract 6th Edition (1991). As above, the employer makes no guarantee that the ground exposed at the tunnel face may be as inferred by the contractor from his Clause 11(2) observations.

The Engineer would formulate a set of reference ground conditions in order to assist him in assessing the contractor's method of working and/or for contractual purposes. These reference conditions would usually comprise a transformation of geological and hydrogeological evidence into a scheme of engineering classification upon which payment for work done would be based. The reference conditions would therefore embody both excavation and support implications. Since water inflows can have a major effect upon tunnel construction, requiring diversionary measures, drainage, or ground treatment to permit construction to proceed, it is essential that detailed information relating to groundwater conditions should be provided in the contract documents and be expressly included in a referencing scheme. It may often be most appropriate to segment the tunnel into a series of zones relating to specific type or types of ground, perceived groundwater conditions, and/or mode of response to a defined tunnelling method, and each requiring to be priced per unit volume of ground removed. The contractor would need to foresee the need for and include in his prices the cost of temporary support. Alternatively, one element of the risk to the contractor could be reduced if the client pays for the support deemed by the contractor to be necessary during the construction but subject to a closely defined set of pre-conditions and at unit rates actually submitted with the tender.

As a simple method of referencing in the context of, for example, rock excavation in a particular designated zone, the tendering contractors might be requested to price for several categories of unconfined compressive strength using machine excavation, culminating with drill and blast for the highest category of strength deemed not to be economically rippable (for example, in excess of, say, 60MPa unconfined compressive strength). However, a contractor may chose not to commit himself to what he regards as excessive refinement in the make-up of his tender and may simply bracket some or all of the categories and enter a composite price.

Uff (1989), for example, has proposed the clear adoption of a system of reference ground conditions as one way of reducing the incidence of Clause 12(1) claims. Cottam (1989), however, doubts the effectiveness of this, feeling that a hypothetical set of reference ground conditions, the very formulation of which means that the conditions are to some extent foreseeable, may simply 'be a re-statement of the borehole conditions'. This latter counter-argument is not entirely correct since the reference conditions would relate a range of possible foreseeable conditions to costs, via required contractor pricing. The problem arises from the range of likely scenarios that would need to be covered by such a system. This point is emphasized by Barber's (1989) comment on Uff (1989) - that the implementation of Clause 12 of the ICE Conditions of Contract (1973, now 1991) depends on objective standards that must be found outside the contract documents, but such standards and their development have not yet been coordinated.

Assuming that a range of possible (uncertain) ground conditions, not revealed by the site investigation, can nevertheless be contained in the bill, or a separate bill, as a schedule of rates, it is most likely that a normal schedule would need to be extended to include specification clauses and techniques for handling that variety of conditions that could be encountered (and even those conditions known with some certainty to be encountered but for which it would not be possible to pre-estimate quantities with the requisite degree of accuracy for satisfactory pricing by the contractor). Because of the uncertainties, it would be sensible to provide for a sliding scale of pricing based on quantities encountered. The unit rate would normally be expected to decrease with increasing quantity as set-up costs are absorbed. Alternatively, these latter costs can more clearly be accommodated using two separate schedule items. The rate for the first part of the work relates to the expected minimum expected quantity of work that the contractor would be required to perform. The relevant specification would require that all the fixed costs be recovered under this part, with some provision for price adjustment if the baseline quantity specified in the schedule is not reached. The rate for the second part relates to quantities exceeding those covered by the first part up to some specified maximum. It would be clearly stated that only non-fixed costs are contained in this rate and that there are no provisions for adjustment.

Just as the tunnel should be segmented into a series of zones in order to be more specific as to the reference ground conditions, so, as an alternative to a schedule of rates encompassing 'unforeseen' but possible ground conditions, each operation within the construction process could be broken down into several parts in order to reduce the spread of residual uncertainty. If neither this nor the schedule of rates approach is acceptable, perhaps on the grounds of complexity and documentation volume, then resort must be made to a daywork (plant and labour) schedule as a provisional item in the bill (*see* Section 5.5) or even a time-related system in the manner of the Norwegian 'equivalent construction time principle'.

An inherent problem related to the scheduling of contingency items concerns the realistic specification of the scale of the quantities. Also, a contractor may over-price or under-price nominal quantities in a schedule of rates without the variation having significant impact on the overall tender price derived by totalling a standard admeasurement bill and a schedule of rates addendum.

It is important when formulating reference conditions that the nomenclature be very carefully chosen and the terms that are used be fully defined. At the Northumbrian Water River Tyne-River Tees water transfer Kielder tunnels, for example, stratigraphical nomenclature was used in the site investigation reports, lithological terms were also named, and a geotechnical classification was used for comparisons between predicted (expected) and actual geology (Berry, 1980). The Kielder tunnel contract was actually drawn up with one method of excavation in mind, but alternative methods were encouraged from the tenderers.

Care must also be taken at all times that the provision of information to the contractor for the purposes of enabling the contractor to submit a tender could not be deemed to amount to a misrepresentation (*see* Section 7.9 below).

5.2.4 Engineer's assessment

About 80% of the clauses in the ICE Conditions of Contract 5th Edition (1973) referred to the Engineer and 27 clauses were 'dedicated' to him. Although not a party to the contract, his contractual responsibilities are heavy. Under the 6th Edition as in the 5th Edition he must act impartially, exercising his professional judgement to decide on such matters under the contract as the contractor's reasonable risks with respect to unforeseen ground conditions (Clause 12), extension of time (Clause 44), ordered variations (Clause 51), valuation of variation orders (Clause 52) and the use of daywork (Clause 52(3)), variation of rates (Clause 56(2)), and differences between the employer and the contractor arising out of the contract (Clause 66). His duties also include certifying the contractor's monthly statement (Clause 60(2)).

The Engineer will usually have been responsible for - or at least heavily involved in - the design of the works and thus owe a duty of care during construction. He will carry insurance and will be able to meet a claim in whole or in part. Although the contractor has a duty under the contract to make good the works, the contractors' all-risks insurance (CAR) together with third party insurance (TP) at a combined cost of 0.5% to 1.5% of the cost of the works is usually limited to where physical damage has occurred. The employer's position will be safeguarded by contract insurances and usually a bond. But the Engineer will increasingly realise that insurers may repudiate a policy on the grounds of non-disclosure of relevant facts and that if this should occur, perhaps by inadvertent omissions, then he could be in trouble.

In addition to negligence with respect to disclosure there is negligence in design which can be a criterion for a claim in tort. In the case of substantial constructions there may be many parties involved as contributors to negligence and this will inevitably lead to litigation complexities. Uff *et al.* (1989) have argued that the concept of strict liability should be adopted, as on the Continent, and that this would lead to less litigation and uncertainty and prove more attractive to the insurance market.

In formulating the contract the Engineer must always be aware of the needs of the employer. The 1992 President of the Institution of Civil Engineers made this point very well (Wilson, 1992):

'Engineers forget at their peril that engineering is a service industry and that engineering expertise should be addressed to satisfying the needs of the promoter in the context of society at large. It is the promoter who initiates the project and it is the investment of the promoter's funds that makes the project possible.'

The Engineer can best achieve this by more resolutely attempting to quantity the risk for the employer, perhaps by carefully probing residual uncertainties identified by an interpretive report and then asking the contractor to price a number of 'what if' ground conditions under a reference ground conditions format.

Following the Abbeystead (Lancaster) enquiry (*see* Section 7.10.5) the Engineer has another important problem to address. Does he carry a continuing (and unpaid) duty to tender advice to an employer where developments in the state of the art suggest that tunnelling works already completed - works that have perhaps been completed many years ago - might be defective? This same question, which has been addressed by Winter (1993), applies also to contractors, and is considered further in Section 8.3.

The Engineer's perception of the ground plus the associated tunnelling risks may veer somewhat towards pessimism. Although this pessimism could incur a 'mark-up' penalty in the tender price if he conveys such pessimism, either directly or by inference, in any report that he writes and decides to issue, this may be no bad thing since the contractor, for competitive reasons, may tend towards an optimistic appraisal of risks. However, the Engineer may not always be correct in veering towards pessimism, especially if he has not fully identified the reasons for this attitude. The main point is that he should commission and receive both a quality site investigation and report, and prepare the tender documents and drawings to a high standard. He must also arrange a final pre-tender meeting with the employer to ensure that the risks involved and the statements that are made in the documents are both understood and accepted.

Example: A contractor claimed extra payment on the grounds that rock strength higher than expected was slowing down the rate of tunnel advance through both excavation difficulty and roof support problems. In most instances there is an inconsistency in such an argument. A 'pessimistic' prediction at the site investigation stage of rock strength being higher than actually proved to be the case may prevent claims related to excavation difficulty and reduced production, but could generate claims based on unforeseen support problems at the tunnel face. Lower strength may therefore correlate with either a lower or higher rate of advance. This actual claim was allowed in part in respect of excavation difficulty but was turned down in respect of support problems. In addition to the intrinsic strength of the rock the influence of discontinuity spacing on both excavation and support were taken into consideration when assessing the claim.

Not only is the Engineer's role in the administration of public contracts being eroded on the grounds that impartiality with respect to decisions affecting the use of public money may not always be in the public interest, but also that ability (indeed requirement) to be impartial in the administration of contract decisions is being increasingly questioned as competitive pressures bear more and more heavily. There are moves on large contracts to put in place a panel of adjudicators to fulfil the role of the Engineer in the resolution of disputes. It is argued that the adjudicator (panel), having no particular relationship with either party to the contract nor any faults in design or decision-making to defend, can referee the work without fear or suspicion of favour. The adjudicator must be willing and able to make a quick decision which is binding on the contracting parties. Reference may be made to the NEC form of contract (*see* Section 2.8) which also goes some way towards overcoming these objections, and to Section 7.8 on the question of alternative dispute resolution and insurance.

5.3   CONTRACTOR'S EVALUATION OF GROUND CONDITIONS

It is not unknown for a contractor to claim or infer that the employer is in effect warranting the ground conditions to be encountered at the tunnel face or even to state that because the conditions were not as expected - conditions which induced the contractor to undertake the Works at inadequate rates and prices - then this amounts to a misrepresentation (*see* Section 7.9 below).

The site investigation report, and particularly the borehole logs, expounds in some detail the material encountered in those boreholes (or trial pits if that be the case) and presents the results of tests on the material. There may not be an expressed - but there is certainly an implied - warranty that the material in the boreholes and the test results are as described. The employer

via the Engineer must accept responsibility for any errors in the descriptions of and tests on the material in the boreholes and/or trial pits. The Engineer can never provide a warranty that the material encountered at the tunnel face will be in accordance with and have the same physical and mechanical properties as that recovered in the boreholes. This means that any expressed or implied warranty with respect to material and material properties does not extend to interpolations and extrapolations from very local and limited borehole or trial pit evidence.

In the same general vein, any claims of misrepresentation must be invalid because the employer should state that the information in the site investigation report concerning the materials in the boreholes or trial pits is related to the materials in those boreholes/trial pits and not to other boreholes/trial pits or other volumes of ground. There is some further comment on misrepresentation in Section 7.9.

Clause 11(2) of the ICE Conditions of Contract (1991) states:

'The contractor shall be deemed to have inspected and examined the site and its surroundings and information available in connection therewith and to have satisfied himself so far as is practicable and reasonable before submitting his Tender as to

(a) the form and nature thereof including the ground and sub-soil

(b) the extent and nature of the work and materials necessary for constructing and completing the Works and

(c) the means of communication with and access to the Site and the accommodation he may require

and in general to have obtained for himself all necessary information as to risks contingencies and all other circumstances which may influence or affect his Tender.'

It is assumed that the contractor (usually his estimator) will have undertaken these pre-tender responsibilities. However, realism requires an acknowledgement that in very many instances this may not be the case. Pressure of work from the sheer number of tenders needing to be compiled forces the contractor into the (risky) position of relying exclusively on the site investigation report(s) as being the *best available* evidence and ignoring the fact that it/they are not the *only available* evidence. Indeed, the contractor may well feel that it is against his interests to seek further information on the ground conditions. At the pre-contract stage a contractor will seek to identify information that promotes the best possible tunnelling conditions but may 'seek to ignore' information that could suggest worse tunnelling conditions, since a conservative stance prompted by the latter approach would price him out of the work. If things go wrong on the contract and 'unforeseen' ground conditions occur, the contractor will claim under Clause 12 of the ICE Conditions of Contract, citing a total reliance under Clause 11(1) on the site investigation report(s) to a degree that it was unnecessary to seek further information and attempting to show, if such further information is available, that the information is not encompassed by Clause 11(2).

A fruitful source of contention lies in the terms 'site' and 'surroundings'. 'The site' is defined in Clause 1(1)(v) of the ICE Conditions of Contract, 6th Edition 1990 as meaning 'the lands and other places on under in or through which the Works are to be executed and any other lands or places provided by the employer for the purposes of the Contract together with such other places as may be designated in the Contract or subsequently agreed by the Engineer as forming part of the site'. It is clear, therefore, that the contract documents should clearly delineate the boundaries of the site and provide for the necessary possessions of it. The 'site' of a tunnel is concentrated in one dimension of length only. Of the other two dimensions, its depth is usually fairly uniform, and its width is negligible. What then are its 'surroundings'? In some instances the site compound may be taken as the reference point for the 'site', but certainly in the case of a long tunnel it seems sensible to adopt as the 'site' for reference purposes the location in the tunnel of the actual cause of a dispute, and in most cases this will be the tunnel face. Definition of 'The Site' in the New Engineering Contract (1993) includes a reference to the works: 'The Site is an area within the *boundaries of the site* and the volumes above and below it which are affected by work included in this contract' (Core Clauses document 1 General, 11.2(7)).

The 'surroundings' are much more difficult to define. Abrahamson (1979) uses the term

'neighbourhood', but this term is no more definitive with respect to distance. It is not difficult to realise that a contractor may claim that a vital piece of evidence does not fall within the surroundings of the site. Distance, therefore, is an essential element of dispute in this context, to be resolved by an arbitrator only with difficulty.

If there is contractually relevant evidence, such as geological exposures, in close proximity (say 200 to 300 metres) to the site, then the sufficiency of that evidence in the setting of the contract for the resolution of the dispute should condition the interpretation of the term 'surroundings'. If there is substantial, vital and easily accessible evidence relevant to a dispute at a greater distance, then the term 'surroundings' should encompass such evidence. The concept of reasonableness will obviously come into the definition of surrounding distance, but it is considered that the concepts of accessibility and importance should be firm determinants of the definition.

Section 4.8.3 includes a form of general notes that might preface a factual site investigation report, released to the contractors for evaluation of ground conditions, in order to cover some of the points made here. Inclusion of such notes should not be interpreted as an attempt by the employer to disclaim responsibility for the conduct of, and report on, the site investigation.

It has been suggested that although tendering contractors should not each be required to make their own sub-surface investigations (because this would be both impractical and uneconomic) there may be some advantage in giving those contractors likely to appear on the select list of tenderers an opportunity to identify any omissions in the site investigation and to suggest any further items that should be undertaken. This procedure has the merit of placing a greater responsibility on the contractors to assess fully the quality and relevance of the site investigation at the pre-tender stage, but it does not provide a licence for unrestrained expenditure. There is added responsibility on the Engineer to weigh up any suggestions with respect to their contractual cost benefit as he sees it.

An alternative and widely used method is to provide only the factual site investigation report to the tendering contractors who would then be required to make their own interpretations of the factual data and to submit details of their evaluations to the Engineer for inspection. This would reveal the thinking behind the build-up of the contractors' rates and, in the case of the successful contractor, would constitute a valid document for reference in the event of disputes. It should be noted that a well-designed investigation is still required in order to provide the contractors with adequate information upon which to base their assessments and prices.

The main danger inherent in this method is that, because of time limitations, the contractors may fail to foresee problems which may have been identified during the course of a relatively lengthy design process. Although these conditions might be judged to be foreseeable, and hence claims could be resisted, considerable costs might still be incurred as a result of delays, disputes and legal arguments. These problems could be reduced, to some extent, by a formal pre-tender meeting to establish the contractors' perceptions of the conditions and indeed provide an opportunity for the contractors to raise queries with respect to both the conduct of the site investigation and the content of the site investigation report(s). Such a meeting would also provide clarifications in the form of the memorandum which forms part of the contract.

With this (factual report only) method, the contractor bears a more substantial part of the contractual risk. There is a consequential risk that many contractors could go out of business and competition could begin to decline, perhaps then leading to higher prices on future contracts. Furthermore, the baseline for the bidding would tend to increase as contractors build into the pricing their own individual assessments of risk. Nonetheless, provided that an adequate site investigation has been performed, this method does provide some attractions for the Engineer in relation to appraisal of financial commitments at the tender stage.

A modified method, which is sometimes used, is to provide both the factual and interpretive site investigation reports to the contractors, but with only the factual report forming part of the contract documents. The interpretive report, presumably unmodified by the Engineer, is thus provided for information purposes only. However, it may be extremely difficult during the course of assessing the validity of a claim to differentiate between conclusions which have

been derived on the one hand from factual data and on the other from interpretive information. The basis for adjudicating possible claims relating to changes in ground conditions may therefore be extremely unclear and so this path of information presentation cannot be recommended. The much preferred route is for full disclosure of all known information and reports related to the site conditions on a full contractual basis. Design calculations leading up to interpretation and opinion given in the reports would not normally be disclosed to tenderers unless there was a specific request to do so, since under the constraints of time and competition it is the final facts, interpretations or opinions that are of immediate concern to the contractors.

Use of reference ground conditions (Section 5.2.3) and the concept of risk sharing is based on the assumption that both parties to the contract will abide by the letter and the spirit of the contract. However, when the economic climate is less than rosy, this assumption is not necessarily valid. It is often the case that the greater the volume of factual and interpretive information relating to ground conditions conveyed to the contractor, the more might he become aware of any optimistic interpretations of the evidence upon which contractual claims for extra payment might ultimately be based.

Both parties to the contract take risks in respect of the site investigation information available to them. Selection of suitable plant and dealing with water are two such areas of risk. Haswell (1989) has noted that 'it is essential for the Engineer to appreciate the difference between contract risk and contract responsibility with only the former at financial risk to the contractor'. He (Haswell, 1986) defines 'contract risks' as including the procurement of labour, materials and plant for the work (and ensuring compliance with requisite standards of materials and workmanship), weather, accidents, and mistakes (such as errors in setting out), and 'contract responsibilities' as including the carrying out of dayworks and works connected with Clause 12 claims where the claim has been successfully invoked by the contractor, provision of pumps for dewatering where such is not billed, and the paying for imponderables such as compressed air. Ultimately, of course, the employer is always at risk because he is the paymaster!

The contractor will place much reliance on the groundwater information - level of water strikes in boreholes, any time-related rise of water, inferred (from an interpretive report) continuity of water supply from silty lenses, and so on - since this will influence his choice of plant, the possible need for compressed air for temporary support of the ground, and the risks to the progress of the works that must be made for him to achieve a satisfactory profit. Based on the site investigation information there might be a bill item on excavation that must include for pumping up to, say, 30 litres/second. The contractor will know that the provision of pumps and their proper function is therefore a contract risk up to this limit, but any need for pumping water at a rate in excess of this is a contract responsibility to be paid separately either through a dayworks item or through a special item in the bill of quantities. (In this latter case, there is some contract risk if the contractor were to be asked to insert a rate-only item.)

From his own assessment, if the Engineer considers that compressed air (*see* Section 5.6) is needed, then an item should be included in a priced bill of quantities for provision of compressed air plant and its operation, and there should be extra over items for working in compressed air. The contractor will do his own experienced assessment and expect the amount of compressed air under the definition of 'contract risk' to be limited by a statement of the quantity that needs to be provided and by an additional item for compressed air quantities over that amount.

Selection of plant as a contractor risk (but sometimes accepted as an employer risk by specification) is often conditioned by intrinsic strength and discontinuities in the case of rock and by the presence of cobbles and boulders (Section 4.3.3) in the case of soil. All this information should be in the site investigation report.

5.4   DEFINITION OF ROCK

A major problem can revolve around the definition of rock. Although it may be convenient to regard material to be tunnelled as comprising rock or soil, certain intermediate materials such as argillaceous 'rocks' (clay shales, mudrocks) tend not to fall conveniently into either category. Contractors may be asked, for example, to price for so many cubic metres of rock, Coal Measures material and clay for a tunnel to be driven partly through till and partly through a cyclothemic sequence of Coal Measures rock. In such a case, the rock would tend to be distinguished for the purposes of the contract from the Coal Measures material on the basis of compressive strength (the Coal Measures material perhaps being defined as amenable to excavation using a light pneumatic pick whereas rock would require a heavy-duty tool for excavation), but if there is no separate billing for boulder rock, or if such rock is not accommodated in the preamble to the bill of quantities, then this could be a likely setting for dispute. In any case, such a style of billing might invite the tendering contractors to bracket items in the bill for pricing purposes, but such an exercise in itself often sets the origins of a later dispute since tunnelling in rock will normally attract a higher bill price than will tunnelling in weaker materials.

It is also most important, when using the term 'rock' in a site investigation report, to define at the outset when the term is being used in its geological sense and when it is used in some contractual and technical sense. The term may have been pre-defined by, for example, some water industry documentation, in which case express reference should be made to that definition. In any event, the geological sense of 'rock' should be restricted to the site investigation reports. The contractual/technical sense of the word will tend to appear in the bill of quantities.

It is very difficult to define rock for tunnelling purposes in a non-petrological manner. A general definition of 'rock', as used in a preamble to a bill of quantities for hand excavation, could be 'that material which, in the opinion of the Engineer, could not reasonably be removed by pick and shovel, light pneumatic pick, or pneumatic clay spade (...in the hands of an experienced operator...) without the use of blasting or pneumatic breakers'. Since any rock could be removed by these implements, given sufficient time, discussion may centre on the definition of reasonable progress rates. An alternative approach would be to define rock as that which could not be dug using a clay spade, thereby accepting a greater volume of priced rock excavation. In any case there will usually be a statement to the effect that the Engineer's definition regarding the classification of rock shall be final and binding on the contractor and only that material which the Engineer actually confirms as rock will be paid for as such.

Rock might occupy all the tunnel face throughout the drive, or it could appear as harder stratifications between weaker soil-like material which could be removed with clay spades. The Engineer might attempt to quantify the percentage of rock in the tunnel face, and possibly even sub-divide the tunnel into zones which contain volumes of rock between upper and lower percentage limits (*see* Section 5.2.3). Although this method of specification is a form of reference ground condition and would appear in the bill of quantities, such detail could incur claims for extra payment should the estimates of rock volume prove not to be reasonably accurate. Indeed, if there are substantial discrepancies, the contractor may request a re-rating (*see also* the comment in Section 7.7). It follows that well-designed investigations are very important for defining the bedrock surface and also the engineering characteristics, such as strength and fracture frequency, of the rock.

> **Example:** If the 'rock' at the tunnel face is stratified and a tunnel is driven by hand excavation along or almost along the strike of the beds, then the stronger bands could be bounded by very weak bands and would present this configuration at the face consistently. The difficulty of excavation is not merely a function of rock strength and discontinuity spacing but it also depends upon the relative strengths of adjacent bands, the thicknesses of the stronger bands (which affect the ease of wedging them down), and the fracture frequency and orientation in the stronger bands.
>
> Attempts were made on a tunnelling contract in Cleveland to characterise the 'excavatability' of such a banded rock sequence by means of a points system which was somewhat similar to Bieniawski's Rock Mass Rating (RMR) system for tunnel support design (Bieniawski, 1973, 1974, 1976, 1979a, 1979b, 1983, 1984, 1988, 1989). This work was only partially completed and a practical compromise system for assessing excavatability was adopted. This simply

consisted of the assistant resident engineer and a representative of the contractor's agent visiting the tunnel face together at frequent intervals and agreeing into which one of five levels of excavation difficulty, each level triggering a specific contract price per unit volume of excavation based on a perceived rate of progress, the particular face should be placed.

Attempts have also been made to define the excavatability of rock in terms of the Q parameter of Barton *et al.*(1974) - *see also* Barton (1976, 1983, 1988) and Barton *et al.*(1980). Reference may also be made to Supplementary Information 14 in Part Two of the book.

Somewhat similarly, because cobbles and boulders are rock there is the implied contention that their excavation should be paid for at the billed rock rate even though the quantities in the bill upon which the tendering contractors had determined their rates had been calculated on the basis of the *stratified* rock at the tunnel face. On the other hand, according to CESMM3 (Institution of Civil Engineers, 1991b, referring to Class T: Tunnels, M3, page 83) 'An isolated volume of rock occurring within other material to be excavated shall not be measured separately unless its volume exceeds $0.25m^3$'. This means that technically for the purposes of the contract individual boulders of volume less than $0.25m^3$ (and all cobbles) are not rock and that the value to the contractor for their excavation is the price that he entered for the excavation of the clay that surrounds them. In other words, the factors controlling 'handlability' of boulders without breakage are not fully addressed, nor is the matter of breakage in limited space. There is some earlier mention of this question in Section 4.3.3. For further discussion on CESMM3 reference may be made to Barnes (1992a, b).

Assuming that the rock of volume $0.25m^3$ is a perfect sphere, then its equivalent diameter is 780mm. A contractor would not relish being reminded that a 780mm boulder in the face of his 1.5m diameter tunnel is soil under the contract. It is likely that a responsible Engineer would adopt a lower limit, perhaps $0.10m^3$ (576mm), for such small diameter tunnels. It would also seem sensible for any future revision of CESMM3 to relate cut-off volumes of this isolated material for separate measurement to tunnel face sizes.

Because this question of boulders can be fraught with contractual difficulty, it is preferable that it be resolved as far as is reasonably possible in the preamble to the bill of quantities, adopting a modification of the CESMM3 recommendation if the tunnel is of small diameter (say 1.5m diameter or less), and by including a specific bill item for boulders. If the reference in the preamble is to boulder volume (as is the case in CESMM3), then it would be sensible also to relate any chosen critical volume to an actual linear measure. Although few boulders will be actually spherical in shape, nevertheless the assumption of a sphere, as above, is perhaps most appropriate using the standard formula (volume of a sphere is $4\pi r^3/3$, where r is its radius). In cases of any dispute, and perhaps as a routine procedure, boulder sizes may have to be measured on exposure at the tunnel face as a joint operation - a representative of the Engineer and a representative of the contractor. Using a tape, the measurement would obviously be around a curved surface of the boulder, and so a value for the radius r would have to be inferred from the circumference formula $2\pi r$.

Example: Notwithstanding the provisions of CESMM3, a typical definition in contract documentation relates to the size of solid 'boulders' being 'not less than $0.1m^3$ in tunnels and headings, requiring blasting, the use of a compressor fitted with a pneumatic breaker, or any similar plant'. A $0.1m^3$ 'boulder' will create a much greater problem (and hence should perhaps be paid for at a higher rate) in a 1m diameter tunnel than in a 3m diameter tunnel. Experience suggests, however, that a contractual specification as to boulder size is rarely varied with tunnel size. On one tunnelling contract a further problem was highlighted. The original intention was to drive a 1.5m (excavated) diameter mini-tunnel but this was changed to a pipejacking operation before the work began. There was some dispute as to whether the change was as a result of the winning contractor's request or whether it was as a result of the Engineer's instruction. The contractor (strictly, the subcontractor) was experienced in both methods of tunnelling and he also had considerable experience of tunnelling in the type of ground (boulder clay and Coal

Measures strata) that would be encountered at the face of the new tunnel. The problems (both contractual and technical) arose not only in respect of the boulders which occupied the area of the tunnel face but also those which individually lay both within the tunnel face area (the payment line) and outside it. The technical problems were most severe with the largest boulders, which either had to be extracted and transported as a whole out of the tunnel or had to be broken up at the perimeter of the tunnel. There was no clear provision in the contract documentation to take account of this out-of-tunnel-section excavation or breaking-up and it was claimed by the contractor that the pipejack operation was especially disadvantaged by the need for this extra work. But the Engineer could forcefully refer not only to the contractor's prior experience in the same type of ground, the nearby boulder clay exposures that would have given him a most realistic impression of the type of ground that he would encounter at the tunnel face, and to his special expertise in pipe jacking operations. This case history and other similar ones indicate that the contract specification and billing for tunnelling in boulder clay and related strata must be written from the perspective of practical experience of the particular problems. They need to be written with great care, and ideally with the benefit of independent advice, if potentially protracted and expensive disputes are to be avoided.

'Measurable' boulders penetrating both the excavated area of the tunnel face and beyond the payment line create particular problems with respect to hand excavation (as a whole or by breakage at the payment line) and back grouting. Such problems need to be foreseen and addressed in the preamble to the bill of quantities. This difficulty is also noted in Section 6.12. The problem posed by 'nests' of boulders and blocks of rock at rockhead has been mentioned in Section 4.3.3. Where such blocks, individually below the critical size for 'rock', together form an obstacle to progress in a soil, it would be reasonable to pay for the obstruction at the billed rock rate, but sensibly this eventuality should be accommodated specifically in the contract documentation.

As a general comment, it is often the small quantities in tunnelling that can create problems for the employer because small quantities in a bill of quantities are prone to loading and so can be taken advantage of by contractors.

5.5 POSSIBLE ITEMS FOR INCLUSION IN THE PREAMBLE TO THE BILL OF QUANTITIES FOR A TUNNEL IN BOTH SOIL AND ROCK

Within the foregoing text there are frequent references to the preamble to a bill of quantities and to items that should be included. Some possible entries, which would be modified and added to in order to suit the particular construction, are given below. The entries relate substantially to hand excavation within the protection of a tunnelling shield but can be changed to accommodate TBM working. The sixth item, concerned with Coal Measures strata, was an attempt on an actual tunnelling contract to address the practical problem of differentiating between geological and technical terms for the contractual purposes of referencing the ground material. An alternative example, referenced elsewhere in the book, is that of the Lower Lias (Jurassic) which comprises clay shale and stratified bands of strong limestone. Reference in the bill of quantities would then be to 'Lower Lias material' (clay shale) and 'rock' (limestone).

- 'The bills of quantities are in accordance with the Civil Engineering Standard Method of Measurement, 3rd Edition 1991 produced by the Institution of Civil Engineers with the following additions and amendments:

- 'The term "rock" means a natural aggregate of mineral particles which would normally require the use of explosives and/or heavy pneumatic breaker to remove.

- 'Measurements of rock volume shall be made and agreed with the Engineer's Representative at the time of excavation. No payment shall be made for rock in any excavation where the Contractor has removed, broken up or buried the same prior to informing the Engineer's Representative and agreeing the dimension.

- 'Boulders shall be deemed to be rock if their dimensions exceed the following limits:

  (a) In Trench Excavation   0.20 (m^3)
  (b) In Manhole Excavation   0.25 (m^3)
  (c) In Tunnel Excavation   0.10 (m^3)

- 'Notwithstanding the fact that the boulders *in situ* may not be spherical, it will be assumed that they are spherical for the purposes of transforming the above volumes into linear dimensions. Any boulders penetrating the ground beyond the payment line of the excavation will be measured for payment as above if removed from the ground without breaking up. No extra payment will be made for breaking such boulders to conform with the payment line nor will payment be made for infilling of voids created by the excavation of boulders beyond the payment line.

- 'To simplify billing, excavation items in the tunnel have been split into three categories - rock, coal measures strata and clay. Rock is defined in Preamble note [.....] (above). Coal measures strata include mudstones, siltstones, sandstones etc. which are weathered so that they do not require the use of explosives or heavy pneumatic breakers to remove. Clay, for the purposes of this contract, is deemed to include all other material which can be excavated by light pneumatic hand tools. This will include sand, gravel, silt etc. as indicated in the site investigation report.

- 'Items have not been included for dealing with groundwater in the tunnel or open excavations. The contractor is required to make provision for dealing with, controlling and removal of accumulations of water throughout the works. The tenderer shall make provision for this requirement in his items detailed in Method Related Charges.

- 'Rates and prices inserted into the bill of quantities are to be the full inclusive value of the work described under the several items, including
  *all costs and expenses
  *all general risks, liabilities and obligations set forth or implied in the documents on which the tender is based.

Where special risks, liabilities and obligations cannot be dealt with as above, then the price thereof is to be separately stated in the item or items provided for the purpose.

• 'A price or rate is to be entered against each item in the bill of quantities, whether or not quantities are stated. Items against which no price is entered are to be considered as covered by other prices or rates in the bill.'

The following notes apply to the above items.

• *General items* in CESMM are to cover elements of the cost of the work which are not considered proportional to the quantities of the permanent works.

• *Method related charges*, to be inserted by the tenderer, are to distinguish between time-related and fixed charges.

• Since 'provisional sum items', 'provisional items', and 'prime cost items' are often included in a bill of quantities, the meanings of these terms should be given.

• *Provisional sum items* are those for which sums of money are provided in the bill of quantities for contingencies, additional or extra works, or for the cost of works envisaged to be carried out on the basis of nominated subcontractors. These sums are only implemented on the direction and at the discretion of the Engineer. Measurement and valuation should be at the rates (or analogous rates) contained in the priced bill of quantities or as prime cost items or on the basis of daywork. Although in a less-than-ideal contractual setting, provisional sum items might typically be used when the works design is incomplete but when there is pressure for the job to go out to tender and for construction to begin. One contract in London (Number 74 Worship Street renovation project), using the Joint Contracts Tribunal 80 form, is quoted as having 85% of an approximate bill of quantities in the form of provisional sums, meaning that only 15% of the work had been measured (Anon, 1990b). Unlike the ICE Conditions of Contract, 6th Edition (1991), the JCT80 form is intended

for works which are 'substantially designed but not completely detailed' and where a 'reasonably accurate forecast of the work to be done' exists. However to include such a high percentage of the work under the heading of provisional sums must be unusual, notwithstanding the fact that it is often not unusual for design work to continue into a contract period.

• A typical *provisional item* would be 'extra over items for excavation' where, in rock, 'making good soft spots at the base of a foundation' require imported fill not accommodated in the bill of quantities for pricing. Neither the 6th Edition of the ICE Conditions of Contract nor CESMM define provisional items, each of which must be noted separately in the preamble to the bill of quantities.

• *Prime cost items* are defined as the net sum entered into the bill of quantities by the Engineer as the sum provided to cover the cost of, or to be paid by the contractor to merchants or others, for specific items or materials to be supplied, or work to be done. Each tenderer is usually required to fill in such separate sub-items as: (a) charges and profit on prime cost items as a profit thereof, and (b) an item for handling and fixing connected with the item for which the prime cost is intended.

• Provisional sums, provisional items, and prime cost sums may also serve to 'pad' the contract, offering greater scope for the final contract cost to be lower than the winning tender price and thereby creating more mutual satisfaction among the participants (employer, Engineer, contractor)!

5.6 SPECIFICATION OF COMPRESSED AIR

Compressed air, the historical use of which in tunnelling has been traced by Hammond (1963), enjoys technical attractions in tunnelling works but also presents obvious physiological drawbacks. It balances the external head of water (the compressed air going into solution with the groundwater over a diffuse zone) and resists its intrusion into the tunnel. It also offers support directly to the ground by resisting the field forces which attempt to move soil particles into the tunnel. There is also the drying out of a narrow 'skin' of (cohesive) soil at the tunnel face, so increasing its effective strength and enabling it more easily to resist the passive pressures in the ground. In sands and silts, compressed air displaces much of the porewater within the range of its effectiveness and induces some cohesion between the grains of the soil as a result of surface tension.

Provided that the tunnel diameter is not large, the air pressure will usually be balanced to the head of water at the tunnel axis (pressure in kPa = 9.81 x water head in metres), being sufficient to leave just a small seepage from the face. These seepages may, however, cause some groundwater lowering and induce early consolidation settlement to be added to ground loss settlement (*see* Supplementary Information 16 in Part Two of the book). Experience is needed to judge the size of plant necessary to accommodate air losses. Wood and Kirkland (1987), quoting Hewett and Johannesson (1922), give a 'rule of thumb' for calculating the volumes of compressed air needed to hold back water in coarse sand as $7.5D^2 \text{m}^3$/minute (equivalent to $24D^2 \text{ft}^3$/minute) where D is the tunnel diameter in metres (or feet). This same relation is said by Dawson (1963) to apply to 'open ground' and the equivalent relation for 'compact ground' to be $3.75D^2 \text{m}^3$/minute (which is equivalent to $12 D^2 \text{ft}^3$/minute), again with D being the tunnel diameter in metres (or feet). There is a UK upper limit of 350kPa (international limits ranging from 350kPa to 400kPa), and if at all possible the compressed air pressure should be maintained below 1 bar (100kPa) in order to eliminate the need for special measures which include the need for medically controlled compression and decompression. Such controls, to restrict the onset of dysbaric osteonecrosis (bone necrosis, or arthritis of the joints), involve strict adherence to the rules and procedures (Work in Compressed Air Regulations, 1958: under revision) laid down in the UK by the Factory Inspectorate (*see* Construction Industry Research and Information Association, 1982). New safety measures have now been drawn up by the Health and Safety Executive's working party on Compressed Air Regulations. These recommendations, prepared for Government consid-

eration, include a reduction in exposure periods from 8 hours to 4 hours, higher levels of training, and stricter medical surveillance which requires workers to have a check-up 12 months after leaving a project.

There have been settlements of compensation claims against the contractors as a result of bone necrosis incurred during the 1970s construction of the second Dartford tunnel under the River Thames. Following one precedent concerning the award in February 1994 of £200 000 to a former Dartford tunnel miner while working for the contractors, Balfour Beatty, it seems that legal actions are planned against contractors Charles Brand and Edmund Nuttall by miners who contracted bone necrosis while working on the River Tyne vehicular tunnel in the 1960s.

Decompression sickness is virtually unknown at pressures below 80kPa and is rare between 80kPa and 100kPa. Timed decompressions, with stops at intermediate pressures, are required at pressures over 100kPa (approximately 1 bar). If there is too quick a reduction from a high pressure, nitrogen dissolved in the body through breathing forms bubbles in the bloodstream. In severe cases these bubbles may travel to the brain to cause death, or form around the optic nerve causing blindness, or around the spinal column to cause a form of meningitis or paralysis. These severe cases are more usually associated with deep-sea diving. Bubbles formed by uncontrolled compressed air tunnel decompression are more likely to cause localised bone death. The effect can be symptomless, with lesions appearing on the long bones, but is debilitating if it occurs near to joints. It is interesting to note that new research involving a perfluorocarbon emulsion blood substitute may overcome the problem should decompression be not properly performed, whatever the reason might be. Perfluorocarbon emulsions, which are in fact combinations of carbon and fluorine suspended in an emulsion, are very stable and are able to absorb and transport much larger quantities of oxygen than can natural red blood cells. Most important, however, in the present context is the fact that they are good solvents for gas (about 10 times better than red blood cells) and quickly re-absorb the nitrogen in bubble form associated by too rapid decompression. The idea is that the emulsion would be administered intravenously within minutes of an erroneous decompression.

This action alone may not be sufficient, but may 'buy time' so that re-compression and then controlled decompression can be applied in a suitable facility.

Too high a compressed air pressure is not only physiologically disadvantageous but it can also prevent groundwater from fully contacting segmental lining sealing gaskets, so inhibiting swelling of the gaskets and allowing compressed air losses.

With interbedded thin layers of sand or silt at the tunnel face, a relatively low ratio between air pressure and external head of water is usually adequate to provide greatly improved stability at the tunnel face (Wood and Kirkland, 1987). In marginal ground conditions, such as may arise with cohesionless soils or in ground containing sand/silt lenses with a water table only 1 to 2 metres above the tunnel crown, the Engineer may be undecided as to whether or not compressed air would be needed. Experience suggests that, where such difficult choices arise, provision *should* normally be made for compressed air. At a fairly low installation cost relative to the total project cost for small diameter tunnels, it is an insurance against subsequent disruption, higher costs and contractual claims.

> **Example:** Because of access problems between the site investigation borehole locations, the holes were sunk at 250m centres for a 1.5m diameter tunnel in boulder clay. Water-bearing sand layers in the boreholes were incorrectly identified as 'small lenses' and no special dewatering measures were expected. In fact, a sand layer extended between the boreholes. There was a 2m head of water above the tunnel and compressed air eventually had to be used at considerable extra cost.

A possible hazard to be foreseen, as noted by Wood and Kirkland (1987), concerns the presence of de-oxygenated air in the face of a tunnel when approached by a second tunnel under construction in compressed air. Another reported problem concerned a tunnel, driven with compressed air support, excavating from clay soil into 'ballast'. Water had been drawn down from the ballast by pumping, the barometric pressure was low, and air was drawn from the tunnel through the clay, so creating an oxygen deficiency at the tunnel face.

In all cases the thickness of cover above the tunnel crown should be checked to make sure that there is no possibility of excessive ground heave from tunnel air pressure. Even more important, the possibility of a blowout when tunnelling in compressed air below water must be avoided. The level in the face at which the air and water pressures balance out relative to the tunnel diameter, and expressed as a distance above the invert, is termed the 'pressure balance level' (PBL). Problems with blowouts can arise when the PBL is low and, with a relatively large diameter tunnel, the air pressure at the crown is sufficient to form a highly permeable cavity for air to escape. Dawson (1963) sketches the use of 'clay pocketing', employed for older hand-driven tunnels, at the tunnel crown just ahead of the shield as a means of inhibiting serious compressed air loss. Blowouts can occur at depths anywhere from 1.5 tunnel diameters for a small, highly pressurised tunnel in clean sand to less than 0.25 of a tunnel diameter for a large tunnel with low air pressures in a very silty sand. For protection, a cover of twice a tunnel diameter is needed in coarse-grained permeable ground and a cover of one tunnel diameter must be specified in a fine-grained soil of low permeability (Schenck and Wagner, 1963). If the site investigation indicates the presence of coarse soils above the tunnel crown, the necessary cover can be reduced to one tunnel diameter by the provision of either a graded filter or an impermeable clay blanket 1 metre thick and six tunnel diameters wide on the surface of the granular soil directly above the tunnel. This is similar to the suggestion made much earlier by Hewett and Johannesson (1922) wherein the thickness of the impermeable clay blanket should be at least one half the tunnel face diameter. For tunnelling under a river, the river bed would need to be dredged over the tunnel centre line and relined with the clay blanket before tunnelling.

Work by Peck (1991) on homogeneous soils has shown that a blowout can propagate from the surface down as well as from the tunnel up, and that although the factor of safety against a blowout can be more sensitive to tunnel depth with greater river depths, the critical depth at which a blowout occurs is independent of the river depth. There seems to be a linear relation between the critical depth of tunnel and the tunnel diameter, with smaller diameter tunnels requiring relatively more cover than larger diameter tunnels. The use of forepoling plates on a hand shield serves to reduce the critical depth by up to 60%, and the importance of good soils investigation is underlined by the fact that soils having 10% or more silt were found to be noticeably less prone to blowouts than was the case with soils having 5% or less silt fraction. A set of nomographs, accommodating tunnel depth, tunnel diameter, internal air pressure, soil type and shield type, and permitting an evaluation of the likely factor of safety against blowouts, serve as a useful preliminary design aid for the project engineer. However, most practising tunnelling engineers will realise that blowouts are most likely to occur in unpredicted, inhomogeneous ground, and so will tend treat such nomographs with some caution and scepticism.

With the advent of new technologies, the attractions of compressed air (such as being able to see the full exposed face of soil and any obstacles to progress) will increasingly be limited to short runs of tunnel. For longer man-entry tunnels, slurry shields and particularly earth pressure balancing shields will tend to be the primary choice. By the year 1985, earth pressure balance machines had taken 62% of tunnelling work otherwise requiring compressed air, with slurry shield tunnelling taking only 23% of the market. These machines, which have been discussed very lightly in Chapter 1 of the book, can be expected to achieve a substantial reduction in surface settlements in comparison with the more conventional methods, and there is also less need for associated ground improvement works, such as chemical grouting, curtain piling, or ground freezing. The slurry shield system is suitable for use in gravels and soft sands. Use of the earth pressure balance shield system, in which the excavated earth is used as the supporting medium and is removed from the face chamber by screw conveyor at the same rate as earth is excavated, is limited to silty and clayey soils. However, there will still be a need for an airlock and compressed air provision with these systems since if large boulders (Section 4.3.3) or strong bands of rock are encountered, de-slurrying and access to the face through the machine head may be required for breaking them up. Access will also be needed for

pick replacement and general maintenance purposes.

5.7   METHODS OF WORKING

If, in the opinion of the Engineer, tunnelling excavation and support hazards exist which could not be overcome by the adoption of conventional techniques such as the use of tunnelling shields, compressed air, dewatering, and so on, then there is considerable merit in billing preventative measures as definite specified requirements rather than leaving them to the contractor's discretion under method related charges. Should this not be done, a responsible contractor may allow in his tender for unspecified use of plant and may well be indirectly penalised by not getting the work. Although an Engineer may bill for a technique he should not attempt to impose engineering decisions (selection of particular plant being one example) that are more appropriately within the province of the contractor and his experience. There may be circumstances, however, when a decision is taken by an employer to provide a tunnelling machine for use by the contractor on a particular contract.

Example: The Engineer may direct that a shield be used for tunnelling in soil, but he should not attempt to specify the type of shield or a particular manufacturer. There may be special circumstances, however, when this rule should not apply. One example relates to tunnelling on Tyneside. On an interceptor sewer contract east of the City of Newcastle upon Tyne, where the tunnelling took place under compressed air in dominantly silty clay deposits below the water table, substantial ground settlements caused damage to a factory above the line of the tunnel. It was suspected that much of this settlement could be attributed to build-up of grout on the extrados of the tunnelling shield, the set grout in effect acting as an oversize bead and creating a substantial over-cut each time the shield shoved off the last completed lining ring. This experience was taken into account when a new contract was designed for the same interceptor sewer, this time in the centre of Newcastle upon Tyne. The sewer was driven beneath the centre of a road, to both sides of which were expensive commercial properties that could be prone to 'hogging' and settlement as a result of any tunnelling induced ground movements. Tunnelling was again in silty clay deposits, with the groundwater at sub-artesian but almost at artesian pressure. Both compressed air and a build-one-ring grout-one-ring construction regime were specified. In an attempt to militate against the problem of set grout adhesion to the extrados of the shield, a special 'teflon' style of coating was prescribed. Whether or not this precautionary move provided the reason, there were few, if any, settlement problems accompanying this particular tunnelling contract.

Example: An example supporting the exception to the rule referred to above concerns the tunnelling for London's water ring main. A 3.8km tunnel, New River Head to Barrow Hill, was driven through mixed ground containing sand lenses filled with water. A Canadian Lovat full-face machine was used, but not in the earth pressure balanced mode. The contract was of ICE 5th Edition form, but the employer, Thames Water, was involved in the choice of machine type.

Example: On many rock tunnelling contracts in urban areas, drill and blast methods of excavation are inappropriate. The associated ground vibrations may be environmentally unpleasant and they could cause structural damage (*see* Supplementary Information 20 in Part Two of the book). The Engineer may direct that the rock be excavated by cutting in order to avoid the vibration problem, but he should usually refrain from dictating the type of cutting machine. Such a choice is within the province of the contractor's experience, although the Engineer will obviously have a professional interest in knowing the potential performance of the machine. Should the machine under-perform in the tunnel, the contractor will then have an additional stimulus to seek geological and geotechnical grounds for explaining that reduced performance, and he must be made aware of the fact that because of the reduced working flexibility, if and when problems arise, the financial costs involved will be substantially

higher than will be the drill and blast plus back-up costs. An alternative strategy to actually proscribing the drill and blast method of excavation, and which was actually adopted for one particular contract on Tyneside, is to specify very low limits of tolerable ground vibration. This should indirectly guide the contractor towards the cutting machine solution without that solution being expressly imposed on the contractor. This is not a favoured solution since the Engineer should really be prepared to take the advice on low-level vibrations, recommending that the employer either accept the reduced production (and higher costs) associated with drill and blast or directly go for a machine drive. It should, however, be noted that it is quite possible using careful, very low level, explosive charges and delayed action detonators to design sensibly vibration-free blasts and still achieve production targets. There is further discussion on vibration in Section 6.10 and in Supplementary Information 20 in Part Two of the book.

For construction in difficult and potentially dangerous geological environments the Engineer may prefer to present the engineering requirements rather than give design details for temporary works, which is becoming a more formal discipline and which is usually the responsibility of the contractor. However, this does represent a greater tilting of the risk towards the contractor and would tend to raise the base level of the bidding for the work, generate more contractual claims, and be generally contrary to the recommendations and spirit of CIRIA Report 79 (CIRIA, 1978). This attitude is likely to be enhanced because contractors do not always demonstrate their acceptance of the principles of Banwell (National Economic Development Council Working Party, 1964), Harris (National Economic Development Organisation, 1968), or the rules of CESMM3 (Institution of Civil Engineers, 1991b, Section 7 p14) in relation to method related charges at the time of tender.

A contractor may offer an alternative method of working from that originally expected, or a different type of construction from that billed. Many of these are clearly of a design and construct nature and would fundamentally change the contract. Before acceptance by the Engineer, any such proposals should receive very careful consideration to establish that the contractor is experienced in, and quite able to complete successfully, the work proposed. However, time will be at a premium for any detailed considerations to be carried out and there will certainly be little or no time for further exploratory works related specifically to the alternative proposals. Consequently, there will tend to be increased risk placed on the employer, for not only will his prior knowledge of the ground now be somewhat less relevant but also he will know less than the contractor about the new design and will need to assimilate a new pricing structure. Before an employer accepts any such bid the contractor should be asked to indicate how risks associated with the original design and contract documents will be changed under the new proposals and a framework for the apportionment of risk arising from the alternative method should be clearly agreed at the outset, with legal assistance in drafting a written agreement if necessary. It is noted that acceptance by the Engineer of an alternative method of working means that the method then becomes his for the purposes of the contract. Notwithstanding these difficulties, alternative tenders can offer good, cost-saving opportunities for the client commissioning the work.

Example: Contractors have proposed, on occasions, to construct larger diameter tunnels in order to suit plant access requirements, or they have offered to construct different types of permanent lining. In the former case, the original payment line would need to be adhered to. An example of the latter case would be the use of a 'one-pass' lining for a sewer tunnel in lieu of, say, a reinforced concrete segmental primary lining and either a concrete permanent lining or a brick invert-to-axis/concrete axis-to-crown permanent lining. The Engineer would need to be satisfied as to the long-term mechanical stability, degradational resistance, and watertightness of the one-pass lining before agreeing to its use.

When alternative forms of construction are being offered by tenderers, the choice of options available should be carefully considered in the light of the available ground investigation information.

Example: In connection with a sewerage scheme on Tyneside, several tunnelling options were offered: conventional hand shield tunnelling, box heading, and liner plate construction. In one instance there was a failure to deduce from the ground investigation that compressed air would have to be used as a temporary works item to stabilise the tunnel face. Under such circumstances a box heading, which was the least expensive construction offer, was clearly inappropriate because of the inevitable air losses. Its choice resulted in significant problems on site and increased construction costs.

Method statements should be encouraged from contractors who would be required to state the major assumptions (dominantly those relating to ground conditions) used in preparing the bid. In most instances it would be helpful for the Engineer to prepare a pro forma for completion by the contractor, the questions in the pro forma being carefully designed to elicit clear replies.

5.8   TENDER LIST AND PERIOD

The short list for the tunnel contract tenderers, usually comprising a minimum of three and a maximum of eight tenderers depending an the size and nature of the contract, should (ideally) be confined to those firms having a substantial record of technical ability, financial standing, managerial capability and a firm organisational structure, a well-founded history of contract performance on comparable construction work, and a reputation for professional integrity. There must be a good health and safety policy and record (the 1994 Construction (Design and Management) Regulations, scheduled to come into force on 31 March 1995, making clients more responsible for health and safety), sufficient insurance cover, suitable quality assurance policies and records, and evidence of good environmental policies, management as per BS 7750:1994 (British Standards Institution, 1994b), and records. It may sometimes be necessary for the client or his agent to visit premises of possible tenderers in order to assess production capability, capacity, methods of operation and historical implementation of attributes noted above. Assessment of suitability will hinge on information proffered in an application for inclusion, and evidence stemming from referees, from the actual client organisation and from other readily available sources. Grounds for exclusion from the list include bankruptcy or receivership, proven professional misconduct, non-payment of tax or statutory contributions, and the provision of misrepresentative information. Firms that are known to have made dubious claims in the past should not be included on the list. In a similar manner, there should be a refusal to allow sub-letting to firms that have previously been involved in dubious claims. Select lists should be continuously updated by client organisations having large volumes of tunnelling works to let, or by their advisers.

It is in the interests of all parties involved in a contract that the tenderers be given sufficient time to examine the tender documents in detail, to seek information from knowledgeable bodies, to visit the site and its environs, to inspect the borehole and other evidence, to price the bill of quantities, and to extract such other information as will be necessary in order to fulfil the contractors' Clause 11(2) obligations under the ICE Conditions of Contract (1991). It may be thought prudent, in the contract documents, clearly to direct the tendering contractors' attention to these matters, especially the need for informed site visits and borehole core inspections. A record should be made on behalf of the Engineer of the dates and times of these latter inspections.

Although contract documentation preparation can be a lengthy operation, but often with a narrow time frame being allowed for this work before the contract goes out to tender, the tender list is usually established early and some of the details, such as the site investigation information and the drawings, could be sent out to the contractors for their evaluation before the full documentation is released. Such early information could add to the value of pre-tender meetings. On the other hand, realism suggests that early information of this nature would not usually prompt contractors into early action; no doubt they would still delay examination of the documentation until the last moment before tender! Another problem could arise from the fact that a great deal of tunnelling work is let to contractors who use subcontractors for the tunnelling, and, notwithstanding the comment above concerning sub-letting, it is very often in the latter that much of the real tunnelling expertise is vested. It is unlikely, in most instances, that a subcontractor would be willing to make himself available for, or would be in a realistic position to offer, a useful contribution at, such a meeting which would need to concentrate on the contractual relations between employer and main contractor.

5.9   EVALUATION OF TENDER

As noted earlier in this chapter, the options for accepting a tender are the *lowest price* tendered or the *most economically advantageous* tender. The criteria upon which decisions will be based typically include factors such as price, operating and maintenance costs, technical and operational characteristics, quality and reliability, service after completion, programme proposed for completion and delivery, and guarantees.

Incorporation of reference ground conditions in the tunnel contract tender documents could stimulate qualifications from the contractors. A contractor may also inject his own perception of ground conditions, in the form of tender qualifications. Acceptance of such a tender could give the qualifications contractual status.

> **Example:** Having perused the site investigation report(s) for a tunnelling contract in an urban area, a particular contractor formed a different interpretation of expected water inflows to the tunnel from those originally formulated by the Engineer and his advisers. The magnitudes of the water inflows influence the control measures needed in a tunnel; for example, forward grouting, compressed air working, or dewatering measures. After discussions it was considered that the contractor's views on quantities carried sufficient weight for his proposals on ground improvement measures to be accepted.

Detailed discussions with tenderers should aim to reveal those contractors who have built into their prices specific assumptions that are unlikely to be achieved. Bidding contractors might be provided with the option of making a specific form of qualification, namely a tender price statement. Such a statement would take the form of an outline reason for any seemingly unusual pricing in the bill of quantities, and would be used only in conjunction with the contractual submissions. It would be used as a basis for further discussion and investigation of the pricing structure, if needed, at a pre-award meeting. A contractor would not be expected to include padding in his tender to meet unquantifiable risks because he recognises the need to bid keenly. In order to detect excesses of over-optimism, particularly in respect of ground conditions, and a failure to understand the complexity of the work, the Engineer must carefully assess the tenders, especially in relation to the method of working. Where difficult or complex working conditions are foreseen, specialist opinions must be sought. Final selection of a contractor should depend to a large extent, and notwithstanding the relative prices of the bids, on the perceived feasibility and flexibility of each tenderer's proposal and on past records for handling difficult tunnelling conditions.

In the absence of a specific strategy to the contrary, a public body in particular may feel under some constraint to accept the lowest unqualified tender. An Engineer must argue the case, if need be, for the employer not necessarily to accept the lowest tender and must, under such conditions, present convincing evidence in a form that is understandable to the general public. His review of the evidence to a wider gathering would include matters such as any qualifications to the tender, the possible methods of working, and the contractor's awareness of health, safety and environmental issues that might affect and be affected by his operations.

All tenderers need to be informed of award decisions and, if the estimated value of the contract has been above the EC threshold, then a Notice of Award will be placed in the Official Journal of the European Community.

6

TUNNEL CONSTRUCTION

6.1 GENERAL

The value of good quality site investigations cannot be over-emphasised for predicting and pre-empting problems which may arise before, during and after tunnel construction. Lack of adequate investigation data upon which to design the contract can lead to increased construction costs, additional costs arising from unforeseen ground conditions, construction periods extended beyond the contract term, and a general atmosphere of ill will on the site.

There is a whole raft of legislation relating to health and safety of which the contractor must be aware. This includes:

- the Construction (General Provisions) Regulations 1961, SI 224;
- the Health and Safety at Work etc Act 1974;
- the Health and Safety (First Aid) Regulations 1981, SI 917;
- the Reporting of Injuries, Diseases and Dangerous Occurrences Regulations 1985, SI 1457;
- the Diseases and Dangerous Occurrences Regulations 1986;
- the Control of Substances Hazardous to Health Regulations 1988, amended by SI 1990/2026, SI 1991/2431, and SI 1992/2382 (which requires that risk assessments be carried out);
- the Noise at Work Regulations 1989, SI 1790;
- the Environmental Protection Act 1990;
- the Workplace (Health, Safety and Welfare) Regulations 1992, SI 3004;
- the Management of Health and Safety at Work Regulations 1992, SI 2051 (under which there is both a legal requirement for risk assessment of work activities and a duty of care responsibility on the contractor's line management to ensure that the workforce is adequately trained for the work in hand, the provision of information being not enough);
- the Provision and Use of Work Equipment Regulations 1992, SI 2966 (which states that equipment must be carefully selected to ensure that it is suitable for its intended use and conditions, that all safety considerations must be taken into account, and that appropriate British, European and International standards should be complied with);
- the Manual Handling Operations Regulations 1992;
- the Personal Protective Equipment at Work Regulations 1992, SI 2966 (which requires that adequate and appropriate protective clothing, such as safety helmets, must be worn by all participants, that it must be checked regularly, maintained in good condition, and worn correctly);
- the Lifting Plant and Equipment (Records of Test and Examination etc) Regulations 1992;
- the Construction (Design and Management) Regulations 1994.

There are statutory provisions of Acts of Parliament, and of Regulations made under them, which remain active until they are replaced by Regulations. The means of enforcing them have, however, been changed by the provisions of the Health and Safety at Work etc Act.

- The Explosives Act 1875 (except Sections 30-32, 80, and 116-121).
- The Explosives Act 1923.
- The Public Health Act 1961 (except Section 151).
- The Employment Medical Advisory Service Act 1972 (except Sections 1 and 6 and Schedule 1).

This list of generally applicable health and safety legislation is by no means exhaustive; for example, additional legislation will apply when working in, or in the general proximity of, coal seams where methane gas could be a hazard. In addition, reference may be made to Davies and Tomasin (1990) and to the following Government publications.

• *Management of Health and Safety at Work - Approved Code of Practice 1992*, HSE Books, ISBN 0 71760412 8;
• *Control of Substances Hazardous to Health Regulations - Approved Codes of Practice* (4th Edition) 1993, HSE Books, ISBN 0 11882085 0.

Of particular interest to tunnelling are the Construction (Design and Management) Regulations 1994, effective 1 April 1995. These allocate responsibilities to the various parties within a construction project with a view to managing health and safety from the feasibility study at the outset to completion of the construction, and beyond. There is a new duty of care on clients to appoint a 'competent planning supervisor' whose duties are to coordinate the management of health and safety issues for the project right through the design stage and to interface with the contractor. The planning supervisor must assess whether the project's financial capacity and timescale are sufficient to allow its completion safely and without risks to health. He must prepare a health and safety plan, assess the design in respect of its impact on health and safety during the construction phase, prepare a health and safety file, and coordinate the activities of the members of the design team by ensuring liaison between them and the main contractor.

In addition to the above statutes there is also, of course, the common law, and particularly the torts of nuisance, negligence and trespass to be considered during the tunnelling works.

In subsequent sections of the book an indication is given of some problems which may arise during the currency of a tunnel contract, all illustrating the requirement for relevant site investigation data. In addition, several general experiences in tunnelling contracts, not all of the experiences being closely related to site investigation problems, are described.

6.2   GROUNDWATER PROBLEMS

6.2.1 General

If the site investigation reveals a water table at or above the tunnel horizon then its implications with respect to tunnelling progress need to be assessed in the context of the ground conditions, particularly the structure of the ground. A high water table in a homogeneous low permeability clay soil may have no effect on the tunnelling works, which may well progress under mainly dry conditions. However, should the clay contain sand/silt lenses, or if the tunnel is driven in cohesionless soils, then water inflows could create severe problems at the tunnel face.

In the case of rock strata, a high water table may not affect tunnelling works if there is a suitably impermeable stratum at or above the tunnel soffit. Because the rock above this stratum may be highly discontinuous, rendering it very permeable, there is an obvious need for the site investigation to prove, or at least imply, a relative lack of discontinuities, along which water could be conducted, in this lower stratum. If, after careful appraisal, there is some doubt as to the eventual watertightness at the tunnel face, but it is decided to progress the tunnel without any special measures being implemented, there could be some merit in arranging at the outset for the construction of a bulkhead and an airlock as a contingency measure should compressed air ultimately be required. A maximum of 1 bar pressure would be adopted these days and the airlock would cost about £10k for a 6ft diameter tunnel. A steel bulkhead incorporating the airlock would be bolted to 75mm steel flanges projecting from tunnelling support rings.

Example: The site investigation for a coastal tunnel revealed a water table 8m above axis level and that the tunnelling would be mainly in sandstone, but with a thin transition bed of silty sand above the sandstone, and till above that. For part of the time the tunnel soffit would be in the till but at no time would the covering of till be breached. Because of concern over the head of water, and an unwillingness to adopt a compressed air solution, it was decided to tunnel full-face using a bentonite tunnelling machine. There were subsequent steering problems with the machine and it proved impossible to maintain line and level. Without considering the reasons for these problems, it is not entirely clear why a closed form of shield needed to be chosen for the work in the first place. Although the high head of water gave rise to some concern, the impression from the site investigation report was that the till would perform as an impermeable or partially impermeable barrier, protecting the tunnel from significant water inflows. On one occasion the shield was stopped and the tunnel face was accessed through the bulkhead by engineering geologists for the purpose of taking samples of the ground. There was little evidence of water. A less expensive tunnelling solution might have been the use of a boom cutter in a short shield, contingency provision having been made for an airlock at the entrance to the tunnel and for pumping ahead of the tunnel face should it have proved ultimately to be necessary, perhaps in conjunction with a reduced compressed air facility.

Careful site investigation is essential to allow a realistic evaluation of possible groundwater control measures which will be appropriate at the site, together with possible consequential effects, for example longer-term consolidation-type settlements (*see* Supplementary Information 16 in Part Two of the book). In addition to special gasketing of lining segments, control measures might include: sump pumping if the inflows are light; well-point pumping (*see* Supplementary Information 15 in Part Two of the book); grouting from ground surface or from face probe holes (*see* Supplementary Information 8 in Part Two of the book); ground freezing (*see* Section 6.5 below); the use of cut-off walls such as sheet steel piling or slurry walls taken down to an aquiclude below the tunnel invert if the tunnel is not too deep; pilot tunnels, and water diversion. In the case of water diversion, plastic sheeting may be supported on panels of steel mesh, with longitudinal french drains at each side of the invert which are grouted up in the final operation after completion of a cast *in situ* lining. Alternatively, instead of plastic sheeting a quick-set mortar applied pneumatically to steel mesh may be used, with internal pipes to drain off the water flow. The pipes can then be closed at the end of the operation, or they can be left open but with the possibility of encouraging longer-term consolidation settlement (*see* Section 6.3.2 and Supplementary Information 16 in Part Two of the book). In the case of a cut-off, even if termination of the cut-off toe in an aquiclude is not possible, flow net sketching may suggest a suitable termination depth that will considerably reduce the inflow volume to a degree that can be

pumped from the tunnel. With low water heads, less than 2m or so, and low or medium permeability soils, consideration should be given to the use of a tunnelling shield equipped with poling plates, possibly accompanied by provision for the adoption of compressed air (*see* Sections 5.6 and 6.4) should it prove to be necessary. Tunnelling has been successfully accomplished at large diameters and under severe ground and groundwater conditions in the United States (Washington Metro) using this technique without resorting to compressed air or extensive dewatering.

If compressed air is an acceptable option, it must be decided, after careful appraisal of the groundwater head and the geology, whether compressed air will be needed, with the consequent expense in additional plant and extra rates of payment related to the chosen working pressures (*see* Section 6.4). Dewatering will often be the preferred option if the ground permeability is high. Evaluation of the viability of a dewatering scheme should be based mainly on *in situ* permeability tests (*see* Supplementary Information 13 in Part Two of the book) because of the danger of loss of fines during boring (*see* Section 4.3.7). If grouting is to be adopted, either alone or in conjunction with compressed air, and the particle size of the ground is sufficiently large to permit the use of a particulate (usually cement) grout, then this is the least expensive option. Chemical/resin grouts, used alone or in association with a particulate grout in a staged manner, will be more costly. On a 275m long tunnel in Stratford High Street, East End of London, chemical grouting was used to reduce the compressed air pressure that would otherwise have been needed. Bentonite-cement grout was injected to infill large fissures and silicate bicarbonate grout was used to tighten up the ground in the path of the proposed tunnel.

Grouting can be from ground surface or, in the case of a large diameter tunnel, from a pilot tunnel. There is comment on grouting in Supplementary Information 8 in Part Two of the book.

6.2.2 Entrained water behind tunnel lining

In some tunnelling methods, the void behind the permanent lining may tend not to be filled until the section of tunnel is completed, but in a three-segment mini-tunnel, for example, pea gravel or Lytag injection should take place after lining but with the grouting of it being delayed until tunnel completion. Some pipejack sections can be up to about 600m long before sealing takes place if indeed sealing is done at all. The presence of such voids not only encourages ground loss settlements (*see* Section 6.3 and Supplementary Information 16 in Part Two of the book) but can also pose water problems at the tunnel face if the workings are to the dip.

> **Example:** One sewage interceptor tunnel in Coal Measures rock on Tyneside was constructed by pipejacking according to the client specification. The contractor chose to excavate oversize in order to use a boom cutter in a short shield. In the area of the tunnel, coal had been extracted in the past, and so a programme of grouting of the old workings was instituted prior to the tunnelling works. The tunnel line was inclined at an angle of 1 in 20. For technical reasons the contractor, with the approval of the Engineer, chose to work to the dip. At a distance well into the section drive water entered the tunnel face in quantities that could not be removed by standard pumping, and there was some concern for the safety of the operatives in the event that a major inundation occurred. The slurry created by the action of the water and the cutters on the mudrock (*see*, for example, Section 4.4.5 and Supplementary Information 6 in Part Two of the book) could not be removed against the incline of the tunnel.
>
> Investigations established that the water was running down the outside of the pipe from water-bearing voids intercepted at a higher level but which had not released water in any quantities when first exposed. (A pipejack cannot be sealed until it is completed.) Where the problem was most severe, it was known that water was not entering directly at the tunnel face. The bulk grouting operation had clearly not been successful and it was necessary to institute a re-grouting programme.   Before that, however, a pumping well was sunk just to the rise of, and outside the section of the tunnel and a small dip heading driven from the tunnel to the well. This served to deflect much of the entrained water and allowed

> tunnelling   to   proceed   under   more manageable conditions.
>
> The tunnel face was affected by the severe inflow at a distance of 90 metres from the shaft towards which it was being driven. Had the pumping not been effective, the only reasonable alternative strategy would have been to have driven, by drill and blast, a small heading from the shaft to the rise in order to intercept the tunnel face, to drain the water away, and so allow the boom cutter and shield to complete the tunnel.   The client would have been at a contractual disadvantage when negotiating the payment for construction of the heading.

There could be circumstances when a tunnelling machine becomes trapped and it is decided that the cost of recovery would exceed the value of the retrieved machine. A bypass tunnel will then need to be constructed, with the cost of this additional work being met by one of the parties to the contract or else shared in some way by the parties   to   the   contract   depending   upon establishment of the causes of the problem. The trapped machine may have been purchased by the client for operation by the contractor. Depending on the terms of the works insurance policy taken out by the contractor jointly in his name and that of the client it may be possible for the client to seek recovery of the value of the lost machine. This will ideally be done under the joint names, but the contractor may well be constrained from joining in the attempted recovery of cost because of its possible effect on his future levels of insurance premium and the fact that he will gain little but goodwill from the client should the exercise be successful. If the contractor declines to become involved, then the client will have to go it alone with the insurer.

6.2.3 Perched water tables

Perched water tables can pose problems for tunnelling projects within urban areas, and so their detection during the ground investigation is most important. It is generally very difficult to assess whether several perched water tables are hydraulically connected. Compressed air, if used, could drive away water within these perched tables, perhaps causing some ground settlement beneath adjacent buildings when the air pressure is taken off and before porewater pressures re-equilibriate. Sometimes it may be possible to deal with perched water tables by allowing the water to drain from the tunnel face and then into a sump for pumping out of the tunnel, thus avoiding the need for compressed air or expensive dewatering measures.

6.3 GROUND MOVEMENTS AND SETTLEMENTS

6.3.1 General movements and surveys

It is important to determine the temporary and permanent components of the unavoidable surface settlement caused by tunnel and shaft construction, particularly in urban areas where the cost of damage to both surface and underground structures could be high. Equations for the estimation of short-term and long-term consequential settlements are given in Supplementary Information 16 in Part Two of the book. Settlement would normally occur above a tunnel in the form of an inverted bell-shape depression as a result of unavoidable loss of ground into the tunnel and due to associated groundwater movements (Norgrove *et al.*, 1979; Attewell and Yeates, 1984; Attewell *et al.*, 1986). It is possible to estimate the three-dimensional character of the ground-loss movements (Attewell and Woodman, 1982; Attewell and Yeates, 1984; Attewell *et al.*, 1986), the possible transfer of those movements to buildings and buried pipes (Attewell and Yeates, 1984; Attewell *et al.*, 1986) and both the magnitude and form of the longer-term consolidation-type movements (Attewell, 1988b). For relatively shallow tunnels in cohesive soil approximately 50% of the final total settlement occurs above the advancing tunnel face. It is axiomatic that good quality site investigation information is critical to allow realistic estimates to be made of these movements.

Although attention in this book is directed to settlement, the vertical component of ground movement, it is important to understand that the induced horizontal displacements, the vertical and horizontal strains, and the induced curvatures, all as a function of depth, are also important parameters affecting above-ground and in-ground structures. Reference should be made to the publications cited above and to the

more recent paper by New and O'Reilly (1991). The question of adjacent and super-adjacent tunnels is addressed in Attewell (1978) and the relevant ground movement equations for twin tunnels can be found in New and O'Reilly (1991).

The term 'unavoidable' is used in the sense of Clause 22(2)(d) of the ICE Conditions of Contract (1991). Some ground movement is the inevitable consequence of tunnelling. Other contributions to ground movement may be a function of tunnelling technique and operator experience, and therefore 'avoidable'. When tunnelling in soil having cohesive characteristics, radially inward ground-loss movement occurs through the overcut annulus created by the bead, which is attached to the leading edge of the shield and hood and covers the upper 180° of the shield circumference or sometimes the whole of the circumference. Inward ground movement continues around the tail skin of the shield and around the newly erected lining until arrested by the gravel and/or grout injected behind the lining to infill the annular void between the lining extrados and the cut surface of the soil. When tunnelling in granular media in free air with a hand shield to which is attached a tail skin, the ground losses at the tunnel may be quite high because there is always the danger of uncontrolled movement of the ground inwards at the face and of collapse on to the newly erected ring before any contact grouting at the back of the lining can take place. Ground losses will also be associated with the cutting of an elliptical section when a shield advances around a curve (major axis horizontal) and if it is driven with a 'look-up' attitude to counteract diving in weak ground (major axis vertical). A non-articulating shield could become 'locked' on a tight transition curve, leading to consequential ground losses associated both with the stand, even when the face is boarded up, and with the overbreak as the shield is dug out. When assessing responsibility for settlement damage, the question of 'unavoidable' and 'avoidable' demands careful consideration.

Restriction of the radial take component of ground loss depends upon a number of factors including early contact grouting behind a pre-cast concrete or cast iron segmental lining (from the base upwards) and minimum bleed during the setting of a cement grout. It also depends upon how many rings are erected before grouting takes place. As noted above, Lytag (a sintered pulverised fuel ash) is often blown into the voids behind the segments of a mini-tunnel after a substantial portion of the tunnel has been built, the Lytag then being injected with a cement grout. It is difficult to check on the pervasiveness and injected efficiency of Lytag. It can, for example, disintegrate during handling and placing. Water under pressure in the cement grout can be forced into porous Lytag particles, so reducing the grout fluidity and its ability fully to permeate the granular infill. Any voids remaining behind the lining will encourage residual ground losses and settlements.

Pipejacking techniques (*see* Chapter 1) are becoming much more prevalent in the tunnelling industry both for non-man-entry (remote control microtunnelling at diameters up to 900mm) and for man-entry size tunnels. A major advantage with a pipejacking system can be the control on face take, and also to some extent on radial-take ground loss when operating with an earth pressure balance or slurry shield at the face. An earth pressure balance machine tends to have no bead but there is usually a difference between the outside diameter of the lining and the cut diameter of the tunnel (40 to 50mm in the case of the Markham EPB machine used on the Bordeaux, France contract, as described by Wallis, 1994). One way of limiting radial ground loss on such a machine, and to compensate for the overcutting that is needed for steering the machine on line and level, is to inject the annular void through the machine tail skin, but using means to ensure that there can be no set-adhesion of grout to the metal of the tail skin. In the case of the Markham EPB machine a thixotropic grout was injected through twelve grout channels spaced at equal intervals around the inside circumference of the tail skin, injection occurring immediately behind the wire brush tail seals which are there to prevent pressurised soil extruding from the tunnel face around the outside circumference of the shield. About 4.5m^3 of grout were injected at a pressure of about 15 bars behind each 1.2m wide lining ring.

Pre-contract structural surveys must be carried out on all buildings likely to be affected by the tunnelling operations. These surveys are particularly important for high value properties, settlement-sensitive structures or listed buildings. Properties which lie within a distance of about

two tunnel depths either side the tunnel centre line should be surveyed and levelling stations (temporary bench marks) established. For tunnelling in firm to stiff clays, a survey distance from the tunnel centre line of 1.5 tunnel depths should be adequate. The surveys should be conducted by a qualified building surveyor, perhaps accompanied by a structural engineer, and property schedules prepared by the quantity surveyor. These schedules comprise detailed expositions of all existing damage, together with colour photographic evidence, against which any post-construction damage can be compared and assessed for compensation payments. A fictitious property schedule is provided in Attewell *et al.* (1986).

It will be realised that such surveys become especially important where a suspect building might be rendered unstable by the tunnelling operation and, as a result, could cause injury to people in the building or outside. The case in tort of *Donoghue* v *Stevenson* [1932] AC 562 suggests that the designer must take reasonable care to avoid any acts or omissions that can be reasonably foreseen which would be likely to injure a neighbour. However, where injury can be foreseen, a defendant would not be charged with an absolute duty to avoid such injury; he is required merely to take reasonable care to avoid it. In assessing this reasonableness, the law takes into account such matters as to the magnitude of the risk (which would involve the input of substantial structural expertise), the cost of avoidance, and the actual nature and benefit (to society) of the defendant's actions which invoked the risk. In the case of tunnelling works, no doubt the latter would hold important sway.

Compensation for structural damage would normally be covered by the works insurance taken out by the contractor on behalf of himself and the client, but there may be a situation where the operation of the contract is under the control of project management carrying its own insurance. Under these particular circumstances the grey areas of insurer liability may be broadened and the time taken for settlement of claims increased as the several insurance companies (including that or those of the building owner) deliberate their individual responsibilities under the policies.

Consideration should also be given to the possible effects of the estimated ground movements on underground services, particularly old cast iron gas mains and branches from mains to individual buildings. Many of these pipes may lie along the centre line of the tunnel, or parallel to the centre line and at a distance of less then 1.5 tunnel depths from it. (This arises because significant lengths of many urban tunnels tend to be driven beneath and parallel to roads.) Pipes above the tunnel centre line would be subjected to direct longitudinal and bending strains caused by the forward (temporary) settlement trough. Those displaced from the centre line are subjected also to bending in a horizontal plane as a result of the transverse (permanent) settlement trough. Pipelines lying at right angles to the tunnel centre line will be subjected to a range of bending and direct strains that will maximize with the passage of the face and will remain as a permanent feature of the pipe. Those pipeline segments towards the transverse extremities of the settlement trough are subjected to the highest levels of tensile strain - superposition of direct and bending tensions - and it is here that the pulling effects on branches which supply individual properties with gas and water are most severe. If possible, very old high pressure gas mains in cast iron should be considered for replacement before tunnelling takes place. High density polyethylene pipe will often be used, a suitable betterment element (*see* the example in Section 4.3.4) of the cost being negotiated between the client body for the tunnel and the owner of the pipe, in this case British Gas. Various methods of estimating upper bound pipe strains and of accommodating ground-pipe interaction effects in the analyses have been considered by Attewell and Yeates (1984) and by Attewell *et al.* (1986).

Apart from the re-laying of pipelines, there are several other measures that can be taken to mitigate settlement effects. Most of these require careful attention to detail at the design stage, good workmanship by the contractor, and attention to detail at the tunnel face by the Engineer's team. One example of the design relates to the specification of a build-one-ring grout-one-ring construction regime, which may be somewhat irksome to the contractor but which will have the effect of offering more immediate ground support at the point of ground loss. London Underground, for example, specifies that the unsupported length should not exceed 1.5m,

and that the maximum distortion of the tunnel should be 15mm.

6.3.2 Groundwater effects

Drawdown of groundwater can affect the development of a settlement trough in several ways, and these can have important implications for surface and sub-surface structures. The use of compressed air can also result in increased settlements (*see* Section 6.4).

Groundwater will drain into unsealed portions of the tunnel. If it drains through open joints in rock then it causes a problem of water containment only. If it flows through infilled joints and filling is washed out, the material is affected and instability could ensue. Flow through soil can cause particle loss and some ground settlement, interbedded coarse and fine-grained layers or sand lenses creating particular problems. In the absence of any significant recharge, dewatering alone of the superadjacent ground without any particle movement will result in longer-term consolidation settlements, organic soils creating a particular problem. In addition, reduced or negative porewater pressures in the ground surrounding the tunnel can promote drainage, increases in effective stress in the drawdown zone, and subsequent longer-term settlements.

If the ground permeability is isotropic, then the effect of such drainage may be to deepen the settlement trough without significantly widening it. This will result in greater ground curvatures and a greater possibility of damage to structures within the limits of the trough. If the ground permeability is highly anisotropic, for example with much greater horizontal than vertical permeability, then the earlier ground loss settlement trough could widen considerably but deepen less than it would under isotropic conditions. More structures may be affected by the wider trough, but angular distortions and ground curvatures (Attewell and Yeates, 1984; Attewell *et al.*, 1986) would be reduced. The problem of estimating longer-term ground settlements has been addressed by Attewell (1988b) and is considered in Supplementary Information 16 in Part Two of the book. For the reasons mentioned above, and for such estimates, it may be important to establish at the site investigation stage the degree of anisotropy of the strata by, for example, careful examination of the soil structure and perhaps by suitable permeability tests on the strata, although the results of such testing could be misleading with respect to the mass properties of the ground.

6.3.3 Whole body tunnel movements

Where tunnels are to be constructed in very weak ground and/or ground that has a high potential for compaction (such as fill material, especially when it contains a major proportion of waste deposited less than, say, a few decades ago), time must always be allowed for settlement to take place prior to placement of any permanent lining. In an infilled valley, increasing whole body settlement may be expected towards the centre of the valley, principally caused by the weight and vibration of construction plant operating within the tunnel. The lined tunnel itself will usually be slightly negatively buoyant. The question of lining-down will not arise in the case of the increasingly popular one-pass lining and also pipejacked lining, nor in the case of non-man-entry machine-driven tunnels, so these systems need to be considered carefully in the context of this type of ground.

It would be usual to calculate any expected whole-body differential (over the tunnel length) settlements, to construct the tunnel over-size and, after allowing sufficient time for the settlement to cease, to accommodate the deformations by longitudinal reinforcement in the lining. Procedure on a sewerage scheme in the north-east of England, even in competent natural ground, has often been in the past to allow a time delay of several months between installation of the primary support and casting of a permanent concrete lining. Actual measurements that were carried out on an interceptor sewer suggested that delays of this length were unduly pessimistic and could be reduced significantly.

The reverse problem in such ground relates to the effects of a rising water table. This is causing concern in several cities including Paris, New York, Tokyo, London, Liverpool and Birmingham. In London, for example, the water level in the Chalk and sands which under lie the Tertiary clays has been lowered by abstractions during the past two centuries. At its lowest point the water level had dropped by up to 70m, but now, because of reduced abstractions, it is rising

at rates typically up to 1m/year. Water levels in London could return almost to their original values within the next 30 years unless specific action is taken to offset the rise. But in any case, the general problem needs to be recognised in the context of tunnel design. If tunnels could become just buoyant, then special measures to prevent this may need to be incorporated at the outset. Other measures, including improved sealing of segments against higher water pressures and associated seepages, may need to be suggested for those tunnels that are already in the ground in areas that are now known to be vulnerable.

6.4   COMPRESSED AIR EFFECTS ON THE GROUND

In cohesionless soils, compressed air may penetrate the ground for considerable distances from the tunnel, sometimes becoming apparent as it escapes adjacent to manholes, cellars, or even at ground level. However, if air at contained high pressure penetrates a wide area of laminated clays and/or silts, then the potential for unacceptable levels of ground heave and even shear failure under low effective stress is quite high. The magnitudes of air pressure in the tunnel should thus be the minimum necessary to control the inflow of water at the face. This implies that the balance should be such as to permit a tolerable seepage at the face.

If the site investigation shows the groundwater at tunnel level to be under artesian or high sub-artesian pressure, then in the case of very shallow tunnels it may be difficult to achieve a suitable balance in compressed air pressure, maintaining the pressure high enough to prevent substantial inflows at invert and low enough to avoid the possibility of a blowout at soffit. It may even be necessary, if conditions permit (surface access, for example), to surcharge the ground above the tunnel.

In soils such as laminated clays and silts the expulsion of water from the ground by compressed air may result in increased settlements. When using compressed air for the support of tunnel excavations in alluvial deposits it should be remembered that settlements can occur over a long period of time and so any monitoring to establish the magnitudes of developing and terminal settlements and the styles of those settlements should be continued for an equivalent period.

Example: Where compressed air was used for a siphon tunnel, part of the north bank interceptor sewer of the Tyneside Sewerage Scheme driven in alluvial silt below the water table, it had the effect of expelling the water for a considerable distance from the tunnel line. As a result, settlement and deformation of the tunnel, together with damage to a factory above the tunnel line, took place over a long period of time until equilibrium conditions were re-established in the groundwater surrounding the tunnel.

6.5   GROUND FREEZING

When freezing is contemplated for the purposes of ground stabilisation, this is often regarded as the last resort and specialist advice is usually sought. Reference on the subject may be made to Czurda (1983). Other alternatives that might be considered comprise claquage grouting, ground pre-treatment, jet grouting, dewatering or even tunnel diversion, if considered possible. Freezing is applicable to a wide range of water-bearing soil types, including mixed ground where grouting may be inappropriate, or when compressed air pressures would be too high. Freezing is normally carried out from ground surface but it can be done from the tunnel face.

Line circulation of refrigerated brines (calcium chloride) as secondary coolants through tubes driven into the ground is usually the 'indirect' method that is adopted. Alternatively it would be possible to circulate the primary refrigerant through tubes in the ground. As a primary refrigerant, freon ($CHClF_2$) is non-toxic, non-flammable, and non-corrosive, and its boiling point at normal temperature and pressure is -40.8°C. Ammonia (NH_3) and CO_2 are not commonly used today. Since the freezing point of calcium chloride at a concentration of 30% is well below that of freon, this will be the preferred coolant for injection into the ground tubes. Where space for plant is restricted in a built-up area and/or where a faster freeze down to about -25°C is needed the more expensive and more direct injection of liquid nitrogen, which boils at a temperature of -196°C under normal pressure conditions may be called for. It is also very

useful to have this N_2 facility on-hand for contingency purposes. In a typical installation, liquid nitrogen freeze tubes might be inserted in 100mm diameter pre-drilled holes at 1m spacing. An 18mm diameter conduit would carry the liquid nitrogen down the centre of 50mm diameter low carbon steel tubes and release it about 1m below the tunnel invert. The gas tends to boil off within about 2 minutes of being released and it soon occupies about 600 times its liquid volume. Its cooling effect is achieved by the nitrogen's latent heat of vaporisation.

Soft ground frozen down to a temperature of about -10°C to -15°C has an unconfined compressive strength in the range 10-20MN/m². Typically, soft ground stiffening by a liquid nitrogen freeze should be accomplished within two to three days compared to four to eight weeks with brine. However, the actual freezing time depends upon the presence of any salt in the soil or if the pore fluid differs from water (oil contamination, for example), the spacing of the freeze pipes for different soils, the water contents of the soils, the ambient ground temperature, and the required cooling temperatures. Flowing water carries a large amount of heat and creates an additional loading on the freeze pipes. Any freezing operation should be carefully designed and closely monitored so that the location of the advancing ice front is known at all times. Its effects on the surrounding ground, buried structures and services may then be reasonably predicted, both at the freezing and thawing stages. When the ground thaws out, its shear strength may be reduced compared to its strength before the freezing operation began. This strength reduction could have implications with respect to the support properties of the tunnel lining.

During tunnelling under property it may not be economically possible to freeze at all. Monitoring of a freezing operation is especially important in urban areas where public utility services could be placed at risk, not only from the temperature effects but also if blasting were to be used to excavate the frozen material. Services can, however, be isolated from the effects of freezing by excavating the surrounding soil to below their level. The strength of frozen soil increases with decreasing pore space, that is, as the soil density increases, and also with water content to a maximum value and then declines as the water content increases. In practice, a soft core of partially frozen ground should normally be left within the frozen zone at the position of the tunnel face (and/or shaft location) in order to facilitate the excavation along the tunnel or shaft alignment.

Freezing may also be used to supplement claquage - a process of fracturing by injection in order to achieve more comprehensive grouting.

There can be in-tunnel dangers associated with the use of liquid nitrogen for ground freezing. No general freezing should be carried out within 4m of the tunnel face when men are working there. Should a freeze pipe be fractured, either by ground movements ahead of the tunnel or by direct interception at the tunnel face, nitrogen gas seepage into the tunnel could cause asphyxiation of the miners. A fast-response O_2 deficiency meter is therefore needed at the tunnel face when such freezing operations are in progress. There is a danger of unintentionally injecting liquid nitrogen into a pipe that has been cut away at the tunnel face. Strict procedures, such as the works contractor clearly informing the freezing (sub)contractor of which pipes have been cut in the preceding two shifts, must be in place to prevent this happening. There should also be an awareness among staff of the possibility of cold burns from inadvertently touching cold metal such as exposed freeze tubes. Only one man should enter a tunnel within frozen ground at the beginning of a shift and he should carry an O_2 deficiency meter. If the O_2 level is below 19%, further tunnel ventilation will be needed to purge the gas and re-entry should be allowed only after a further 15 minutes. Of course, all men should evacuate a tunnel if an O_2 deficiency alarm rings.

Example: 128 tonnes of liquid nitrogen were used each day on a 55m length of the 6km Three Valleys tunnel, part of the London Water Ring Main. This 2.44m internal diameter concrete-lined tunnel is an alternative water supply route for Iver treatment works should river pollution force closure of the key Thames river intake for London at Sunneymeads. At the freezing location artesian pressures from the Woolwich and Reading sands meet downward intrusion of gravel in a region where Woolwich and Reading clays underlie firm London Clay. The frozen ground had a permeability of about 10^{-7}m/s.

6.6   TUNNELLING IN FILLED GROUND

Special problems may arise in tunnelling through fill (*see* Sections 3.4 and 4.3.4), particularly in non-geological waste disposal areas (industrial waste and domestic waste). A method statement covering health and safety matters must be prepared by the Engineer, its provisions to be implemented as part of the contractor's responsibilities. Alternatively, the contractor can prepare such a statement as part of his broader quality assurance system. Artificial ventilation must be provided for any tunnel driven through, and shaft sunk in proximity to, municipal and/or chemical wastes, and the requirements of The Control of Substances Hazardous to Health Regulations (Health and Safety Executive, 1988a, b) taken fully into account. Reference should also be made to BS 6164: 1990 (British Standards Institution, 1990a). Air quantities commensurate with safe working conditions should be defined at the outset in the contract documentation. Both methane and oxygen concentrations should be continuously monitored at the tunnel face since it may be necessary for flameproof equipment to be used, provision for such equipment having been made in the contract documentation. If continuous shift working is not in progress, gas monitors should be in place half an hour before shift working begins, a klaxon should be linked to the monitors, and fire extinguishers should be to hand in sufficient quantities. Leading tunnellers should be required to sign registers to confirm that they have tested for gas. In the case of domestic waste, gases are produced as a function of temperature and moisture content under anaerobic conditions (*see* County Surveyor's Society, 1987; Her Majesty's Inspectorate of Pollution, 1989). The range of gases present in the ground and their likely release volumes should be determined at the ground investigation stage and steps taken to ensure that the proposed air quantities forced into the tunnel will be sufficient to dilute them adequately. Flammable gases can, of course, also occur naturally from organic alluvial deposits or from Coal Measures rocks especially where discontinuities in the rock encourage ducting of the gases to the tunnel face and through any loose seals in the primary lining. In the case of one stretch of Los Angeles downtown metro (Red Line) tunnel the possibility of methane gas from old oil workings (and perhaps from other decomposed organic matter) had to be prevented by sandwiching 2mm thick high density polyethylene sheeting between the initial 230mm thick pre-cast concrete segmental primary lining and 300mm thick slipformed 5.4m internal diameter reinforced concrete permanent lining. All the joints and membrane seams were heat welded and vacuum tested for leakage and the protection was enhanced by a system of gas monitors and ventilation fans capable of coping with a gas flow of $1.40m^3/m$ for a 1525m run of tunnel.

Problems may arise in the case of hand-shield excavated tunnels as a result of water inflows at the face at points where the tunnel face crosses the boundary between natural sloping ground and fill material at the edge on an infilled valley (*see* Section 4.3.4), although efforts will normally be made to re-route such tunnels or if feasible excavate full-face by remote control. Special consideration will have to be given to protection of both the tunnel workers and the tunnel lining from any leachate entering the tunnel and affecting the extrados of the tunnel, particularly from fills such as chemical wastes. In the former case the miners would need to be provided with protective clothing and gloves, and there would need to be forced ventilation equipment and monitoring equipment for methane and carbon monoxide gas. In the latter case an epoxy resin could be applied to the extrados of the lining, taking care to ensure that the coating extended undamaged along the full thickness of the lining. It follows that the site investigation should also be required to provide suitable information on the chemical characteristics of the soils and groundwater (*see also* Section 4.3.8).

Example: Tunnelling through filled ground, with its attendant environmental problems, may not be the only solution. For one sewer to be driven through fill, the client was faced with several construction options but with the primary choice of tunnelling or construction from ground surface. The latter was chosen, and involved a concrete box culvert within which was built the circular cross-section sewer. The fill was at first stabilised by vibrocompaction, with 38mm crushed stone or gravel being used to form pillars at the compaction points. The specification permitted use of either water or compressed air for forming the compaction points but compressed air was not allowed to be used beneath the groundwater table. The exact spacing of compaction points, normally 2m, was decided on site by the specialist subcontractor in order to achieve the specified degree of support. Stabilisation was carried out from a level 500mm above culvert founding level and the compaction points terminated in clay 500mm below the base of the fill which was 6m maximum thickness.

When the requisite depth had been reached by the vibrating probe, stone was placed in the hole to build up a column to working level, each hole being filled and compacted before moving to the next position. The bearing capacity of the ground, after stabilisation, with respect to a uniformly distributed load was specified as 240kPa at founding level in the open trench and, with this load acting, the overall settlement at founding level was not to exceed a maximum of 12mm over a width of approximately 3.6m. In order to check that the specification was being satisfied, machine and test data had to be recorded. This information comprised: machine details; depth of penetration of probes tested; ground conditions; load, time, deflection curves; meter reading of work done by the probes when reaching the required compaction (measured in watts or lbf/in^2 hydraulic pressure); quantity and type of aggregate used in the probes tested; if water was used for forming compaction points, then full details of maximum flows and quantities had to be recorded. SPTs in the fill and the stone were required to be performed before and during the work in order to monitor the compaction progress. The tests were carried out after excavating to culvert founding level in positions chosen by the Engineer. There was an option to substitute, subject to the Engineer's approval, dynamic penetrometer or other types of test for some or all of the SPTs. The Engineer was also allowed to specify one or more zone tests on the treated ground to supplement the SPTs. Spacing of the compaction centres determined the size of the reinforced concrete base necessary to assess the settlement of a representative area of ground. The bearing area was expected to be 2m square. The reinforced concrete base was specified at 600mm or more deep and to be cast with its invert at culvert founding level. A total load of 240kN multiplied by the base area was applied in four increments at a rate of one increment per hour or until settlement did not exceed 0.1mm/hour, whichever was the longer. The maximum load was held until the settlement did not exceed 0.1mm/hour. After release of load, the recovery was observed for not less than 2 hours. Settlement of the test base area was measured by four dial gauges mounted on suitably stiff rails supported not less than 2m distance on either side of the test base.

Subsequently, settlement up to 40mm was monitored in the culvert following completion of the vibrocompaction works, and settlement was observed in both the fill and the clay. The vibrocompaction contractor considered that he was responsible only for settlements within the made ground and that the probability of compression of the laminated and boulder clay below the fill had been pointed out earlier by him. This case history thus indicates that the assignment of responsibilities and the detailed specification of what is implied by a geotechnical parameter - in this case the apportionment of potential movement within materials - should be carefully set down.

6.7   SELECTION OF THE TUNNELLING SYSTEM FOR ROCK EXCAVATION

The site investigation may not always identify the full range of rock strengths likely to be encountered during the works. Rock which proves to be stronger or weaker than predicted is often found during tunnel construction. It is the former that tends to lead to contractual claims on the grounds of excavation difficulty but the latter could also create problems with respect to tunnel support (*see* the Example in Section 5.2.4). In those situations where experience and knowledge of an area suggest that rocks have greater strengths than those indicated by the test data it would not be wise for a contractor to choose an excavation system such as a shielded boom cutter which could only just cope with the rock strengths actually quantified. Instead it would be sensible to select a machine having a greater cutting capacity (*see also* Section 7.6) or to choose a system which could easily be replaced, either temporarily by retraction from the face or permanently by removal from the tunnel, by drill and blast methods.

Successful and efficient blasting depends upon drill hole patterns and spacings, selection of 'cuts' ('burn', 'wedge', 'fan'), and the correct choice of explosive and quantities of explosive at the right locations so that the free surface into which the fractured rock can displace is maximised. These variables are a function of the rock material strength, its stratification and its discontinuous nature. Contractor experience and specialist advice from explosives manufacturers is of paramount importance, but Wilbur (1982), for example, has given some guidance (Table 6.1).

To maximise blast excavation efficiency (and also to reduce vibration - *see* Section 6.10) millisecond delayed action detonators will be used, perhaps also combined with the burn cut technique. If the compressive stresses in the ground are not too high to prevent cracks developing between holes, these objectives may also be more easily attained by the use of presplitting whereby small diameter, lightly charged and closely spaced holes (of the order of ten diameters apart) around the perimeter of the excavation are fired first on zero delay, thereby

Table 6.1 Guidance on blasting (after Wilbur, 1982)

Tunnel cross-sectional area (m²)	Required number of blastholes per round		Quantity of explosive per round (kg/m³)	
	Highly fractured rock	Massive rock	Highly fractured rock	Massive rock
10	23-27	35-50	1-4	5-7
25	45-50	60-70	1-2.5	4-5
50	75-85	95-110	1-2	3-4

forming the unlined tunnel walls with minimum overbreak and to some extent isolating the surrounding rock mass from the heavier excavation charges. In general, however, smooth wall blasting is the norm in tunnels.

Both in the site investigation report and any interpretive or design report, if it is considered that the rock test results might not completely reflect the upper strength of the rock to be tunnelled, it would be entirely sensible for the reports to carry warnings that stronger rocks could not be ruled out and that appropriate heavier duty excavation plant might be required. It is also necessary to use words carefully, realising that such words as 'strong', 'very strong' and so on are immediately quantified by implied reference to one of the accepted classifications (*see* Section 4.4.2). To avoid any doubt as to this quantification, the reports should carry a table relating the adjectives to the numbers and referencing those relations to published source.

One such source, often quoted, is the Geological Society of London (1977). Because of the relationship between weathering and strength, a table defining degree of weathering should also be included (*see* Section 4.4.2).

As noted earlier in this book, the tendering contractors have a duty to inspect very carefully the rock cores, descriptions on the borehole logs (in the context of the visual core evidence), and the test data. The times and dates of those inspections should be noted by a responsible person authorised to do so by the Engineer. This is the time also when the contractors should raise questions about any matters on which they are uncertain and need further guidance in order to fulfil their Clause 11(2) responsibilities under the ICE Conditions of Contract (1991) - *see* Section 5.2. The Engineer, when judging a claim, should confirm that the cores were properly inspected by appropriately qualified personnel engaged by the contractor at the pre-tender stage. He will also bear in mind that the contractor is, by virtue of his appearance on the short list, 'experienced' in such tunnelling work and should have suitably performed his inspection work on the cores, at the site of the proposed tunnelling work, and on any features in the general vicinity of the site that could affect his decisions before pricing the bill of quantities.

Example: On one tunnelling contract in the north-east of England, tests as part of the ground investigation revealed Coal Measures rocks at tunnel horizon having unconfined compressive strengths not generally exceeding 50MPa (the upper limit of 'moderately strong'). Contractors would usually consider such rocks to be cuttable at acceptable production rates. However, some very strong bands were encountered at the tunnel face and, using a boom cutter for excavation, these resulted in locally reduced rates of advance and a substantial contractual claim. In the site investigation report there were many references to 'strong' rocks, meaning that the unconfined compressive strength could be up to 100MPa as per the Geological Society of London (1977) classification tabulated in notes prefacing the site investigation report. A careful comparison of rock cores from the site investigation, and both cores and blocks of rock taken from the tunnel face, confirmed that, had the contractor carefully examined the cores at the pre-tender stage, he should have perceived the potential problems posed by the strong rocks and adjusted his proposed method of excavation and tender price accordingly.

6.8   TUNNEL SUPPORT

Until the widespread adoption of numerical (mainly finite element and boundary element) methods, mechanical analyses of supported and unsupported tunnels have usually been confined to circular or elliptical cross-sections deforming in a linear elastic manner. A starting point has been the unsupported circular hole in an isotropic, homogeneous, elastically deforming plate (plane strain) subjected to orthogonal field stresses applied at infinity. The well-known Kirsch (1898) equations (*see* also Savin, 1961) express the radial, tangential (hoop) and shear stresses in the plate at any radial distance from the hole, the interesting stresses being, of course, those on the boundary of the hole. In a civil engineering and tunnelling setting these equations are available in a number of books, such as the one by Obert and Duvall (1967). Pender (1980) has analysed the equivalent deformations corresponding to the states of pre-ground stress and post-ground stress perforation.

Of course, neither soils nor rock are homogeneous, isotropic and elastically deforming. However, if the varying states of stress in mechanically disturbed soils are to be reasonably analysed for practical purposes these assumptions may need to be tolerated, but account must be taken of the three-phase (solids, water, gas) character of the material. In the case of engineering operations in rocks, conditions of anisotropy, and inhomogeneity in the form of discontinuity spatial and orientation density, will be routinely tolerated in analyses, but the presence of water will tend to be ignored.

An early analysis of the support offered to the ground by a stiffer circular annulus is attributed to Spangler (1948), but the theoretical work by Morgan is one of the first analytical contributions specifically directed to tunnel linings. There was subsequent comment on that work by Engelbreth (1961) and Hsieh (1961), and these were followed later by Muir Wood's (1975) important paper. Other significant papers on this subject include those by Peck *et al.* (1972), Ladanyi (1974), Muir Wood (1979) and Krakowski (1979). In the more practical setting of lining design, reference may be made to the report by O'Rourke (1984). The book by Széchy (1973) should also be consulted.

Theoretical analysis of stresses and bending moments in linings with and without joints which provide a direct means of articulation need to be accompanied by measurements on actual tunnel linings. The early work of Ward and Thomas (1965) and Ward (1970) on linings in the London Clay is notable in this respect.

Most existing civil engineering tunnels are of circular cross-section and are supported usually by reinforced pre-cast concrete segments or by cast iron segments. Cast iron segments were first used in the London Tower subway in 1869, later becoming the standard lining for the London metro tunnels until the 1930s when concrete segments became established. Now grey cast iron has been almost phased out. Pre-cast concrete segments now take about 70% of the market, jacking pipes take about 15%, and other materials, including spheroidal graphite ductile iron, account for the rest.

The segments are usually bolted together through the flanges on the longitudinal and circumferential joints to create a stiff ground support ring that will tend to articulate around the longitudinal (cross) joints under high differences between horizontal and vertical ground deformations as the field stresses relax locally. Any such articulation reduces the bending moments and consequential tensile stresses in the lining.

When in place, the segment joints are (or have in the past) sealed (caulked) by a patent compound (for example, PC4, now without asbestos) inserted by means of a mechanical (pneumatic or hydraulic) hammer. The joints now tend to be sealed with elastomeric or hydrophilic gaskets rather than the traditional caulking and pointing, these new gaskets providing higher levels of waterproofing. Segments need to be thicker and smoother internally, with provision for the fixings to be hidden. Their standard width has increased from 610mm to 1m. In the context of the loadings that they are generally called upon to support, the segments themselves have tended to be somewhat over-designed in that except for the most extreme local conditions of ground contact stress they are unlikely to fail in service, but they must be capable of withstanding the rigours of transportation to and handling at the tunnel face.

In contrast to the stiffness of a bolted segmental lining, which is erected within the protection provided by a tunnelling machine or hand shield as the case may be, the segments (usually pre-cast concrete) may be designed for articulation, without bolted restraint, along the horizontal joints, in which case the cross-sectional shape of the lining changes according to the local relaxation state of the ground around the radius of the tunnel when it is initially erected in direct contact with the cut surface of the tunnel. Bending moments and tensile stresses are substantially reduced, except for the contact points between the segments which are strongly reinforced. This type of lining has been used to support sections of the London Underground Jubilee Line driven in the London Clay. During erection, the segments must be expanded directly against the ground, without the presence of any protective canopy, in order to introduce a degree of pre-stress in the lining ring, but this does mean that the geotechnical properties of the ground must be well understood. The ground must be self-supporting during the time of erection and if, for example, the tunnel carries close-tolerance rail carriages there must be an assurance that any longer-term ovaloiding of the cross-section as the ground stresses equilibriate with the lining will not reduce those tolerances to unacceptable levels. The benefit of the expanded, flexible lining is that by more readily accommodating, through deformation, the local states of stress, the self-supporting properties of the ground around the lining are more readily activated. On the other hand, the extrados of this type of lining cannot be grouted up and unless there is the possibility of fixing a plastic membrane, hot-welded into place, before the segments are expanded it will not be suitable for wet ground conditions.

In some vehicular tunnels and metro running tunnels, and also sewer and water transfer tunnels, the primary support serves also as the permanent lining. A special form of lining for the latter two utilities is the bolted, segmental, pre-cast concrete, smooth internal finish, patent one-pass lining which is noted in Supplementary Information 21 in Part Two of the book. In other tunnels the permanent lining takes the form of cast *in situ* or slip-formed concrete placed directly against the primary support.

An ultimate expression of the enhanced self-support principle, characterised earlier by the flexible expanded lining, is incorporated in the New Austrian Tunnelling Method (NATM), more information on which is given in Supplementary Information 17 in Part Two of the book. Reference should also be made to the papers by Kovári (1994a, b). In this system, once the ground is exposed it is subjected to a combination of dowelling, latticework around the exposed perimeter with supporting girders, and shotcreting. The aim again is for this somewhat insubstantial primary lining to sustain the external loading by utilising the bearing capacity of the ground in the short term until such time as a stiff permanent concrete lining can be installed. The NATM also often assumes, for design purposes, that the initial primary support behind the permanent lining deteriorates in the medium term and contributes nothing to the long-term support of the tunnel. (Steel dowels/anchors should really be pre-coated with epoxy resin to ensure that no corrosion occurs if air gaps are left during insertion of the resin grout.) The NATM technique is most appropriately applied to rock tunnelling but it is increasingly used for the primary support of weak chalks and marls and of stiff to hard overconsolidated clays.

Example: The English cross-over cavern in the Channel Tunnel is 7km from the English coast line, 36m below the sea bed and 68m below mean sea level. It is 164m long, 21.2m wide and 15.4m high (excavated), and provides a height of 9.5m above the rail track and an internal width of 18m. Excavation for the large diameter cross-over was in stages: sidewall drift headings, then sidewall drift invert construction, the crown heading, and finally shotcrete removal, bench and invert excavation. Primary lining was by dowelling, lattice girders and shotcreting. The permanent lining was 600mm thickness of unreinforced cast *in situ* concrete. A composite waterproof membrane and geotextile drainage fabric were installed between the primary and permanent lining.

Of course, spray concrete may be used independently of the NATM. It will generally comprise up to 12mm aggregate, mixed in a gun, with or without steel fibres or alkali-resistant glass fibres, together with cement and water having a low water-cement ratio, and sprayed on to the exposed rock surface at the end of the tunnel. The thickness of the concrete layer will usually be up to about 75mm, but can be more, and there will be a need to keep the rebound material to as low as 10% as the material is fired at a velocity of about 100m/s in order to drive it into discontinuities and fissures.

Before the NATM is implemented, a high quality and much more extensive site investigation than is normally adopted for tunnels must be implemented. Some or all of the following elements should be put in place: continuous core sampling, with total core recovery better than 99%, using triple tube core barrels with wireline techniques where necessary and suitable flushing fluids; thin wall samples (push-in) in clays which are overconsolidated; pressuremeter or dilatometer testing in order to derive shear modulus and deformation modulus values for the ground (*refer*, for example, to Bellotti *et al.*, 1989; Clarke *et al.*, 1989; Fahey and Jewell, 1990; Jardine, 1991; Jing *et al.*, 1994; Ortigao, 1994); high and low flow packer permeability tests for determinations of permeability; closed circuit television or acoustic profiling down boreholes particularly in order to evaluate discontinuity characteristics; inclined core drilling, especially in portal areas.

Shear modulus (G) information is provided by the pressuremeter, usually for adoption in finite element analyses. For an (assumed) elastic material, G is related to the deformation (Young's) modulus (E) by the expression

$$E = 2G(1 + \nu)$$

where ν is Poisson's ratio. If ν is taken as 0.25 for rock, then $E = 2.5G$, approximately. It can also be shown that the bulk (volumetric) modulus K is approximately equal to $5G/3$, in which case $K = 2E/3$, approximately.

Particularly because of the presence of discontinuities, a rock mass will not deform elastically at civil engineering depths, in which case any design in the material may need to resort also to empirical estimates of the deformation parameters. One method of estimating the elastic modulus in discontinuous rock uses Bieniawski's Rock Mass Rating (RMR) value:

$$E \text{ (gigapascals)} = 2(\text{RMR} - 50)$$

after Bieniawski (1978), or

$$E \text{ (gigapascals)} = 10^{(\text{RMR} - 10)/40}$$

after Serafim and Pereira (1983).

The above two equations become almost equivalent at RMR = 55. However, the first equation is valid only for RMR > 50 whereas the second equation gives a rapid increase in the value of E for RMR > 85. The elastic modulus E can also be derived indirectly from a Barton Q parameter via the expression

$$\text{RMR} = 9 \ln Q + 44$$

after Barton (1983), where $\ln Q$ is the natural logarithm of Q. Alternatively, the *in situ* deformation modulus may be determined directly from a derivation of Q via the expressions

$$E_{\text{min}} = 10 \log Q$$

$$E_{\text{mean}} = 25 \log Q$$
$$E_{\text{max}} = 40 \log Q$$

where there is quite a significant range in the minimum to maximum E-values.

Hoek and Brown (1980) have provided approximate equations for the principal stress relationships and Mohr envelopes for different rock and jointed rock masses. The general form of these equations is

$$\sigma_{1n} = \sigma_{3n} + (m\sigma_{3n} + s)^{1/2}$$

where
σ_{1n} is the major principal stress normalised to the unconfined compressive strength σ_c of the rock,
σ_{3n} is the minor principal stress similarly normalised,
m and s are constants dependent on the properties of the rock and the extent to which it has been fractured by being subjected to σ_1 and σ_3.

For intact rock, $m = m_i$, which is determined from a fit of the above equation to triaxial test data from laboratory specimens, taking $s = 1$ for rock material. For rock masses, the constants m and s are related to the basic (unadjusted) RMR as indicated below after Hoek and Brown (1980).

$$m = m_i \exp[(\text{RMR} - 100)/28]$$
$$s = \exp[(\text{RMR} - 100)/9]$$

for undisturbed rock masses (smooth-blasted or machine-bored excavations).

$$m = m_i \exp[(\text{RMR} - 100)/14]$$
$$s = \exp[(\text{RMR} - 100)/6]$$

for disturbed rock masses (slopes or blast-damaged excavations).

As before, Q can be used in these equations by adopting the relationship RMR = 9 ln Q + 44.

The NATM always relies heavily upon a spread of instrumentation and post-construction measurements in order to assess the competence of the installation and to 'tune' the working plan if this is shown to be necessary. There is a need for measurements of convergence at the perimeter of the excavation in order to assess the overall stability of the tunnel. Convergences can be measured by invar tape with suitable tensioning and a dial gauge accurate to $\pm 10\mu$m. In-ground deformations can be measured from the tunnel by means of multi-position borehole extensometers, magnetic probe extensometers or resistance wire extensometers. Stresses in the ground can be measured by suitable hydraulic pressure cells installed in boreholes or by stress gauges set in a triaxial configuration. It would also be necessary to monitor reinforcement loads at the heads of the dowels and to use wire resistance strain gauges mounted on the dowels themselves. Wire resistance strain gauges can also be embedded at various locations in the shotcrete, and pressure cells cast in the permanent lining will be monitored for the long-term performance of the support.

The observational method of proceeding (Powderham, 1994) is fundamental to the use of the NATM (Wood, 1994). For this reason alone the form of a NATM contract needs to be sufficiently flexible to accommodate a substantial range of variations. Also, because of the amount of the instrumentation required to be installed for the NATM, and the time required for its monitoring, it would be prudent to place these responsibilities onto the contractor so that claims for disruptions to the tunnel progress can be avoided. In any case NATM tunnels advance much more slowly than do TBM or shield-driven tunnels - of the order of, say, 2m per day as compared with, say, 10 to 100m per day (Wood, 1994). Assessment of the monitoring results and any associated ordering of variations would be the responsibility of the Engineer and his staff. Management of a NATM project is likely to be more demanding than on a conventionally driven and supported tunnel, but it is often claimed that for large projects the construction costs are often significantly less.

Some tunnels in rock that is strong and not prone to squeezing pressures may stand without the need for a lining. Some blocks, defined by the joint system, may be prone to detachment, in which case they can be secured by bolting. Where tunnelling takes place in highly disturbed or fractured strata that occurs, for example, adjacent to faults, or in generally squeezing ground, the tunnel supports such as p.c. concrete segments could be deformed on the horizontal joint articulation lines, sometimes requiring the

tunnel section to be re-dressed. Squeeze caused by stress relaxation is likely to occur when the tunnel depth is sufficiently great for the stresses on the tunnel wall to exceed the unconfined compressive strength (UCS) of the rock. This can happen when the depth of the tunnel is greater than about half the ratio of UCS to unit weight γ_r for the rock. Inserting typical unit weights and UCSs suggests that most near-surface tunnels should not be affected and that only civil engineering tunnels in mountainous areas could be vulnerable. For deep tunnels it may be necessary to determine the *in situ* state of stress using, for example, borehole inclusion stress meters. Tunnel deformation in some rocks such as expansive shales and mudstones (*see* Supplementary Information 6 in Part Two of the book) is a function of the mineralogy and, to a degree, the presence of moisture. Mineralogical composition can be determined by X-ray methods (*see* Supplementary Information 7 in Part Two of the book).

A particular problem of support design arises with the case of pressure tunnels. These are tunnels which carry water under pressure, typically being head-race tunnels in hydroelectric power schemes. A monolithic concrete lining would be designed to support inward deformations as the surrounding radial ground stresses partially relax onto the extrados of the lining, and these deformations place the lining in hoop (circumferential) compression. (For an analytical solution it is simplest to assume that the external pre-existing ground stress field is hydrostatic and that no bending stresses are introduced into the lining.) The internal water pressure compresses the lining radially and, through Poisson's ratio effects, serves to reduce the hoop compression, but the overall effect of the internal pressure in resisting the external inward pressure is to indicate that the lining could be designed thinner for operational conditions. However, since the lining must support the ground during construction when there is no internal support, and since the internal pressure is a variable quantity anyway, it would be normal practice to design the lining to support the ground and to discount any additional support provided by the internal pressure. This problem is overviewed in Supplementary Information 19 in Part Two of the book, but in practice numerical solutions would be sought within the design exercise.

Instances have been reported of local cracking of the lining of tunnels driven in strong rock. If if can be shown by measurements on lining segments that the cracking can be related to stress concentrations in the ground, then before the broken segments are replaced attempts should be made to resist any possible recurrences of the same problem. For example, the rock in the area of the problem could be excavated back and the overbreak injected with compressible granular fill. In sensitive areas, typically below rivers, where at the outset it is considered that stronger primary support is needed, cast iron segments are often specified in place of pre-cast concrete segments. They are of relatively low tensile strength, brittle, and when originating from a foundry process can contain shrinkage cracks. But they can also generally be caulked more securely to resist moisture inflow to the tunnel. Nevertheless, it is not always certain that the extra cost of cast iron segments is justified.

> **Example:** A sewage siphon tunnel in the north of England was constructed throughout with a cast iron segmental primary lining to resist movements caused by ground pressures on the lining. Several segments cracked at one place in the tunnel at a fault zone location, so requiring replacement. In this case, cast iron segments were no more successful in resisting cracking than were the cheaper alternative of pre-cast concrete segments.

Blasting may sometimes be permitted for excavation in rock. Since blasting from within a shield could prove to be hazardous, colliery arch support might be specified by the contractor in preference to pre-cast concrete segments where the tunnel size permits it and where the grout infill quantities to conform, say, to a circular permanent lining are acceptable. With such primary support it will be necessary to provide additional struts at footblock level to withstand disturbance of the colliery arch supports as a result of the blasting.

Deterioration of Coal Measures rocks, for example mudstones (mudrocks) and seatearths (underclays), is common in tunnel floors (*refer to* the subjects of mudrocks and mineralogy in Supplementary Information 6 and 7 in Part Two of the book). Evaluation of this degradational (slaking) factor is therefore important at the site investigation stage. Stability problems can occur at the footings of the arches, especially where there is a time lag between completing the primary support and commencing the permanent lining. Provision should be made for larger pad footings, even to the extent of concreting the entire floor to protect the footblocks.

The potential for slaking will also condition the choice of excavation system. A track mounted roadheader could be inappropriate because it would dig itself into wet, slaking ground and so become operationally inefficient. Weakening of the lining support at the tunnel invert could also reduce the stability of the lining. In these circumstances a boom cutter mounted in a short shield might be the plant chosen by the contractor.

Sewer tunnels carry aggressive effluents and so the linings must be designed to resist deterioration over their design life. Old man-entry sewers were often driven to an elliptical or egg shape cross-section in order to reduce the width at invert and thereby increase the solids retention of the low-volume dry weather flows by speeding them up. Modern sewer tunnels are usually driven to a circular cross-section and, if necessary, the dry weather flows can be accommodated in a smaller diameter invert gully. The subject of sewer tunnel linings is considered in Supplementary Information 21 in Part Two of the book.

Outside the coal mining industry horseshoe-shaped tunnels are not common but there is an interesting example of their use in Seattle, Washington State, USA. As part of an 11km freeway system of Interstate 90 a two-tier, 19.2m internal diameter tunnel was driven through a slip-prone suburban hill. Construction by the contractor Guy F. Atkinson consisted of 24 contiguous small diameter perimeter bores of horseshoe configuration, backfilled with concrete to form a 'corduroy' profile concrete tube from which the internal soil was able to be removed with safety.

6.9   ABANDONED MINEWORKINGS

Site investigations may indicate the possibility of abandoned mine workings and shafts located on a proposed tunnel route, a specialist report having been commissioned in many cases from a mining consultant. The Engineer will have to decide whether the presence of possible workings would require special investigation/treatment works during the tunnel contract period. Probing ahead of the tunnel face may have to be specified in order to establish the possible presence of voids and to determine water levels in the workings.

Example: On one Tyneside contract, probing ahead for a distance of 10 to 15 metres was carried out where there was a possibility of encountering old disused shafts, although none was found.

It may be possible to backfill workings when they are encountered during the tunnel construction or, alternatively, voids can be filled prior to tunnel driving using pulverised fuel ash/ordinary Portland cement mixtures. The pfa must be suitably bound by the cement-water mix. This infilling process will stabilise the ground and allow the tunnel to be excavated through the zone of the workings. The grout mix should be designed to form a stable tunnel face but should also be sufficiently weak for it to be excavated by light pneumatic picks and clay spades. A 10:1 pfa/opc grout mix may, in general, be the most suitable, satisfying the criteria of fluidity, bulk-set stability, and excavatability.

Example: On a contract for a 2m diameter interceptor sewer tunnel at a depth of 7m in Newcastle upon Tyne, and following an extensive video camera pre-contract examination of abandoned mine workings which showed that a partially worked coal seam would be intercepted at the tunnel face, boreholes were drilled at 3m centres at about two tunnel diameters either side the future tunnel centre line and curtain walls were formed using a 10mm pea gravel/opc mixture injected down these holes. Further holes were then drilled between these original holes and also at 5m centres along the centre line of the future tunnel, and the cavities both along and between the curtain walls filled with a 7:1 pfa/opc grout mix injected under pressure. This stabilisation proved to be a complete success. When exposed at the tunnel face the

old mined-out coal workings, the remaining broken coal on the floor of the seam, and broken rock adjacent to the seam were comprehensively grouted in a most impressive manner. The grout itself presented the appearance of a weak, laminated, calcareous rock. It had not only achieved its primary objectives of infill and stability but it was also sufficiently weak for it to be excavated with ease using light pneumatic picks and clay spades.

Grout holes should be suitably sleeved in order to prevent any buried services in close proximity from being subjected to grouting pressures.

Voids immediately below a tunnel could also affect the long-term stability of a tunnel lining. In such instances, probing would be carried out from the tunnel invert at regular intervals, say 2 to 3 metre centres, with a view to filling any voids with pfa/opc mixtures.

Old mine workings will have created particular patterns of ground strain distribution above them. There have been several occasions in north-east England when the distribution of tensile and compressive strains caused by the working of different coal seams and adjacent coal seams by longwall methods have been analysed in order to assess their effects on either future tunnelling activities and the structural integrity of tunnel linings or upon buried pipelines. An early and notable example was the analysis by Boden (1969) of the coal mining-induced ground strains below the River Tyne. Drawing on practical experience of work on and general knowledge of the Tyne vehicular tunnel ground environment, Boden analysed superimposed strains in connection with the future (at that time) River Tyne sewage siphon tunnel which was designed to carry south bank interceptor tunnel sewage to the north bank Howdon sewage treatment works (*see* Supplementary Information 18 in Part Two of the book concerning water pollution). Old coal workings project beyond both north and south river banks beneath the line of the sewage siphon tunnel and it was feared that the superimposed tensile strains projecting forward from old longwall faces on both sides of the river could have opened up pre-existing joint systems in the rock, so increasing the mass permeability of the ground above the tunnel. The urgent concern was to avoid the problems that beset the earlier Tyne vehicular tunnel, just over 15m below the river bed and 1.676km long. This tunnel, which was designed by Mott Hay and Anderson and constructed by Edmund Nuttall, was begun in 1961 and opened by Queen Elizabeth on 19 October 1967. At one stage during construction it was lost to the river through a collapse at the crown and was only retrieved through a combination of infilling from the river itself, grouting from the tunnel, and the use of high compressed air pressures which led to well-publicised incidences of bone necrosis and thereby triggered the development of rigorous controls in the form of the 'Blackpool Tables' for compressed air working

When excavating through Coal Measures strata, additional project costs may be incurred due to the need to follow mine safety regulations. As noted above, it is necessary to carry out air quality monitoring and to ensure that adequate safeguards are taken with respect to ventilation and the use of flameproof equipment. The recommendations with respect to gas that have been mentioned in Section 6.6 above apply equally if not more to the conduct of tunnelling in this type of strata. All these matters concerning health and safety need to be suitably emphasised in the contract documents.

6.10   VIBRATIONS

Blasting may be required (and permitted) in rock tunnels and installation of driven piles may be necessary for open cut, chamber, or shaft construction. The vibration levels at adjacent properties due to these operations must be estimated and assessed in the context of both property damage and environmental impact. In the UK, reference needs to be made to the Control of Pollution Act 1974 in respect of noise/vibration and also the Environmental Protection Act 1990 for other environmental implications. Reasonable estimation of vibration levels requires an understanding of the input energy to the ground, the transmission properties of the ground and the structural response of the buildings to be affected. Heavy blasting in a tunnel may also affect the integrity of the lining and/or any rock anchorages in close proximity to the blast. In the case of TBMs, resultant peak particle velocities are generally low (less than 10mm/s) with attenuation rates, according to

Flanagan (1993), similar to those found for various other construction equipment. At a distance of 45 to 50m, vibrations from tunnelling machines will probably not be felt and structural damage is not really an issue.

Monitoring of blast vibrations at ground surface would normally take the form of measurement of peak particle velocities (or accelerations) in three orthogonal directions on the ground surface at known radial distances from the source, the dominant frequencies of vibration, and the amplitudes of vibration in buildings likely to be affected. From this information a resultant peak particle velocity (or acceleration) can be derived. The records must be retained for later scrutiny and for evidence in the event of dispute and/or building damage. Details of drill hole configurations and lengths, explosive weight and type and the weight of explosive used per delay, and date and time should be recorded. When monitoring driven pile vibrations, the type and manufacturer of the driver must be recorded (hydraulic, air hammer, single- or double-acting, bottom-driving, vibro-driver, and so on), the notional energy input as stated by the plant manufacturer, the distance from the pile to the recording point, and the depth of the pile toe at the time of the recording.

Empirically derived information is readily available on tolerable peak particle velocities in relation to induced building damage (Attewell, 1986). Unfortunately, but not unexpectedly, different countries have adopted different criteria, and because all buildings tend to be constructed differently there is no reasonable assurance that these criteria are valid for particular cases. In terms of permissible peak particle velocities, there is no current UK specification relevant to all vibration, although British Standard BS 5228 Part 4 has been revised (British Standards Institution, 1992a) to accommodate vibrations from driven piling. Some extracts from this Standard are included in Supplementary Information 20 in Part Two of the book. In Sweden, Australia and Austria, for example, the respective vibration velocity figures are 30, 25 and 15mm/s independent of frequency. For blasting, the Swiss specify 12mm/s up to 60Hz; thereafter the permissible level rises by an additional 0.45mm/s per Hz up to 70Hz. For machinery vibrations, the Swiss specify a limit of 5mm/s up to 30Hz, with an additional 0.12mm/s per Hz up to 60Hz. The German DIN 4150 code specifies 5mm/s up to 10Hz, thereafter with an additional 0.4mm/s per Hz up to 50Hz. These figures simply serve to demonstrate the range of national recommendations.

Driven piling operations could subject a building in close proximity to many high amplitude vibrations. Those standard building damage criteria which are based on single transients take no account of fatigue effects and should be used only with caution, but they often have to be adopted because there are no suitable fatigue-based criteria available. Attention should clearly be paid to the type and function of the structure (Skipp, 1984; Attewell, 1986).

Example: On one tunnelling contract, a rock tunnel was planned to be driven quite close to a glass-fronted swimming pool. Such a swimming pool would be vulnerable to cracking at edges and corners, and would have presented an even more obvious and unacceptable hazard had it also been roofed in glass sheeting. No sensible engineer would be likely to permit blasting in the vicinity of such a structure at times of occupation. On the same contract, the tunnel was to pass close to a retirement home. In the contract documentation the Engineer had the option of either expressly banning drill and blast methods of excavation or of setting such very low levels of tolerable vibration that the contractor would feel constrained to adopt an alternative method of excavation. It should be noted, however, that it is possible to excavate a tunnel face in moderately strong rock and generate only very low ground vibrations if low charge weights per delay are used in a carefully designed delayed action blasting pattern and incorporating the burn cut technique.

Vibration can be perceived at very low amplitudes and may prove to be extremely disturbing to inhabitants of buildings close to the vibration source. BS 6472 (British Standards Institution, 1984; now revised as British Standards Institution, 1992b) addresses this problem by specifying upper level vibration criteria in terms of ground velocity and acceleration. In addition to its amplitude, tolerance of vibration is to some extent dependent upon forewarning, the age of the recipient and whether he/she is standing, sitting or lying down, how long it persists, the time of day, and whether the person owns the property. In urban areas, full disclosure of information on blasting/pile driving times is essential. The public relations expertise of the resident engineer and his staff should be an asset in this respect. Older people need to be reassured when, for example, rattling of china in a glass cabinet might suggest much higher vibration levels than actually occur. In these circumstances the resident engineer might suggest that his staff be allowed to package such items until the construction is completed. Noise levels need also to be measured and assessed. There should be early discussions and continuous liaison with the local environmental health officer whenever blasting or pile driving operations are to be undertaken. Consent to work agreements with the local authority, in accordance with Section 61 of the Control of Pollution Act (1974) and based on acceptable noise and vibration levels, should be sought from the local authority environmental health officer. The approach is best made by the employer at the outset if according to the conditions of the contract noise and vibration are deemed to be an issue. Alternatively, the responsibility can be placed on the contractor.

Both vibration and noise matters are covered more fully in Supplementary Information 20 in Part Two of the book.

6.11 GROUT ADHESION TO THE TUNNELLING SHIELD

The rough metal and any poorly ground-down welding on the outside of a tunnelling hand shield and tail skin tend to encourage the adhesion of grout. In addition, grout may migrate forward into the circular void around the shield. Unless restrained, such grout adhesion can lead to problems in shield advancement as the necessary jacking pressures have to be increased, and also to increased ground settlements which could not reasonably be termed 'unavoidable' (referring to Clause 22(2)(d) of the ICE Conditions of Contract, 6th Edition, 1991). Attention should be paid by the contractor to the packing around the leading edge of the lining ring ('fluffing-up') behind the shield before grouting operations begin. Better control of ground movement could be attempted by the use of a tail seal or a smooth-treated shield and tail skin extrados. Tail seals will be incorporated on TBMs.

> **Example:** On one tunnelling contract on Tyneside, grout adhering to the outside of the shield and tail skin gradually built up a wedge on the outside of the tail section. When the shield was shoved forward in weak alluvial clay, the solid grout around the tail gouged a larger void in the soil and so promoted greater than necessary ground settlements (causing structural damage to a building overhead) and a wider settlement trough.

6.12   PAYMENT FOR OVERSIZE CONSTRUCTION

Primary tolerances for a tunnel are typically ±20mm on level and ±50mm on line when the tunnel is to be lined down to size with a secondary lining. On a one-pass sewer tunnel lining they may be ±50mm on both line and level.

On small cross-section tunnels in rock, say up to 3m diameter, a contractor using a rock cutting machine may decide to construct with an oversize primary lining to ease tolerance of the specified internal lining. The contractor will then use rather more grout behind the lining rings, and additional concrete will be needed for lining the tunnel down to its permanent size. A larger cross-section TBM cutting squeezing rock may be equipped with adjustable gauge cutters which allow overcutting to compensate for the inward squeeze. The same decision on overcutting may be made for a tunnel in soil in order to compensate for squeeze and also to ease shield movement round curves.

It must be made clear before the winning contractor's tender is accepted that the bill of quantities has been priced for excavation in rock (or soil) at the specified diameter. The payment line (*refer to* CESMM3: Tunnels T pp82-87) would normally be defined as the outside perimeter of the lining. If the contractor proposes an oversize tunnel, than the basis of payment for any unforeseen variations, such as an excessive overbreak from blasting, should be settled at either the tender-acceptance stage or at the time the contractor produces his Clause 14(1)

programme under the ICE Conditions of Contract (1991).

An alternative approach for, say, a sewer tunnel would be to question why primary tolerances for a tunnel which is to have a secondary lining need to be specified at all. The idea would be to define a minimum thickness of concrete secondary lining and tolerances for its construction, leaving the choice of segment diameter and primary driving tolerances to the contractor.

Care should be taken when defining and discussing the term 'overbreak' in the preamble to a bill of quantities. The term has rock (brittle) fracture connotations, and the development of overbreak is often related to the contractor's choice of excavation method and the discontinuity spacing. Its use should therefore be restricted to rock, since although soil can deform it cannot 'fracture' or 'break' *sensu stricto*. There might, for example, be pockets of soil or zones of rock degraded to soil consistency within a rock tunnel, and for purposes of stability the contract might require these to be excavated beyond the payment line and infilled with concrete or cement grout at the contractor's expense when intercepted. Information on the possibility of such occurrences could be available in the literature and should be identified in the site investigation report, but their frequency and locations cannot be known with any precision. In constructing his preamble to the bill of quantities the Engineer must accommodate these unknown factors by a suitable form of wording. He must also be very careful to attribute the term 'overbreak' to rock and to use a term such as 'over-excavation' for soil or degraded rock.

A particular source of dispute might arise from tunnelling in boulder clay (*see* Section 4.3.3), especially if pipejacking methods are used. Not only does the basis of payment need to be settled in the contract documents for those boulders above measurable size that protrude partially into the tunnel section and would need to be cut, but there would also need to be clear reference to the method of payment for the back-filling of cavities created by the removal of those boulders that were reasonably amenable to such action. There should be an item for boulders in the bill, and a rock overbreak item, the contractor being paid on a square metre basis.

6.13 ACCESS SHAFTS AND MANHOLES

The most common methods of construction of access shafts and manholes have been segmental (underpinning), sheet pile enclosures, and caisson construction with kentledge used to force the lining into the ground (using a suitable lubricant, such as a bentonite paste behind the lining ring, to ease its progress). Secant piling is now becoming more common for shaft construction because it does offer some advantages in bad ground. The main problems to be overcome are boiling/piping in waterlogged granular soils, base heave in soft clays, or uplift caused by artesian/sub-artesian water pressures below the excavation. If such pressures are suspected before undertaking the ground investigation then, rather than sinking a borehole at the centre of each shaft position, the hole should be offset half a shaft diameter from the shaft wall.

Construction in water-bearing granular soils can be undertaken using a segmental method in conjunction with conventional dewatering such as well-points or wells. Alternatively, sheet piles could be driven below the base of the excavation to cut off or minimise flows into the excavation base, or caisson construction could be used with a compressed air facility. Base heave can be prevented by driving sheet piles into firm strata below the excavation base. Uplift can be minimised by the installation of pressure relief wells.

Other, more sophisticated, methods that could be used for shaft construction include diaphragm wall installation, grouting of cohesionless soils (*see* Supplementary Information 8 in Part Two of the book), and ground freezing (*see* Section 6.5). Although in some cases being overtaken for this purpose by secant piling, the technique of jet grouting may have wider future application, not only for shaft construction but also for such specialist purposes as sealing off when a machine breaks out into waterlogged ground. An alternative solution for this type of ground is to use compressed air below a blister lock on the shaft.

Example: For a sewer tunnel at Stockton-on-Tees, a 3m diameter shaft was sunk to a depth of 7.5m through made ground and soft organic clay into loose, fine-to-medium sands. The water table was located at a depth of 4m, so that water inflows and boiling were expected to occur in the sands. Jet grouting was used to form an enclosure around the shaft, extending for several metres below the shaft base and also along the adjoining tunnel alignments. The tunnels were constructed in compressed air.

6.14 SEWER TUNNEL PERMANENT LINING - ENVIRONMENTAL EFFECTS

Many tunnels are constructed as part of sewerage renewal schemes. Although a materials consideration rather than one of geotechnics or engineering geology, it is useful to note the effects of raw sewage on a sewer tunnel lining.

The permanent lining of a sewer tunnel constructed at the present time is usually designed for a 100-year maintenance-free life and must be able to withstand continuous immersion in sewage having the following characteristics:

- Normally slightly alkaline, but with a pH varying between 6 and 10.
- Sulphates (as SO_4) not normally greater than 1200mg/litre.
- Cyanides and all compounds capable of liberating hydrocyanic acid gas on acidification, present in quantities not usually greater than 2mg/litre.
- Sulphides, hydrosulphides and/or polysulphides, which are not normally greater than 2mg/litre.
- Phenols, not greater than 20mg/litre.
- A temperature within the range of 5°C to 20°C, with an average of, say, 15°C. The maximum possible temperature from an industrial liquid waste source would be about 43°C, but this temperature could fall quite quickly in the sewer when mixed in with a cooler main flow, particularly under winter conditions.

Particularly aggressive industrial discharges can, of course, be refused by the tunnel owner, or accepted only after pre-treatment to an acceptable level. Under the operation of Integrated Pollution Control as per the Environmental Protection Act 1990, Her

Majesty's Inspectorate of Pollution has statutory powers to restrict such discharges. For further information on this aspect of pollution, reference may be made to Supplementary Information 18 in Part Two of the book.

It should be quite feasible to achieve a nominal 100-year relatively maintenance-free life with most domestic effluents using concrete linings, but if greater chemical resistance is considered necessary then Class A brick linings to the relevant requirements of British Standard 3921, set in either cement or epoxy resin mortar, should suffice. Where industrial discharges or high temperatures occur, special measures would be needed and would involve special tiles or epoxy resin linings. However, UK tunnels usually contain a high proportion of domestic sewage and surface water which serve the purposes of dilution.

Mortar quality is obviously a crucial factor in the long-term performance of brickwork but there are few extrapolated durability figures for polyester resin mortars exposed to both acid and alkaline environments other than in refractory kilns. Brick pointing with an epoxy mortar, although more costly than with a conventional cement mortar, has been shown to withstand acid and alkaline reactions rather better. On the other hand, deterioration of brickwork in cement mortar under neutral or slightly alkaline conditions has been found to be minimal. CIRIA Research Report Number 14 (Vickers et al., 1968) shows that 8:1 mortars have a relatively good resistance to abrasion. Mortar for engineering brickwork at 3:1 should have a much better resistance. In practice, for a sewer tunnel permanent lining constructed entirely in brick, if such mortars fail to maintain their bond the bricks will still remain in place because of their location below springing.

An additional form of protection involves the use of 1.65mm thick polyvinylchloride sheeting keyed into an *in situ* concrete permanent lining during pouring by means of notched ribs in the plastic which serve to hold it in place. The pvc is placed above design water level only, on the basis

that the bottom need not be lined since it is always submerged and it is the hydrogen sulphide gas that could attack the concrete above water level. Other protection measures, as outlined above, against chemical and abrasive attack on the lining between invert and springing would still need to be incorporated. There is some further comment on sewer linings, and tunnel linings in general, in Supplementary Information 21 in Part Two of the book.

6.15   BUILDING OVER TUNNELS

Requests to build over tunnels are received by tunnel owners from time to time. Such requests should be treated sympathetically. Considerable analytical effort, represented in the form of several reports to the former Northumbrian Water Authority, has been expended by the present writer to establish criteria which would permit sensible technical judgements to be made as to the merits of each case. The inverse problem of post-construction *excavation* is equally relevant, and cases of sequential excavation and surcharge loading have been analysed. Statutory and local authorities have been questioned on this problem with respect to their formal responses when faced with requests for overbuilding. A rule-of-thumb maximum allowable overbuilding surcharge pressure of 25kPa tends to have been quoted in many instances irrespective of ground conditions, tunnel depth and actual lining construction. Heavy superstructures crossing a relatively shallow tunnel would need to be supported on piles lateral to the tunnel sidewalls. Any analytically based design criteria would normally relate to circular cross-section tunnels. Because of the unpredictable durability of the timbering, and the possibility of poor backfilling, it is usually undesirable to allow building over box headings. It is also good construction practice to specify that all headings should be back-grouted to refusal on completion.

A useful starting point for the analysis is the paper by Einstein and Schwartz (1979).

7

CLAIMS AND RECORDS

7.1 GENERAL

As noted by Powell-Smith and Stephenson (1989, 1993), there are four bases on which a claim for (extra) payment may be made in English law:

- Under the conditions of contract themselves.
- For breach of contract, when the contractor's remedy will be damages calculated in accordance with common law principles.
- For breach of duty arising at common law in tort. This is a general liability, and in principle liability often depends on the defendant having, by act or omission, acted in breach of a legal duty imposed on him by law, so infringing a legal right vested in the claimant and causing him foreseeable damage. The remedy is usually an award of money damages as compensation for the damage done.
- On a quasi-contractual or restitutionary basis, often called a *quantum meruit* claim.

This book is concerned, in the main, with contractual claims. It is noted, however, that claims for breach of contract do not hinge entirely on the printed conditions, which are the *expressed terms*, agreed by the contracting parties. Courts may 'breathe' another or other terms, *implied terms,* into the contract to render it commercially effective. The breach of an implied term can also give rise to a claim for damages. There is explicit reference in Clause 49 of the ICE Conditions of Contract to 'any obligation expressed or *implied* on the contractor's part under the contract'. Terms may also be implied by the employer.

The contractor may also invoke the law of tort by alleging breach of a duty arising at common law other than in contract, and such a claim in tort will normally lie, if it lies at all, against the Engineer and not the employer. In many, if not most, standard forms of construction the

administrator of the contract, in this case the Engineer and in others perhaps the architect, is not party to the contract between employer and contractor. His contract is with the employer, and so any action by the contractor against the Engineer for unfair administration of the contract (failure to act impartially), or interference with the contract, would fall under the law of tort. There seems to be no statement in either the ICE Design and Construct Conditions of Contract (1992) or the IChemE 'Green Book' (*see* Section 2.7) to the effect that the Engineer must act impartially in the administration of the contract.

A particular case of interference with a contract arose in the case of *John Mowlem and Co plc* v *Eagle Star Insurance Co Ltd*. Carlton Gate, a company created for the purposes of carrying out a redevelopment in London and 25% of which was owned by Eagle Star Insurance, entered into a management contract with Mowlem to do the development. There were claims by the contractor for extensions of time and the contract was terminated by the employer, who subsequently went into receivership. It was found under arbitration that the actions by the employer had been wrongful and that the contractor was entitled to financial redress (£9.4 millions plus interest of £2.9 millions). The contractor received only £0.5 millions in retention monies and then began proceedings to recover in tort from Eagle Star, the defendants, and the architect under the management contract, alleging wrongful suppression of extensions of time by the architect and claiming, among other things, (i) knowing and wrongful interference with the management contract, and (ii) conspiracy with another to injure the contractor's interest under the contract by unlawful means, viz. interference with the performance of the contract. Although there was an application by the architect that substantial passages of the Mowlem statement of claim should be struck out, the judge rejected the application in respect of these two. The judge listed the elements of the tort of actionable interference as

- knowledge of the existence of the contract
- intention to interfere with the performance of the contract
- the doing of unlawful acts with such knowledge and intention
- causing a breach of the contract or the non-performance of a primary obligation in consequence.

From this judgment it follows that the tort is actionable even if the consequence is not a breach of contract by the contracting party interfered with, and the party is not liable in damages. It is sufficient that the party is prevented or hindered from performing its contract (that is, interference with the execution of the contract - point (i) above). These are the circumstances under which, when he cannot seek redress for breach of contract from the employer, the contractor will explore means of seeking relief from the architect or Engineer to the contract. The judge admitted the claim under (ii) above subject to an indication by the contractor of a preparedness to have 'intention' pleaded on terms. As noted by Atkinson (1992), this is an interesting development of the law, because conspiracy does not require the intended result actually to be achieved, rather requiring only agreement and intention.

Claims for misrepresentation (*see* Section 6.9), either at common law or under the Misrepresentation Act 1967, may also arise under this particular setting.

A *quantum meruit* (meaning 'as much as it is worth' or 'as much as he deserves') claim is a claim for the value of services rendered or work performed where there is no contractual entitlement to payment. Such a claim may arise when work is done on the basis of a letter of intent and without contractual liability. This type of claim can also arise when work has been performed by a contractor without there being any express agreement as to price. The term can also be used loosely to describe a contractual claim for a reasonable sum of money where, for example, the contract rates have ceased to be applicable. *Quantum meruit* is appropriate where a breach of contract by the other party prevents performance. There is a *quantum meruit* basis to

many contractual claims, contrary to the intentions of the ICE Conditions of Contract. As noted by Powell-Smith and Stephenson (1989), acceptance of this principle is to be found expressed in Clause 6.1 of the Minor Works Conditions:

'if the contractor carries out additional works or incurs additional cost....the Engineer shall certify and the employer shall pay to the contractor such additional sum as the Engineer after consultation with the contractor considers *fair and reasonable*'.

The Engineer and his representative must respond immediately, positively and directly, and in writing, to all notices of claim, including those mentioned casually at meetings or orally on site. The response must detail the full history of the events and point out, in particular, any perceived shortcomings of the contractor.

It is most important to maintain complete and authenticated records of the work on site. Under the pressure of site operations there will always be a temptation to defer committing observations and decisions to paper. However, it is essential that good and detailed records should be started at the beginning of the work and should include the following:

- plant, equipment and labour
- rate of tunnelling progress over pre-determined increments of time
- quality of workmanship
- ground conditions, including the geology, rock/soil strengths
- groundwater seepages and estimates of inflow rates
- records of discussions, agreements, disagreements, and claims.

The content of such records should be agreed by the representatives of the Engineer and the contractor although, realistically, there are likely to be some disagreements with respect to the quality of workmanship. It is also sensible for all on-site observations and logging to be conducted jointly by these representatives, but it must be recognised that there may be circumstances on particular contracts when the Engineer's representative may not always be available at the requisite time. For example, on a smaller contract there may not be a clerk of works on one or several particular night shifts at a time of both rapid advance and quickly changing tunnel face geology. Under such circumstances, the Engineer's record of the tunnel face geology and the excavation difficulties that the geology may have imposed will be incomplete and, in the event of a contractual claim, the contractor will almost certainly present his own records as authentic for the purposes of filling in lacunae (gaps). The Engineer should wisely foresee such eventualities and so institute a firm set of procedures for dealing with such matters. It should be made clear in the contract documents that it is the Engineer's records that carry legal force.

In practice, on any contract there will usually have to be a compromise between the amount of site information that should ideally be recorded and the availability of site staff to do the work. In deciding what can be achieved with the resources available the Engineer must determine at the outset what records will constitute essential evidence in the event of claims and what will be important for carrying out an effective post-construction technical audit (*see* Section 8.1).

There may also be records of a less formal type. From an Employer's point of view it will also be an advantage for the resident engineer, any assistant resident engineers and clerks of works to use their site diaries to record their own day-to-day feelings on the progress of a contract, together, perhaps, with any rumours bearing on the contract that they may hear. For example, records of plant breakdowns taken together with verbal complaints from the labour force about the plant might suggest that the contractor is under-resourcing the contract, and this in turn could indicate an under-priced tender.

During the currency of a contract, the Engineer and his representative should be administratively strong and firm with the contractor. There should be little or no ground given to appeals for favours or on hard-luck stories; there should be strict adherence to the contractual line. At the end of the contract it may often be possible to adopt an easier 'tack' in order to achieve a settlement of any point or points of difference between the contracting parties. If a satisfactory outcome to any difference is not possible, then the employer should be advised to make an offer in settlement

as soon as possible after careful consideration of the issues in order to secure a protection on costs (*see* Section 7.8 below).

Notwithstanding these comments, there is always room for a managed, conciliatory approach to the resolution of contractual conflict. However the ultimate costs of a dispute are decided, there will always be unrecovered hidden costs incurred by both sides, such as professional staff time devoted to the problem. Whitfield (1994) notes that even on the most successful claim the professional is unlikely to recover more than the contractor's normal or tendered overheads, and that the costs of a conflict could represent as much as 20% of the contract value on a contentious project. The book by Whitfield is recommended for its thoughtful distillation of the management of construction conflicts.

7.2 FACE LOGGING AND TESTING

For the purposes of assessing contractual claims, and of course only in the case of man-entry size tunnels, it is recommended that, whenever reasonable and possible, face logging and rock/soil testing be carried out at intervals laid down in the contract documents, with provision for more frequent assessments should the Engineer so decide. If the contractor requests more frequent assessments then the Engineer should normally agree, subject to the cost being covered by the contractor. In the case of uncomplicated geologies the work of face observation could be carried out by a clerk of works trained to perform the duties. It would be in the contractor's interests for him to agree the assessments at the time they are made, the form that the agreement should take to be stipulated in the contract documents. In the event of disagreement the Engineer's records must normally be accepted as authentic. This should be specified in the contract.

On larger contracts, say in excess of one million pounds, it may be worthwhile for the Engineer to engage, as an assistant resident engineer, a geotechnical engineer or engineering geologist to perform the logging duties, oversee any testing, and be responsible for reporting the results and keeping the records. The comments made above in Section 7.1 concerning possible gaps in the Engineer's records would still apply, but on such a contract a clerks of works would most likely have been engaged and would thus provide cover on those shifts for which the geotechnical engineer/engineering geologist was not available. It must also be recognised that, in some tunnelling situations, access to the face may not be easy for the purposes outlined above. This problem will arise when machine tunnelling methods are used.

In addition to careful sketches, photographs of the tunnel face geology frequently prove to be useful in the event of a dispute. It is recommended that, whenever possible, polaroid style colour photography should be used. Experience shows that stratifications not readily apparent to the naked eye under artificial lighting are much better defined by flash photography. On the other hand, if one purpose of the photography is to identify the presence of cobble and boulder densities in a face of till material, shadows may define indentations in the face caused by tool action, the indentations having the appearance of cobbles and/or boulders and perhaps being prone to misinterpretation during investigations into the status of a contractual claim. When tunnelling in areas where methane gas might be present, FLP Regulations will preclude the use of flash photography.

In respect of rock strength assessments at the tunnel face, there are certain quick *in situ* tests that can be performed during break periods and at a change of shift in order indirectly to quantify the difficulty of excavation. These are mentioned in Section 7.3 below.

7.3   ROCK TUNNELLING

Contractual claims during rock tunnelling frequently relate to rock strength and the difficulty of excavation. At each stage of tunnel advance, the rock strength and its variability over the area of the face should be recorded in the manner noted in Section 7.2 above. Rapid monitoring of strength is perhaps best done indirectly using a Schmidt hammer. The results of the tests can be correlated with unconfined compressive strength and tensile strength values derived from laboratory tests on cores or block samples. The advantage of such tests is that they do not involve the retrieval of rock material from the face and can be carried out rapidly at break times or between shifts.

Example: In the case of a tunnel driven along the strike of low angle beds, Schmidt hammer tests were carried out on a 200mm grid at the tunnel face. The points of intersection of the template grid lines marked the Schmidt hammer test locations and these were rigidly adhered to even in those instances where there was a discontinuity (which promoted zero rebound) passing beneath the location. Two methods of sampling at each point were used: in one case a continuing series of blows until such time as a constant rebound reading was achieved, and in the other an average of five rebound values. An average rebound hardness number was then defined for the rock at the face and this was then able to be assessed in the context of a changing geological (structural and lithological) profile over the area of the tunnel face.

In addition to - or even in the absence of - strength tests, the Engineer's representative, jointly with the contractor's representative, should also record his perception of excavation difficulty on a shift by shift basis. At its simplest, there could be five levels of perceived excavation difficulty: very easy; easy; average; difficult; very difficult. Many factors, including the geology and equipment used, contribute to the placement at a particular level. A check list should therefore be designed to take account of all the relevant parameters, suitably weighted by the Engineer on the basis of his experience and the particular circumstances of the job, in order to facilitate the classification.

Example: At one tunnel face of large diameter, thin, stronger bands of rock were in juxtaposition with bands of clay shale. Using heavier duty hand-held pneumatic picks, the excavation was placed in the 'easy' category. At another tunnel face of the same diameter the bands of rock were thicker and light pneumatic picks were used. This excavation was described as 'difficult'.

7.4   BOREHOLES AND/OR TRIAL PITS

There may be instances when 'unexpected' ground conditions are encountered and it is considered by the Engineer that it would be advisable to investigate the ground ahead of the tunnel face. (The reasons for doing this, as related to how unexpected the conditions are and the contractual implications, need to be very carefully thought through by the Engineer.) Sometimes the contractor may request further advance information if he is proposing to change his method of tunnelling to cope with what he perceives as significantly changed ground conditions, and which would therefore almost certainly be the subject of a contractual claim for extra payment. To assist in these decisions, it may be agreed that a borehole or boreholes and/or trial pit(s) should be sunk ahead of and just off the line of the tunnel face.

7.5   SUPERVISION

It is essential that the Engineer and contractor provide experienced and competent personnel to control the contract on site. If a resident engineer and the contractor's agent are obviously incompatible, then it should either be accepted that both be changed or other members of staff could be appointed to maintain the necessary liaison by acting as 'buffers'. The site staff may not be able to prevent a dispute developing but they must be capable of establishing and maintaining harmonious working relationships throughout the period of a claim.

7.6   PROGRESS

A key to the actual quantification of any claim is productivity - actual and expected. Payment is related to the rate of advance and it is in the contractor's interests to base any calculations for loss of production on the most favourable rates that have been achieved on the contract. The greatest rate of progress often occurs beyond the beginning of a contract following an initial learning period. In the absence of evidence to the contrary, an average of the rates immediately before and after the incident or incidents that generated the claim would normally be adopted, provided that the method of working had not been changed.

Whilst reduction in the rate of advance may be a major element in the evaluation of a claim for extra payment, it should not be used as the sole method of determining costs. Down-time during tunnel construction may often be more

than 50%. Poor progress may be due to inefficient working caused by such factors as poor spoil and material handling, lack of resources on site to meet short-term emergencies, old, unsuitable or low-capacity plant and equipment, and so on. The contractor may fail to fix a suitable bonus system relating to his expected rate of advance. This situation can indeed be exacerbated by a failure to fix bonus rates while tunnelling through what he considers to be changed ground conditions, so resulting in slower progress and an incentive to claim for extra payment. The problem for the Engineer is then to establish a method of assessing the loss that can reasonably be deemed attributable to these various factors.

Clause 12 of the ICE Conditions of Contract (1991) provides for reimbursement for the cost of work done and additional constructional plant used that would not otherwise have been required, together with a percentage in respect of profit. Under such circumstances the contractor should be required to open his books and show the costs that he has incurred over an agreed period. This action should identify any elements of cost which ought to be borne by the contractor but, in practice, there may be some difficulty in substantiating its authenticity in detail.

Any liquidated damages specified in the contract documents must be carefully and realistically formulated since there could be reference to their make-up in the event of a dispute concerned with unforeseen ground conditions and their effect upon progress. At a pre-tender meeting it should be made clear on behalf of the employer that such damages will be rigorously applied.

7.7   PLANT AND CLAIMS

A contractor may decide at the beginning of a contract to over-invest in plant to a degree that would not be necessary if the ground conditions proved to be exactly as predicted (strictly, estimated) by the site investigation. One example would be the choice of a roadheader that had the capability of cutting rock stronger than that suggested by the results of the site investigation. By improving his working margins in this responsible way the contractor might reasonably argue that his actions should be recognised within the setting of the contract.

Suppose in this example that ground conditions as revealed would provoke a Clause 12 claim but they are overcome without delay by virtue of the fact that heavier duty plant was immediately available. There is a case for logging such incidents in order to quantify the effective gain in productivity for balancing against any conditions that might generate future contractual claims for extra payment stemming from unforeseen ground conditions and/or delays to progress. A contractor might claim for a proportion of his additional costs, but the Engineer would argue that the contractor had used his experience properly and should expect no extra payment.

Also in respect of plant it must always be remembered that when reduced or increased quantities above those billed are encountered, and the contractor requires an increase in rate, that change in rate *may* be based on the use of equipment and/or resources that would not have been required had the conditions not changed from those upon which the tender was priced (*see also* Section 4.3). Employers who are often surprised to receive requests for re-rating have failed to recognise this factor.

7.8   ARBITRATION

Pre-contract under the ICE Conditions of Contract 6th Edition 1991, the Engineer is the employer's professional adviser. But once the contract comes into force he assumes additional and quite separate functions - specifically, he is then required to administer the contract independently of both contracting parties. During the currency of the contract he will have made decisions on claims, and one that he has rejected may be returned to him for a Clause 66 decision. Under Clauses 66(1) and 66(3) of the ICE Conditions of Contract 6th Edition (1991), the Engineer must re-assess the claim on its merits and come to a fair decision that balances all the circumstances of the case. Although he is sometimes referred to as a 'primary arbitrator', and his duties at this stage of the contract are often seen as being 'quasi-arbitral' (Hawker *et al.*, 1986), he is not an arbitrator because he is not bound by the rules of natural justice nor, strictly, is he bound to hear or receive submissions from the parties to the dispute before reaching his decisions, although he may

often be well advised to do so. If possible he should arrange for his representative (the resident engineer on site) to put the employer's case, and for the contractor to put his, so that a clear impartial judgement can be made of the issues even when his own liability may sometimes be on the line. If either the employer or the contractor are dissatisfied with the Engineer's decision, under Clause 66(5) the matter of the dispute may be referred first to a conciliator under Rule 4 of the Institution of Civil Engineers' Conciliation Procedure (1988). Each party is jointly and severally liable for the conciliator's account under Rule 13 which provides that each party pay an equal share. Then, if either party to the dispute remains dissatisfied with the recommendation of the conciliator, the dispute is referred to an independent arbitrator (Clause 66(6)) and the arbitration is conducted in accordance with the Institution of Civil Engineers' Arbitration Procedure (1983) which falls under the Arbitration Acts 1950 and 1979 (*see* Hawker *et al.*, 1986). Details of both the Conciliation Procedure (1988) and the Arbitration Procedure (England and Wales) (1983) are included as separate inserts to the ICE Conditions of Contract (1991).

The question of when a dispute actually begins under the main contract has exercised, and will no doubt continue to exercise, legal minds. Clarke (1993) responding to Stephenson (1993), in considering the position of a subcontractor when in dispute with a main contractor, and consolidation of arbitrations under the main and subcontracts, has referred to the case of *Erith Contractors Ltd* v *Costain Civil Engineering Limited* in which counsel on behalf of the claimants submitted three possible answers: when a contractor has sought from an employer payment or some other contractual right and been refused; when the contractor or the employer has invoked the provisions of Clause 66 of the main contract; and when an arbitration under the main contract is actively commenced. Notwithstanding the fact that a 'dispute', *sensu stricto*, may be said to have arisen when an Engineer rejects a contractual claim, the judge in the Erith Contractors case held that the earliest date upon which a dispute has arisen under a main contract is the date of issuance of a Clause 66 notice, and this is, in fact, the sense of Clause 66(2), the 'notice of dispute clause', of the ICE

Conditions of Contract, 6th Edition. The definition in Clause 66(2) also applies to the Federation of Civil Engineering Contractors Form of Subcontract (September 1991). A contractor can only require a subcontract dispute to be dealt with jointly under a main contract when he has requested a Clause 66 decision, but once that has been done he can still require, under the FCEC Form of Subcontract (Clause 18(8)), that the subcontract dispute be dealt with jointly under the main contract. The time period within which this should take place could be expressed as an addition to this clause. The subcontractor must be bound by any award made under the main contract arbitration whether or not an arbitrator has been appointed under the subcontract. This situation differs from that which applied under the old (1984) version of the FCEC Blue Form in which Clause 18(8) was then Clause 18(2).

Some authorities delete Clause 66, so forcing the contractor to go to the courts for his protection. As noted by Armstrong (1991), such a forced action is not generally in the contractor's interests; his case is better dealt with by an experienced and capable arbitrator at a pre-decided date for a hearing in private, without the formality and inconvenience of a court, but with the back-up of the High Court, and without certain problems related to the Court's powers.

The Arbitration Acts operate in England and Wales under English law. The 1950 Act was a consolidation of previous Acts of 1889 and 1934, and the 1979 Act is an amendment of the 1950 Act. In Scotland the main Act governing arbitration is the Arbitration (Scotland) Act of 1894 and that applicable to Northern Ireland is the Arbitration Act (Northern Ireland) 1937. The 1979 Arbitration Act considerably restricts involvement of the courts in the process of arbitration compared with the provisions of the 1950 Act. Reference may be made to Mustill and Boyd (1989).

As determined by the case of *Northern Regional Health Authority* v *Crouch* [1984] QB 644, the role of an arbitrator under a typical construction contract differs from that of a court, the latter having fewer powers than the former. However, any procedural rules laid down in a standard form of contract must be followed by the arbitrator unless the parties to the dispute make a subsequent agreement altering the

procedure. Not only must the arbitrator act fairly in his handling of the procedures but he must also adopt an adversarial rather than an inquisitorial style of operation (Mustill and Boyd, 1989). It is not his role to seek out the truth by engaging in speculation, pursuing enquiries or calling for and examining witnesses but rather to choose between two alternative versions of the truth as presented to him by the parties in dispute.

Clause 66 of the ICE Conditions of Contract gives the parties to the contract the right to appoint the arbitrator. Failure to make an appointment within one calendar month allows either party to apply to the President (or a Vice-President) of the Institution of Civil Engineers to arrange an appointment. The Arbitration Procedure (England and Wales) (1983) provides for a similar course of action. If for whatever reason Clause 66 does not apply, the appointment of an arbitrator is dealt with by Section 10 of the Arbitration Act 1950. This allows the High Court to appoint an arbitrator when the parties have failed to agree. This same power may also be used when an arbitrator refuses to act, or is incapable of acting, or dies. Arbitrators appointed by the High Court have the same powers to act and to make an award as if they had been appointed by the parties to the dispute. Grounds for disqualification and removal of an arbitrator are given by Manson (1993).

The 1983 ICE Arbitration Procedure gives the disputing parties and particularly the arbitrator various choices as to how a dispute should be investigated. Clause 66(8) deals with the ICE Arbitration Procedure or any amendment or modification thereof at the time of the appointment of the arbitrator. The procedure for England and Wales is provided in the back folder of the ICE contract document and the very different procedure for Scotland is also available (*refer to* Clause 67(2)(c) and also to Hunter, 1987). Under Clause 66(8)(a) the arbitrator is given 'full power to open up review and revise any decision, opinion, instruction, direction, certificate or valuation of the Engineer'. According to Clause 66(8)(b) neither party to the dispute is limited in the proceedings before such arbitrator to the evidence or arguments put before the Engineer for the purpose of obtaining his decision under Clause 66(3). It should be noted, however, that to proceed to such an arbitration at all is, in itself, an admission of failure in the professional conduct of contractual procedures. In tunnelling there are usually no absolutes; no one is absolutely right or absolutely wrong. The answer invariably lies in between. Arbitration could perhaps be more readily avoided if the employer was prepared to demonstrate to the contractor that the Engineer had been given full power to act as an independent arbitrator within the meaning and spirit of the ICE Conditions of Contract. The problem is perceived as being most acute with local and public bodies which not only act in the role of employer but which also provide the Engineer from within their staff of salaried employees. Such a difficulty may be alleviated to a degree by adopting an open policy whereby the most senior engineering staff of the employer, having taken part in the project at the overall planning stage, then withdrew completely from any involvement in the detailed design and the implementation of the contract, leaving a middle-ranking engineer to act as the Engineer for the contract free from any interference from above when fulfilling his role as 'primary arbitrator'. Another solution, perhaps having completed all the design and pre-contract work in-house, would be to engage an independent project manager to oversee the implementation of the tunnelling contract, the Engineer being appointed from within the company. The appointment of an independent Engineer for the currency of the contract is indeed one of the provisions in the Institution of Civil Engineer's New Engineering Contract (1993). It has also been suggested that all Engineers, because they are required to adopt an arbitration role, should be trained in arbitration.

Under a procedure analogous to Order 14 of the Rules of the Supreme Court one of the parties to the arbitration may apply at any time for a summary award. After taking account of the sworn affidavits and accompanying exhibits submitted on behalf of the parties and being addressed, often by learned counsel for the parties, the arbitrator has the power to award payment by one party to another of a sum representing a reasonable proportion of the final net amount which, in his opinion, that party is likely to be ordered to pay after determination of all the issues in the arbitration. The arbitrator

will also make an order as to the costs of the summary award hearing. These costs will usually be a small proportion of the total disputed claim, so there may be great incentives and considerable advantages for the claimant - usually the contractor - to test the 'temperature of the water' in this way by 'winkling out' more details of the respondent's case and perhaps bearing the risk of holding back some expert evidence for later exposure as proof of evidence prior to the main hearing. The evidence upon which Rule 14 judgments have been made may be re-examined at the main hearing.

A summary award should not be lightly made. Hawker *et al.* (1986) state that a summary award is appropriate: 'Where it appears from the evidence that the creditor is bound in the end to obtain some award. However, the power will not be exercisable where liability is genuinely in issue....' (page 72); and 'Where there is no, or no serious defence' (page 72). Also, from Hawker *et al.* (1986 page 2), this power (of summary award) 'is analogous to the High Court's power of summary judgment under Order 14 of the Rules of the Supreme Court'. Reference to the White Book page 141, paragraph 14/3 - 4/8 indicates relief only where 'there is no reasonable doubt that a plaintiff is entitled to judgment'.....'no real substantial question to be raised'.....'no dispute as to facts or law which raises a reasonable doubt that the plaintiff is entitled to judgment'. Reference may be made also to *Smallman* v *R.D.L.*(1988), 5 C.L.J. 1 page 62 at 71 (C.A.) per Bingham, L.J. (dealing with the Order 29 interim payment burden): 'Order 29 deals with a situation in which a final enforceable judgment cannot be given for the plaintiff without full trial, but in which (a) it is either certain or appears overwhelmingly likely that the plaintiff will win when the trial takes place, and (b) it appears overwhelmingly likely that he will recover at least a substantial sum by way of debt or damages. The object is to prevent a defendant, in the interval before a trial takes place, keeping a plaintiff out of money which the plaintiff is likely to recover at the trial, however the trial goes.' Further reference may be made to *Breeze* v *McKennon* (1985) 32 BLR 41 at 49 (C.A.) per Croome-Johnson, L.J. (dealing with the Order 29 burden): 'The onus of proof to "satisfy" the Court on liability under 11(1)(c) is high. It is equivalent to being sure that the plaintiffs will recover. A mere *prima facie* case is not enough.'

Offers may be made during the currency of the dispute. *Open offers* are of particular value in, for example, matrimonial cases concerning financial provision on divorce where the behaviour of either or both parties *in relation to the conduct of the litigation* can be a factor relevant to the quantum of the award. However, matrimonial cases may be *sui generis* in this respect and that in the context of other litigation open offers may most frequently be made by accident (a solicitor omitting to insert the words 'without prejudice' at the top of his letter!). An 'open offer' can be referred to by either party to the dispute at any stage of the hearing. Conversely, a 'without prejudice offer' cannot be referred to by either party at any stage. The privilege conferred by the words 'without prejudice' cannot be waived by the offerer without the consent of the offeree. For this reason the accepted wisdom seems to be that without prejudice offers have only limited advantages; they may, for example, be useful at a very early stage of negotiations, perhaps even before proceedings are formally issued, where the 'delicate' step of an informal offer without a Calderbank (*see* below) threat may be conducive to an early settlement. A respondent may make a 'without prejudice' or open offer, to include interest charges and perhaps some or all of the claimant's costs, as a full and final settlement of all matters in the dispute, and to be open for acceptance for a fixed period of time. The offer must be based on a very careful calculation of the claimant's real loss (which will usually be less than the claimant's own calculated loss) and will be pitched at a judicious level above this figure in order to extract a reaction from the claimant (who, if the contractor, may also be under pressure - perhaps the threat of another arbitration, with the contractor this time as respondent - for extra money claimed by a subcontractor). If, in fact, the two contracting parties are not too far apart in their calculations, or if other factors intervene to press the claimant (such as lack of staff time to pursue the dispute in the face of resolute action from the respondent or the need to move on to another contract and maintain goodwill), then he will eagerly accept the offer.

If the open offer is not accepted, either by direct refusal or by non-reply, then a 'sealed offer' mechanism - the 'Calderbank' offer format - may operate within the arbitration in order to place the claimant under some risk with respect to costs. The *Calderbank* v *Calderbank* [1975] 3 WLR 586 (also [1975] 3 All ER 333 and [1976] Fam 93) case concerned a husband and wife dispute. A Calderbank offer represents what can be termed a 'half-way house' between an 'open' and 'without prejudice' offer (the contents of which cannot be referred to during a hearing with reference to liability or quantum) but in which the maker of the offer expressly reserves the right to refer to the offer once the substantive findings have been made and when the question of costs is being considered (by the arbitrator). This form of offer thus bears many similarities to a payment into court except that no money is actually deposited. Under this system the parties to the dispute may hand to the arbitrator, usually before or even during the hearing, a sealed envelope containing an offer sum in settlement, if there is such an offer, or otherwise a sheet of plain paper. The arbitrator opens the envelope, which is marked 'without prejudice save as to costs', after making the award but before settling costs, so giving the letter privilege except as to costs which are decided by the arbitrator having regard to the offer. Such a sealed offer made before the hearing has the advantage to the respondent of fully protecting his position on costs should the claimant proceed to a full hearing. If the claimant refuses to accept the offer, and if the arbitrator's judgment for the claimant is for a sum less than or equal to the sealed offer sum, then the claimant will usually be required to pay both his own and the respondent's costs incurred from the time of the offer up to the date of judgment. The claimant will, however, still be entitled to receive a payment in respect of those costs that he has incurred prior to the date of the offer. In addition to this payment, the respondent must also cover his own costs up to the offer date. The offer remains open for acceptance until the arbitrator has made his award, unless it is formally withdrawn. However, withdrawal would remove the protection on costs afforded to the respondent. If the offer is accepted, all the other party's (respondent's) taxed costs after the date of the offer must be paid by the accepting party (the claimant).

The reasoning behind this offer mechanism is that if the arbitrator's award is for less than the offer, and had the claimant accepted the amount of the offer, then he would have been no worse off, and would have avoided the need for either party to the arbitration to incur costs beyond the date of the formal offer. In practice, a sealed offer must normally exceed an earlier open offer, and a great deal of technical, legal and contractually psychological skill would go into the determination of its magnitude. This offer must be high enough to present the claimant with a real risk that he would not beat the figure in the end. If a claimant does not accept this offer, the clear indication then is that he does not consider his chances of recovering a sum of that order to be non-existent.

In awarding costs, the arbitrator has to indicate which of two scales will apply. The first scale is 'Standard Basis', which is a reasonable amount in respect of all costs reasonably incurred, and the second scale is 'Indemnity Basis', which is all the costs except insofar as they are of unreasonable amount or have been unreasonably incurred. With indemnity, payment is made unless it can be shown that something is unreasonable. With standard, payment is limited to a reasonable amount provided that those costs have been reasonably incurred. The indemnity basis is obviously more generous, but it is usually applied only when the standard scale is not, in the judgment of the arbitrator, an appropriate award to make.

In practice, when considering the actual costs to be paid over, it is usually held that approximately one-third of all costs incurred by a party in civil proceedings are solicitor and own client costs and as such are not recoverable from the other party to the action. Thus, only two-thirds of the costs are potentially recoverable as a result of standard basis taxation.

A respondent may be induced to increase an earlier, time-barred open offer, but he should do this only with very great caution. If a later and higher offer is accepted, then the claimant party can recover its costs from the date of the first offer to the date of the second offer, and those costs could be substantial.

A claimant will provide detailed calculations in support of the sum that he is requesting for payment. The respondent will perform his own calculations in support of his offer, and with the benefit of having the claimant's calculations to hand. Notwithstanding the technical detail of his own calculations, and in addition to the reasons noted above, the respondent, for his protection, will usually be advised to make a sealed offer of a sum greater than his calculated assessment on the basis that it is advisable to have a 'cushion of protection' in the event that the arbitrator sees things differently from the respondent and awards a sum greater than that carefully calculated by him. With such an uplifted offer, the respondent will inevitably feel frustrated if the claimant accepts it simply because the claimant will then have extracted more than the respondent's firm (and, to him, fair) valuation of the claim. If a claimant feels that his case is going badly during the hearing, and it is unlikely that the arbitrator's award would reach the offer level, then he can accept the offer at that stage. The respondent's frustration would then be magnified.

The timing of an offer, if made, is important. Ideally, as noted above, it should be made early to fully protect the respondent from payment of the costs of the action. Outside objective advice should normally be sought to ensure that the value of the offer does not exceed a realistic valuation of the claim plus, if it is felt to be necessary, any 'psychological probing element' and protective cushion referred to separately above. This latter point, however, is important because an early, high offer could be quickly accepted by the claimant, with the respondent being in the position of having to justify his actions. On the other hand, an offer that the claimant readily sees as being far too low may be taken as a signal that the respondent concedes some validity in the claim and that there is a preparedness for negotiation towards a higher offer.

The successful party to a dispute could encounter some difficulty in obtaining payment by the other party. Under Section 26 of the 1950 Act, with leave of the High Court or a judge of the High Court an award can be enforced in the same way as can a High Court judgment or order. If the High Court does give leave, then judgment may be entered in the terms of the award. This effectively means that the debtor's goods may be seized, his land may be charged with payment of the debt, and money may be obtained from his bank account by means of garnishee proceedings.

A respondent would be wise, if in dispute with a small plaintiff company, to check the financial standing of that company. Unlike payment into a court, the sealed offer procedure can be criticised in that the firm or person making the offer is not compelled to deposit money to back up the offer. So, if the offer is accepted, there is no guarantee that the resources are available to pay the money. Further, if the value of a company is discovered to be less than its potential liabilities should its case be lost, then consideration should be given by the respondent to obtaining an Order for security against the company just in case it should go into liquidation. However, notwithstanding the size of the company, and even a low value placed on it, if it is seen to be trading successfully such an action may not succeed.

Much of the discussion above concerns the assignment of costs, but the British practice whereby costs follow the event is not universal. In almost half the countries of the world each party to a dispute pays its own costs irrespective of the outcome of the proceedings. In many instances in Britain the legal costs can greatly exceed the quantum of the award, and in such cases it would push the disputants to an early settlement if they knew that they were to be responsible for all their own costs from the outset. However, Section 18(3) of the 1950 Arbitration Act invalidates any such agreement made before a dispute has arisen.

One suggestion is for the implementation of a hybrid form of 'flip-flop arbitration' as practised in America, the arbitrator coming down firmly on the side of one of the parties, with no intermediate award, but only in respect of costs and not of the action itself. The parties to the action would be paid the amount of the arbitrator's award but the costs of the action would be awarded to whichever party's pleaded position was closest to the award in the end. This style of judgment would require an amendment before tender to the arbitration clause in the particular contract being used.

Arbitrators on civil engineering matters in general and tunnelling problems in particular are required to exhibit legal expertise. They may not always be engineers by training and thereby have specialist knowledge of the subject in dispute. It

might well be claimed that such a lack of specialist knowledge could be an advantage when handling adversarial proceedings, as are required under the arbitration rules. However, explicit engineering specification requirements may not always be contractually weighted by a non-engineering arbitrator, or even by an engineering arbitrator, because of the amount of practical engineering evidence presented to him for his consideration. This problem can be alleviated to some extent by the provisions of Rule 22 in the ICE Arbitration Procedure (1983) which allows meetings of specialists to discuss and agree technical and/or measurement matters prior to a main hearing. The arbitrator may request that such a meeting or meetings take place, notwithstanding the fact that one of the parties may be reluctant to participate. This procedure is based on documents, a site visit and oral submissions or questions. The actual meeting is of a formal nature between the arbitrator and experts, but without professional advocates taking a direct part although they may advise the experts on what questions to ask.

Example: A contractor experienced problems constructing a sewer tunnel in soil and requested the use of compressed air at the expense of the employer to enable him to complete the tunnel in accordance with the Specification. A claim was upheld by an arbitrator on the basis that the tolerances specified in the contract documentation required good ground conditions for their implementation and that, where ground conditions proved to be poor, it would not then be reasonable to expect the tolerances to be achieved without introducing special measures.

It may be argued that a contractor bids for a contract on the basis of the tolerances as specified. Therefore, provided that the tolerances are not impossible to achieve, a claim under Clause 12(1) of the ICE Conditions of Contract (1973, now 1991) should be no excuse for non-compliance with the Specification. However, it is suggested that if relatively tight tolerances are specified, it would be prudent to state in the contract documents that the nature of the ground should in no way be inferred from the specified tolerances.

Outcomes of arbitration cases sometimes hinge on the lack of time given to an Engineer adequately to prepare pre-tender documents, the feasibility study, the economic design and planning of the works, the site investigation, and the contract documents and drawings. Provision for continuity of construction during arbitration must be expressly written into the conditions of contract. At the core of the Engineer's thinking should be a perceived differentiation between the contractor's contract risk and contract responsibility (Haswell and de Silva, 1989). Arbitration awards are usually binding without appeal unless some degree of invalidity can be firmly demonstrated.

Subcontracts may be different in respect of arbitration. Although the amount of money in dispute may be less than in the case of a main contract dispute the effect on the subcontractor could be greater. Such cases may tend to be settled by writs in the courts rather than by arbitration, or even threatened arbitration. A contractor could experience problems where a main contract is sealed and a subcontract is arranged 'under hand' due to different time bars (Armstrong, 1991, and *see* Section 2.2.2). Reference should be made to the earlier comment on subcontracts.

As indicated in the discussions above, the arbitrator is constrained on the matter of costs. There is an argument that arbitrators should be allowed more freedom to allocate costs and that the parties to an arbitration should be allowed to agree beforehand to share the costs of the action. It is expected that such matters will be addressed when a new arbitration act is on the statute book, but although a quarter of all arbitrations arise in construction and these account for approximately one-third of the value of all claims that go to arbitration, there will not be a separate bill for the industry.

A committee was sitting in 1991 under Mr Justice Steyn to consider reform of arbitration law by consolidating several different statutes of various ages, and there is the draft Bill noted above also under examination. The new Act will embody important provisions brought about by the Courts and Legal Services Act 1990, which was effective from 1 April 1991. Under the first provision the arbitrator is given full power over the process of discovery of documents, with the courts no longer having the authority to order the

parties. In another provision the arbitrator has the power to strike out a claim if a party to the action fails to progress the arbitration properly. Further, arbitrators might no longer be bound by strict laws of evidence. Many arbitrations operate under rules that include a contrary agreement to that effect, but in other cases the Act may end the risk of appeals being reinforced by the argument that a particular finding of fact relied upon inadmissible evidence. A proposal in the draft Bill allows an arbitrator to appoint a legal or technical assessor to assist in weighing the evidence presented to the court. There is also the possibility that an arbitrator with technical expertise would be able to limit the number of expert witnesses, in either a direct manner or by refusing to award costs to cover the fees of expert witnesses. There might also be some advantage in arbitrators having powers to make interim awards in advance of a full hearing, the sums awarded to be recovered later if shown to be unjustified. On the whole it seems that arbitrators will be granted greater discretionary powers in the awarding of costs, and to order security for the costs a party will incur in defending a claim or counter-claim. Finally, arbitrators could enjoy greater freedom to adopt an inquisitorial style of procedure rather than resort to the normal adversarial methods, the argument being that a more business-like hearing is then likely to result.

Referring back to Section 5.3, it is useful to note that breaches of warranty and/or questions of possible misrepresentation may not be considered to fall within the provisions of Clause 66 of the ICE Conditions of Contract and are therefore not matters to be dealt with by an arbitrator. Powell-Smith and Stephenson (1989) state, however, that claims for misrepresentation may be raised in arbitration or litigation. Further comment on this matter is in Section 7.9.

There is reference above to the possibility of a respondent's 'without prejudice' or 'open' offer and the inclusion of interest charges. Official recognition of the interest charge factor has come from an arbitration judgment, upheld by the Official Referee in September 1990, to the effect that an ICE 5th Edition (and presumably now 6th Edition) contract entitles a contractor to interest on the money when a disputed claim is resolved in the contractor's favour irrespective of whether the Engineer had failed to certify at all the sums claimed to be payable, irrespective also of his reasons, or had failed to certify an adequate amount, and that the contractor is entitled to compound interest on the sums of money owed. The case of *Morgan Grenfell (Local Authority Finance) Ltd* v *Sunderland Borough Council and Seven Seas Dredging Ltd* (No 2) 51 BLR 85 has also shown that an Engineer acting in good faith when applying judgment on certification will have to operate with even greater prudence in the future, since although under-certification will entitle a contractor to compound interest, over-certification will not allow an employer to recover interest from a contractor. The employer might therefore attempt to recover that interest from the Engineer.

As indicated above, reference of a dispute to arbitration does not preclude reference to the civil court, although it is usually expected that there should be a degree of finality in an arbitrator's award since the parties to an arbitration bind themselves to accept the decision of the arbitrator. Since the 1979 Arbitration Act, with its much reduced means of appeal against an arbitration award, the decision of the arbitrator has been much more difficult to challenge. Only when a matter of law arises in connection with a standard form of contract do the courts tend to involve themselves in the problem. However, as noted by Armstrong (1991), resort to the courts may be needed when several parties are involved in an action.

The legal forum for the construction industry is the Official Referees' Court which was created by the Judicature Acts of 1873-75 on the recommendation of a Royal Commission following criticism of the manner in which building and other complex cases involving technical issues were handled by juries. At the outset the Official Referee's function was to investigate and report on complex issues of fact as referred to the Court by High Court judges. The Official Referee could not try cases or make judgments. In 1884 a new Act allowed cases to be *referred* to Referees for trial, and they acquired all the powers of High Court judges including the ability to make judgments and orders for costs. It was only in 1982 that cases could actually *start* in the Official Referees' Court. With a heavy workload, full-time Official Referees are joined at the Court by part-time QC Recorders (part-time judges) who hear smaller

cases. Notwithstanding the powers assumed by the Official Referees, their status is still only that of circuit judge.

As is the case with arbitration, the Official Referees' Court has provision for a 'without prejudice' meeting of professional experts in order to identify core issues at dispute and to agree on matters of fact; in other words to simplify and streamline the subjects for the Court's consideration. There is also the feature of statements, or 'proofs of evidence' by key witnesses, to be exchanged before the trial. Again, these set out the facts that would otherwise need to be elicited through a process of time-consuming questioning by counsel for the expert witness's own side.

In response to the rising cost of insurance cover, some firms, especially in the USA and Australia, are offering self-insured 'layers of coverage' (the policy deductible) and are encouraging a more widespread reliance on the idea of 'alternative dispute resolution' (ADR) procedures. The American Society of Civil Engineers (1991) of 345 East 47th Street, New York 10017-2398, USA has published an important work on ADR, claiming that it can lower the cost of claims dramatically by expediting their resolution, often without extensive resort to or reliance on lawyers (attorneys). It will also offset to some extent the current trend against innovatory design. In Britain the ADR system is being operated on a limited basis, but has been used in connection with the Channel Tunnel construction. The Centre for Dispute Resolution (CEDR) offers ADR advice and facilities, and both the Commercial and the Official Referees' Courts are beginning to encourage the use of ADR.

According to Bradshaw (1994) the term 'alternative dispute resolution' will include conciliation, mediation, concilio-arbitration, adjudication, dispute review board and executive tribunal. A mediator or conciliator need not be restricted to the legal or even technical issues. The duty of a conciliator (Whitfield, 1994) is to be investigative, to correlate the facts, to try to reconcile the opposing views, to prompt the parties into proposing settlement offers, and to highlight the possible consequences of failing to settle while indicating strong and weak points in the cases of both parties. He does not propose his own settlement position. Mediation is broader, and while still helping the parties to decide on the issue allows the mediator to propose settlement terms of his own. The difference between conciliation and mediation on the one hand and adjudication on the other is the lack of legal procedure and the fact that the adjudicator decides on the issue himself. This relative informality in conciliation and mediation, and because the hearings are usually without prejudice, can be to the advantage of the parties, but when there is no legally binding result considerable measures of good faith are required from them. Although the proceedings are informal they can be given a degree of formality by including documents-only hearings or oral evidence at which both parties would be present.

ADR (Hibberd and Newman, 1994) can be a staged process. It is basically conciliation with the assistance of a neutral party appointed by the two disputants. This is the first step. If this fails then the next step is mediation. Such a procedural eventuality should be incorporated into the contract, since in the UK a recommendation by a mediator can be made binding. A mediator in the USA will assist negotiation by moving between the parties, acting as a catalyst to assist them in reaching an agreed settlement. In many states in the USA, a judge will not accept a case in the courts until there has been an attempt at mediation. An alternative to mediation is a mini-trial in which a top executive, usually a senior manager, from each of the disputing parties sits with a neutral chairman acting as mediator. Each side receives the full strength of the opposition case, in which there is reference to the 'core bundle' of documents, which lawyers and experts may present. In the USA the chairman of a mini-trial, often a senior lawyer, listens to the evidence from both sides, delivers a summing-up, and then encourages the panel to reach a settlement. The panel may ask questions, but otherwise there is no cross-examination as such. Failing agreement, the chairman is empowered to provide a non-binding opinion as to the possible outcome if the case should proceed to a full court hearing or to arbitration. This opinion then assists the parties in their subsequent negotiations. After an agreement is reached, a joint written statement is prepared and signed by the parties, the signatures being witnessed by the neutral party, and the agreement is legally binding. Another alternative

is to use expert appraisal. If a cause of the dispute can be suitably identified, then liability can be apportioned. Adjudication is a final and binding ruling unless and until the matter is referred to arbitration.

It has been argued that such a system would help to preserve better business relations, would provide a closer involvement in and more control over the settlement process, and would also avoid much unwanted publicity that could otherwise be quite harmful to the parties in dispute. The attractiveness of the system grows with the increase in litigation and litigation costs being experienced in the civil courts of Western countries, and also, as is becoming increasingly common, when the technicalities of a case are so complex as to test the understanding of a jury.

It needs to be re-stressed that for the successful implementation of ADR it is essential that the parties really do desire to achieve a settlement. The procedures must be fully understood, properly administered, and operated by properly qualified people under, in the UK, the Institution of Civil Engineers. It may also be necessary to suspend the timetable to Clause 66 of the ICE Conditions of Contract 1991 during a period of conciliation.

On the plus side, ADR seems to offer speed of conflict resolution and cost effectiveness, mitigating to some extent the hardening and entrenchment of attitudes that can occur with arbitrations, particularly when protracted. On the other hand, the speed of the ADR method could militate against a proper level of document research by a neutral mediator into dispute detail.

7.9   MISREPRESENTATION

A misrepresentation is a false statement of fact (and not of law) made during pre-contractual negotiations and which is one of the inducing causes of the contract. The maker of the statement of fact is the person who has the knowledge, whereas a statement of law is a matter of open knowledge because the statement can be checked by referring to appropriate sources of information. Expressions of opinion or belief are not misrepresentations unless the person making them has some special knowledge or skill. Misrepresentations may be (a) innocent, (b) negligent, or (c) fraudulent. Prior to 1967 there were only innocent and fraudulent misrepresentations, but the position was changed by the case of *Hedley Byrne & Co. Ltd* v *Heller & Partners Ltd* [1964] AC 465 which allowed a party to recover damages when the representation had been made negligently. However, this was available to the representee only when the two parties were in a 'special relationship' and it did not cover the general case of negligent misrepresentation. But under the 1967 Misrepresentation Act, unlike an action under the rule in *Hedley Byrne* v *Heller,* there is now no need for a representee to show that the elements of the tort of negligence are present; rather, all that he now has to show is that if the representation had been fraudulent it would then be actionable, which, in effect, widens the scope of the tort of deceit to encompass even careless misrepresentations. Misrepresentations now begin from a legal position of fraudulence, without imputing to the representor any fraudulent action unless the representor can prove that they were made entirely innocently. By 'entirely innocent' is meant that the representor really did believe the statement both at the time that he made it and at the time that the contract was entered into, and that he had reasonable grounds for that belief.

If a court finds that there has been a misrepresentation it may award rescission to the innocent party. This means that he is restored to the position that he was in before the contract was made. However, the Misrepresentation Act 1967 does allow a court or an arbitrator to declare that the contract shall continue to exist even if the representee has declared the contract to have been rescinded. Damages may be awarded to the innocent party in lieu of rescission when a contract is declared to subsist, and this will apply to almost all consultancy and construction actions.

Any misrepresentations which do not become part of the contract may give rise to liability both at common law and under the Misrepresentation Act 1967.

Typically, a contractor may claim against the employer on the grounds of misrepresentation about site conditions and other risks made during negotiations preceding the actual contract, notwithstanding Clause 11(2) of the ICE Conditions of Contract 6th Edition 1991 which specifies the responsibilities of the contractor in respect of his (the contractor's) site inspections and acquisition of information pertinent to the prosecution of the works. Powell-Smith and Stephenson (1989) note one relevant example concerning boulders, a geological feature which receives attention in different sections of this book, and particularly in Section 3.3.3.

Powell-Smith and Stephenson (1989, p123) quote the Australian case of *Morrison-Knudsen International Co. Inc.* v *Commonwealth of Australia* (1972) in which the contract contained a clause similar to the ICE Clause 11(2):

'The contractor acknowledges that he has satisfied himself as to the nature and location of the work, the general and local conditions, including......the structure and conditions of the ground......Any failure by the contractor to acquaint himself with the available information will not release him from estimating properly any difficulty or the cost of successfully performing the work. The [employer] assumes no responsibility for any conclusions or interpretations made by the contractor on the basis of information made available by the [employer].'

The contractor was provided by the employer at the pre-tender stage with a document called 'Engineering Site Information', which provided basic information on the soil conditions and which also contained a disclaimer of liability. The contractor claimed that the information so provided 'was false, inaccurate and misleading [and] the clays at the site, contrary to that information, contained large quantities of cobbles'.

The High Court of Australia held that the documents did not disclose that the contractor had no cause of action, the words of the Chief Justice being as follows:

'The basic information in the site investigation document appears to have been the result of much highly technical investigation on the part of [the employer]. It was information which the [contractors] had neither the time nor the opportunity to obtain by themselves. It might even be doubted whether they could be expected to obtain it by their own efforts as a potential or actual tenderer. But it was indispensable information if a judgment were to be formed as to the extent of the work to be done...'

Misinformation given to the contractor on behalf of the employer could also give rise to a claim for breach of an implied warranty. In the case of *Bacal Construction (Midlands) Ltd* v *Northampton Development Corporation* (1975), contractors had submitted as part of their tenders foundation (sub-structure) designs and detailed priced bills of quantities for six selected blocks of dwellings in selected foundation conditions. These formed part of the contract documents and had been prepared on the basis that the soil conditions were as shown on the relevant borehole data provided by the employer. The contractor's design was adequate in respect of those soil conditions.

The employer's tender documents stated that the ground conditions at the site comprised a mixture of Northamptonshire sand and Upper Lias clay. Tufa was found in areas of the site as work progressed, and as a result the foundations had to be re-designed and additional work carried out. The Court of Appeal held that the contractors were entitled to recover some compensation for breach of an implied warranty by the employer that the ground conditions would accord with the hypothesis upon which they had been instructed to design.

A civil engineering example (*Howard Marine and Dredging Co. Ltd* v *A. Ogden and Sons (Excavations) Ltd* [1978] QB 5740) of negligent misrepresentation is given by Manson (1993).

7.10   COLLATERAL WARRANTIES, LATENT DAMAGE, INSURANCE AND NEGLIGENCE

7.10.1 Collateral warranties

The expression 'collateral warranties' relates to documents that are given or are intended to be given by one party, acting individually or for or through partnerships or companies, for the purpose of creating legal relationships. They are a supplement to an actual civil engineering works contract. The warranties may take the form of contracts (and therefore will need to conform with the requirements of a contract, which includes a consideration or execution under seal - *see* Section 2.2.2). Alternatively, they may be in

the form of acknowledgements of other duties, that is, in tort or delict in Scotland. The idea behind collateral warranties is to create legal relationships, particularly duties, that would not exist in their absence. There is also the possibility that they could change the character and range of current valid legal relationships. The persons receiving or benefiting from such warranties may be actual parties to construction or other civils works, or they may be third parties, such as purchasers or future tenants of a building, and their identity(ies) may not be known at the time that the warranty is issued. If warranties are issued to unknown parties, then the risks underpinning the legal undertaking are usually increased. An example of a warranty between parties known to each other is one given to the employer by a design engineer employed by a contractor, the employer and the contractor only being initially in contract. A subcontractor may also give to an employer a warranty covering, for example, design of civil engineering works.

A deed of covenant, a memorandum of agreement, and a duty of care formal agreement are all the same as a collateral warranty.

An engineer, if asked to give a warranty, would charge a fee, not only to compensate for expenses directly incurred but also because under English law this action enables the warranty to be under signature rather than under seal, in which case the limitation period for claims is six years rather than twelve years (*see* Section 2.2.2) and the engineer would accordingly be less exposed..

In a quite general setting, and perhaps not so applicable to tunnels, the difficulty that consultants/designers find themselves faced with in the form of collateral warranties is causing concern in the industry. There is evidence that developers are squeezing tenants and consultants into situations where the tenants are forced to demand, and the latter to sign, collateral warranties to the effect that the design is 'fit and suitable for its intended purpose', or that it is such that it will ' meet the stated requirements' of the client. Consultants would argue that they can never guarantee the results of their design; all that they can be required to do is to exercise 'reasonable skill and care' in the performance of their duties under the contract of their appointments. But a consultant's professional indemnity insurance would not normally cover him for consequential losses that might accrue to a tenant as a result of a failure that is deemed (by the tenant) to have been caused by a fault in the design. Such insurance cover could only be secured by vastly increased premiums, which would naturally be a charge on the cost of the construction.

This tripartite problem arises because the tenant has signed a form of lease that binds him to rectify anything that might go wrong with the structure during his tenancy. He is then bound to demand, via the developer, a collateral warranty with the consultants (for the project) with whom he would otherwise have no contractual redress (but could claim liability in tort). Had the consultant not entered into the warranty arrangement at the outset then he would not have been appointed for the design work.

As a possible example of such a problem that could perhaps arise with respect to tunnels, consider the case of a venture capitalist financing the design and construction of a high-speed rail tunnel which is then let on a fixed-term lease to an independent operator. A few years after construction is completed and rail traffic has been using the tunnel it is found that the clearances over one stretch of the tunnel have reduced to unacceptable levels. Rail traffic through the tunnel has to cease until the deformed span of lining has been re-built. The consequential (and no doubt substantial) loss to the operator, who under the lease has accepted responsibility for repair, of revenue from tunnel traffic is then laid at the door of the consultant under the provision of a collateral warranty. The consultant may then wish to investigate whether the terms of the traffic flow - the operating speeds and the loads in particular - have remained within the terms of the design brief, and, if they have not, there might then be grounds for a counterclaim against the tenant/operator. Notwithstanding the potential scenario for legal conflict, the consultant would have foreseen at the outset, when he bid for the design work, the consequences of such a tunnel failure, and particularly the fact that his indemnity insurance would not protect him from those consequences under the arrangement of a collateral warranty.

The Confederation of Construction Specialists (CCS) has developed a model form of collateral warranty in which there has been criticism from the British Property Federation of Clause 6 which allows the warranty to be reassigned only once. Subsequent occupiers of a property would lose benefit if the building were to be re-sold or re-let. There is also some unease that there is no disputes resolution clause in the CCS model form. The Joint Contracts Tribunal (JCT) has been negotiating with British insurance companies in an attempt to cover all parties' responsibilities in project insurance and to produce a draft version of a warranty in 1991, but there is general acknowledgement that if conditions are made too onerous then insurance could prove to be impossible.

7.10.2 Consumer services and latent damage

Also bearing on these problems is the Latent Damage Act (1986) which relates to damage, not involving personal injuries, due to negligence in tort arising from the construction of buildings to the manufacture of goods and to the provision of services.

Professional liability, under the law which operates in England and Wales, arises generally in *contract*, in *tort*, and under *some statute* particular to the problem in hand. A *contract* will usually be recorded in writing or by agreement under seal, but it can also come into force orally, or by the unconditional acceptance of an offer, or by performance such as when one party begins work offered by the other party irrespective of whether that offer has been accepted in some other way. In the case of a service contract, in addition to the terms that are expressed there are implied obligations (which carry legal force) that the service will be carried out with reasonable skill and care within a reasonable time and for a reasonable price. In the case of the sale of goods, an example being that of supplies for constructional purposes, it is implied under common law, increasingly stiffened by statute law, that those goods must be of merchantable quality and reasonably fit for purpose.

The basic principle underlying the Latent Damage Act, which applies in England and Wales, is that an action in negligence accrues *on the date that damage occurs*. From this date there is a 6-year limitation period (3 years for personal injury) within which an action must be brought. (In contract the period is 6 years, or 12 years in the case of a contract under seal.) However, in addition to this 6-year period there is a further 3-year period running *from the date of discovery*, or from the date on which the plaintiff ought, with reasonable diligence, to have discovered that he or she had cause for an action. Notwithstanding the 6-year and 3-year periods, the plaintiff's action is statute-barred after the expiry of a *15-year 'long-stop' period* which runs from the date of the defendant's breach of duty. It is the view of the construction industry that the 15-year period should be reduced to 10 years, and that the commencement date for this period in construction cases should be the date of completion of the works. This 10-year period is the period that applies to manufactured and similar articles under the Consumer Protection Act (1987).

Neither the limitation periods not the 'long-stop' provision will apply if there has been fraud, mistake or deliberate concealment of material facts by a defendant and if it/they may be suspended in respect of any period during which the plaintiff is under any legal incapacity. A deliberate concealment could have the effect of preventing the limitation period from running, as provided under Section 32 of the Limitation Act 1980, even when it occurred after the plaintiff's cause of action arose (*Sheldon and Others* v *R.H.M. Outhwaite (Underwriting Agencies Ltd)* as reported in *The Times* 8 December 1993 in relation to Lloyd's underwriting syndicate problems).

If the ownership of the structure changes, there is provision (Section 3 of the Act) for the second or subsequent owner to 'inherit' the first owner's right of action. Where the defect was not discoverable before the structure changed hands the 3-year limitation period runs from the date on which the later owner discovers or could have discovered it, but again subject to the same 'long-stop' period that would have applied to the first owner.

Under a proposed new European Community Directive, produced by the EU's Consumer Policy Services Department and published in January 1991, clients in the UK would no longer have to prove that consultants were responsible for defective construction work. The onus would be on consultants to disprove claims that their designs were faulty. This change, which would bring the UK into line with consumer law which prevails in other Member States such as Germany, Spain, Denmark, Greece and Belgium, could lead to an escalation of claims for damage from clients, who now only have to prove that damage has occurred. Lengthy multi-party court cases could ensue, and the Directive would almost certainly lead to consultants having to take out higher insurance premiums to protect themselves. It would also be likely to have the effect of consultants attempting to distance themselves from the decision-making process by providing clients with heavily qualified advice and placing the responsibility on the client to decide on a particular course of action. There is no definition of the particular types of construction to which the Directive refers.

In contrast to the 15-year liability period noted above, the EU Directive calls for liability periods of up to 20 years for defective design, with clients entitled to lodge claims for damages up to 10 years after they discover a defect if discovery is within the 20-year liability period. Consultants would not be able to exclude responsibility for defects in contracts with clients. One result of this 20-year long-stop on liability would be that litigation could still be in progress up to 42 years after the end of construction, and consultants would find it very difficult to defend themselves. They would be compelled to maintain even more substantial archival records on past work in order to contest any claim for damages, but computer files with adequate back-up would to a great extent overcome any paper-weight problem. Liability might continue into and through retirement, the risk extending to the heirs of the consultant, so there is a need to maintain professional indemnity insurance albeit without tax relief on the premiums.

The Construction Industry Council has submitted a response to the Department of Trade and Industry with a series of proposed amendments which include reducing the limitation periods on liability in line with those for other services, a clearer definition of the duty of care, and ensuring that the product of the service is excluded from the liability. It says that the industry should be included under a separate construction-specific Directive to be prepared by the Commission's directorate general. If that route is not acceptable, the CIC would like a radical overhaul of the proposals for the services Directive to bring consultancy work into line with other services. This would see a reduction of defects liability periods from 20 years to 5 years and reducing the time limit for the lodging of damages claims from 10 years to 3 years.

7.10.3 Building insurance

There have been moves (late 1990) by Lloyd's insurance syndicates to offer clients a comprehensive building policy called 'Building Line' aimed at offsetting the high cost of traditional indemnity, latent defect, and contract insurance. The idea is for cover to begin at the design stage in the form of a professional indemnity policy, but to be taken up by the developer rather than by the consultant. Contractor indemnity begins with construction, the liabilities of the professional team staying with the policy. On completion of the work, the policy then covers latent defects, with the benefits being available for transfer to tenants or purchasers of the structure. Such a style of policy will require changes to the forms of contract, but with adequate support within the industry could have been operational in 1992. Just before Christmas 1990 the Wren, the architects' mutual insurer for professional indemnity, launched its latent defects insurance packages. Under such a policy, and as is usual with continental practice, building owners and occupiers can insure themselves against costs resulting from structural, weather-proofing and services faults after a building has been completed. This is a move away from reliance on contractual agreements, such as the collateral warranty, between developers, builders and owners, and it serves to overcome the problem of consulting and contracting firms not having adequate assets or insurance cover to support such contracts. As the basis is one of strict liability, it should not then be necessary for the owner of a building, in order for him to be able to recoup the costs of failure or other damage, to have to prove negligence on the part of the designer/consultant and thereby be able to draw recompense from the designer's insurers (*see* discussion on the Abbeystead case in Section 7.10.5 below). The Wren, working with Lloyd's brokers the Miller Group, has developed a scheme that provides automatic cover for UK and Irish contracts up to a value of £30m, each policy being tailored to an individual building and there being provision for insuring larger projects and overseas projects. In line with

the continental (French) system the policies are for 10 years, with a start point on completion of the building. They are not amenable to cancellation, and they cover the building, not the contracting parties involved. The policies are therefore transferable to future owners and occupiers and should be particularly attractive to those people burdened with a full repairing lease. A structural survey is needed so that the insurer can monitor first the design and then the construction quality of the project. Although variable, the costs are of the order of 0.5% to 1.5% of the sum insured. A similar facility is available for building occupiers to provide cover for business interruption losses resulting specifically from a latent defect. Seemingly an advantage of this form of cover is that buildings are easier to let or sell.

There have been two fairly recent House of Lords cases concerning contractors' liability for defects. The cases were *Linden Gardens Trust Limited* v *Lenesta Sludge Disposals Limited and Others*, and *St Martin's Property Corporation Limited and Others* v *Sir Robert McAlpine & Sons Limited.* The cases concerned assignment clauses in contracts. In each case the original owner, who was also the employer under the construction contract, sold the structure and assigned the benefits of the contract to the new owner. The contractor argued that the original owner should not be able to recover substantial damages for defects because he had suffered no loss due to his having disposed of the structure for its full value. The new owner was prevented from recovering because the assignment of the right was invalid, and the House of Lords decided that prohibition of assignment without consent was effective in preventing the new owner from suing the contractor. On the other hand, an original owner is still able to sue even though he no longer has an interest in the property.

What the judgment amounts to is that once a structure has been disposed of, and the contractor has not given a collateral warranty to the new owner, the contractor can no longer assume that he is protected from prosecution. It would appear that the new owner can ask the original owner to sue for any defects that might be revealed after the structure has been sold, and because the new owner's losses could be greater than the original owner's losses then questions of foreseeability could be raised.

Typical of most construction contracts, the ICE Conditions of Contract 6th Edition contains the words 'Neither the Employer nor the Contractor shall assign the Contract or any part thereof or any benefit or interest therein without the prior written consent of the other party which consent shall not unreasonably be withheld'. In fact, it is benefits, not contracts themselves, which can be assigned, the benefit in the cases given above being the right of a future owner to sue a contractor for defective work.

7.10.4 Design and construction

A claimant seeking redress in the form of damages for injury or loss arising from construction works will look to the professional indemnity policy held by the consulting engineer, or, if he has been engaged for all or part of the design of the works, by the contractor. Under the normal terms of such a policy, payment will be made only in respect of a liability due to a negligent act, error or omission by the person, firm or his or its employees. Liability due to faults which do not constitute negligence are not covered, and so, in order to have access to this source of funds, the claimant must succeed in proving negligence on the part of the professional person, firm or his or its employees.

Unless a contract is let for both design and construction the two functions are subject to separate contracts let to different companies. Design of the permanent works is usually the responsibility of a specialist consultant firm. Design and implementation of the temporary works, and construction of the permanent works, are normally the responsibility of a contractor. However, the works contract may provide for part of the permanent works to be designed by the contractor. This eventuality is taken into account by the ICE Conditions of Contract 1991, Clause 7(6) and, more specifically, by the ICE Design and Construct Conditions of Contract 1992. Reference should be made to Chapter 2 of the book.

Clause 8(2) of the ICE Conditions of Contract 1991 requires the contractor to 'exercise all reasonable skill care and diligence in designing any part of the Permanent Works for which he is responsible'. Clause 8(3) states that 'The Contractor shall take full responsibility for the adequacy stability and safety of all site operations and methods of construction'. Under Clause 20(2)(b) the contractor is not liable for loss or damage to the extent that it is due to 'any fault defect error or omission in the design of the Works (other than a design provided by the Contractor pursuant to his obligations under the Contract)'. The design of the permanent works relates to those works during and after construction.

If a contractor's remit is for *construction only*, he then builds the works in accordance with the designs given to him. This is an *absolute duty in contract* and not one that depends on the exercise of *due skill and care*. The issue of *negligence* does not then normally apply between him and the employer. Claims of negligence could arise, however, in respect of a duty of care to *third parties*.

If the contractor engages in design as part of his operations his duty then will *normally* be to provide works that are reasonably *fit for the purpose* for which they are intended (but see the comments in Section 2.2). Such a requirement is *absolute* in the sense that the contractor is under strict liability to ensure that the works must be so, regardless of whether due skill and care were taken in their design and construction. This absolute requirement is really unsuited to construction work because the contractor must unreasonably commit himself to the specification in advance without being sure of suitable recompense for additional costs and delays that might arise. In addition, certain unfavourable conditions might arise, critical to the design, for which the contractor must carry responsibility and the responsibility for the consequences of not taking them into account.

Agreement between the employer and contractor can remove this fitness for purpose standard and replace it by a duty on the contractor to design the works using due skill and care and to construct in accordance with the design. The contractor would still be subject to claims in negligence from the employer for defects in the design (as would a consulting engineer) although his responsibilities might be greater because he exercises wider control of the work. Nonetheless the contractor might prefer this route as being more favourable to him than the fitness for purpose requirement, and in any case the cost of his insurance would most likely be less.

It follows that as contractors take on more design work threats of claims for negligence between consultant and employer will be transferred to the contractor. Even where a fitness for purpose standard is used, negligence can still apply in regard to third-party claims, and where the due-skill-and-care standard is used negligence can apply in regard to claims by the employer. Should a failure of the works occur and the client is unable to prove in contract either negligence or that the design does not achieve fitness for purpose, then no recompense will be forthcoming from the insurers and he must stand the loss himself. As a general point, proof of negligence or lack of fitness for purpose would generally be much easier without separation of the design and construction functions; that is, on a turnkey design and construct type of contract where the contractor takes on responsibility for both sides of the work.

Another variant on this theme arises on a design-and-construct contract when a contractor employs a consultant for the design element on work for which he, the contractor, is responsible. Where the contract has been let on a due-skill-and-care basis then it would be reasonable to expect the engagement between the contractor and consultant to be on that same basis. However, if the contract has been let on a fit for purpose basis, and the contractor has an absolute duty to the employer, then the question arises as to the type of duty that should apply between the contractor and the consultant. The contractor might well require that the duty to him should reflect his own duty to the employer, in which case the consultant would need to guarantee his design which might well have been developed on the basis of insubstantial site investigation information. The level of commercial risk then laid at the door of the consultant would be excessive and unlikely to be covered by his professional indemnity insurance. He should not accept such an engagement. He might, however, be prepared to undertake the design work using due skill and care to achieve the requisite standard, his liability then being that which would apply were he to be negligent. The contractor would then need to be prepared to carry the balance of responsibility (and risk) between that which he owed to the employer and that which the consultant owed to him. To offset this risk to some extent, the consultant might normally be required to work to a fit-for-purpose criterion, without the express contractual requirement to do so. How exactly this standard would be monitored and achieved is not entirely clear. Nor is it entirely clear how the courts would view this general position of consultant design to a skill and care standard where agreement between contractor and consultant is not entirely explicit on the matter. It might be considered that the consultant is under an implied warranty to design the works to be reasonably fit-for-purpose.

Culverwell (1989) has pointed out that the fit-for-purpose standard need not be associated with strict liability, and that although the two are linked in the Sale of Goods Act (1893) and the Consumer Protection Act (1987) the concepts are quite separate. For a condition of strict liability to apply it must necessarily be related to factual and ascertainable requirements, or it becomes unenforceable.

'When applied to manufacture or construction, or to their design, it has normally to be related to performance of the product or works and, where this has to be expressed in generalised form, the term fit-for-purpose is a convenient expression to use. Looked at in the reverse sense, however, there is no reason why the task of manufacturing or constructing something, or designing it, so that it meets some factual and ascertainable requirement, i.e. is fit-for-purpose, should be linked to strict liability. It could as readily be linked to the sanctions that apply to failure to take due skill and care and such a combination could well have been adopted instead. Such a combination .... represents the proper role of a consulting engineer in a design-and-construct contract, i.e. he has a duty to use due skill and care in preparing a design such that the works will be fit-for-purpose, but he does not guarantee that they will achieve this standard'.

The question of negligence is one that fully exercises the skills of the legal profession and the courts. If general support for this statement were even needed, it can readily be found in the three-to-two Law Lords majority judgment in a case involving negligence of a solicitor in failing to carry out instructions in preparing a new will and thereby being liable in damages to the intended beneficiaries (*White and Another* v *Jones and*

Another, Law Report, *The Times*, 17 February 1995). The case, in probing the borderlines between contractual liability and tortious liability and raising consideration of the *Hedley Byrne* principle, does have relevance to the subject of the present book and does merit study by civil engineers.

What may be regarded as negligence could be quite out of proportion to the magnitude of the fault or the degree of carelessness supposedly involved. Also, when a claim is made in negligence against a third party it is often the practice to join the consulting (design) engineer as a defendant even in circumstances where he may not obviously be at fault. (In the case where a consultant works for a contractor, as addressed a little earlier, the ordinary criteria of negligence in respect of the duty owed to third parties will apply; he will have a normal duty of care towards anyone who may use or be affected by the works.) In such circumstances, and under the joint tortfeasor rule of English law, the defendants would be jointly and severally liable and, should the other defendants be unable to pay, the single defendant would himself be required to bear the full amount of any damages awarded. This places the consultant (and contractors engaging in design work) at considerable risk, since small companies (perhaps subcontractors) may go out of business but he will tend to remain. Suppose that on the basis of circumstantial evidence the consultant was found, many years earlier, liable for, say, 5% of damages granted in a court case, he could end up having to pay the full 100% of those damages. There are two logical corollaries to these problems. In the face of liability for negligence there is a real danger that innovatory design will be stifled and a defensive attitude to construction will prevail. Second, in the USA there is an increasing tendency for some firms to 'go bare', that is, not to carry professional indemnity insurance. Without express contractual terms to counteract it, this tendency could conceivably spread in the UK.

In addition to a defendant carrying no indemnity insurance or insufficient insurance from the outset, compensation may not be realisable because what was adequate cover originally has been eroded by successful claims during the period of the contract. The insurer may then be moved to argue that the policy is invalid and refuse to pay, and will also look closely to see if there was any non-disclosure of material facts when the policy was taken out. Under those circumstances the claimant may be prepared to settle for less than the full award. Again, a claimant may, in appropriate circumstances, look to a local authority for recovery of damages if it is felt that the authority could be shown to have been negligent in the conduct of its duties under the Building Regulations 1991 and that safety and health issues might have been relevant.

7.10.5 Abbeystead case

The Abbeystead case has illustrated some of the problems associated with current British legal practice concerning insurance and compensatory redress for construction failure. Further reference may be made to the Great Britain Factory Inspectorate (1985) and Jefferis and Wood (1990).

In May 1984, 31 people from a party of visitors on an outing to the Abbeystead valve house, located adjacent to the River Wyre and below ground for environmental reasons, were killed and injured by a methane gas explosion. Water had been standing in the transfer tunnel connecting the Rivers Lune and Wyre in Lancashire, and the pump was switched on for demonstration purposes. A plug of methane gas, thought to have been released by a drop in air pressure from groundwater infiltrating the tunnel was pushed into the pump house by the water movement and ignited, probably by a cigarette. Compensation in tort for the victims could be forthcoming only by proof of negligence on the part of one or more of the bodies involved in the scheme - the owner (North West Water Authority), the consultant/designer (Binnie and Partners), and the contractor (Edmund Nuttall). Assignment of negligence depended on whether the occurrence of methane gas could have been foreseen, and this in turn depended upon reasonable proof of its genesis. The firm feeling in the civil engineering profession was and is that negligence was not proved, but the original judgment by Mr Justice Rose in Lancaster Castle (*T.E. Eckersley and Others* v *Binnie & Partners and Others* (1988) CILL 388) apportioned the blame for negligence between the three parties to the contract, with the greatest burden falling on the insurers of the consultant. At the Court of Appeal in early 1987, Lord Justices Russell and Fox decided that Binnie's liability had been properly established but that neither Nuttall nor the Water Authority was liable. Testing for methane by the contractor 'was for the benefit of the workforce' and 'they were never requested by the [consultants] to test for the safety of the permanent works'. The Water Authority was not liable because it 'had been lulled into a false sense of security by the consultants'. By concentrating on the specific risk of the particular form of methane leaking into the tunnel at the particular time from the particular place Lord Justice Bingham considered none of the defendants to have been negligent, but of course the majority decision prevailed. Leave to appeal to the House of Lords was refused. Binnie's insurers were left to pay all the claims and all the costs of the action. The costs for 31 victims and their relatives, who were represented by twelve lawyers, were of the order of £1 million at the time.

It is interesting to take note of Lord Justice Bingham's erudite statement on the duties owed by consultants and designers to their employers.

'...a professional man should command the corpus of knowledge which forms part of the professional equipment of the ordinary member of his profession. He should not lag behind other ordinarily assiduous members of his profession in knowledge of new advances, discoveries and developments in his field. He should have such awareness as an ordinarily competent practitioner would have of the deficiencies in his knowledge and the limitations in his skill. He should be alert to the hazards and risks inherent in the professional task he undertakes to the extent that other ordinarily competent members of his profession would be alert. He must bring to any professional task he undertakes no less expertise, skill and care than other ordinarily competent members of his profession would bring, but need bring no more. The standard is that of the reasonable average. The law does not require of a professional man that he be a paragon, combining the qualities of polymath and prophet.

In deciding whether a professional man has fallen short of the standards observed by ordinarily skilled and competent members of his profession, it is the standard prevailing at the time of the acts or omissions which provides the relevant yardstick. He is not to be judged by the wisdom of hindsight.'

There was then a subsequent statement of claim served against Binnie by the Water Authority alleging not only that Binnie was negligent and/or in breach of its contract of engagement, but also that the firm could not deny liability for damage to the Water Authority's property because it had been found negligent under the earlier court action (*North West Water Authority* v *Binnie & Partners* QBD (unreported December 1989)). It was also alleged by the authority that the attempt by Binnie to defend itself on the grounds that this new action raised questions which were entirely different from those which prevailed in the main case was 'frivolous, vexatious and an abuse of the process of the court'. Mr Justice Drake held Binnie liable to the North West Water Authority for damage to the valve house. He said

'So I gave notice to [counsel for Binnie & Partners] at an early stage that I would like him to tell me what practical as opposed to theoretical differences there would be in the two sets of litigation. He pointed out that the issue between the [personal injury] plaintiffs and Binnie was in tort, whereas the action between Binnie and the Water Authority lies in contract.....But I asked to be shown the contract and to have pointed out to me any way in which the contract in practice would modify or give rise to different issues from liability in tort. I was given no satisfactory answer.'

The point that arises forcefully from Lord Justice Bingham's comments is that what constitutes reasonable care in the conduct of professional duties is not to be judged in the light of the standard that a reasonable person would expect from such a professional but rather the standard that other members of the profession would consider to be appropriate. It is therefore up to a plaintiff to establish that what a defendant has or has not done falls below this standard. Establishment of what is a 'common professional practice' does present difficulties. One answer is that a defendant would not be negligent if in his professional practice he had followed a recognised school of thought. Adherence to established codes of practice would be a defence, but this does lead to the charge that by so doing innovation can be stifled. Codes of practice such as the British Standards have no force of law and they can lag behind innovation. If an engineer does decide to go beyond a code or codes of practice then he should carefully record his reasons for so doing. The courts may then apply a test by asking what level of skill did that professional engineer actually deem himself to possess, and he should then be judged against that particular standard as to whether 'common, professional practice' has, by extrapolation, been attained.

Another point that comes across from Lord Justice Bingham's comments is that there is an obligation on the professional to keep abreast of developments in his profession. In the case of *Crawford* v *Charing Cross Hospital* (1953), reported in *The Times*, 8 December 1953, the Court of Appeal held that a professional does not have to read every article published in his professional literature and furthermore he does not have to adopt particular techniques until they have become accepted practice in the discipline appropriate to his profession.

The considered opinion of the industry seemed to be that there were 'socio-political' elements to the original main trial Abbeystead judgment. The dependants of victims and those that had been injured *had* to be compensated and the source of that compensation could only realistically be found from professional indemnity insurance by attribution of negligence.

With respect to the second judgment against Binnie, it is noted that the same issues, once decided by a court, cannot be litigated twice. However, it is not always clear what are 'the same issues'. Binnie claimed that there were three main distinctions between the issues in the two cases. First, in the original case, the victims of the explosion sued in tort for personal injury, whereas the Water Authority was suing in contract for property damage. Second, although there had been no question of Binnie arguing a case of contributory negligence on the part of the victims, such a case would be pertinent in respect of the Water Authority since the Authority carried technically qualified staff. Third, there

was claimed to be geophysical evidence (*see* the note that is attached to the end of this discussion) supporting the company's case, the evidence not having been allowed to be presented at the original court hearing because Mr Justice Rose felt that at a late stage in the proceedings it would have been unfair to the plaintiffs. Binnie argued that there should be no such constraints, since because the report was of obvious importance in relation to the question of alleged negligent design and supervision in respect of the Water Authority the consultant to the Authority should be allowed to submit it in evidence.

Notwithstanding these arguments, the trial judge, Mr Justice Drake, stated that

'In my judgment the proceedings between these two parties have reached the stage where it can emphatically be said that it is in the public interest that there should be a finish to this litigation'.

While excessive litigation and its associated cost are clearly not in the public interest, justice both done and seen to be done most certainly is, and, in cases where there is conflict between the two, it does seem entirely reasonable to require that justice should always prevail.

7.10.6 Abbeystead post-litigation note

The tunnel became operational in 1977, seven years before the disaster at Abbeystead. The region through which the tunnel was driven is underlain by a succession of gently folded and locally faulted Namurian sandstones and shales. It was deduced from investigations that the methane gas was mainly of geological origin and was entering the tunnel through cracks in certain parts of the lining at a rate of about $40m^3$ per day. No methane gas had been identified during construction of the tunnel.

The geophysical evidence referred to above took the form of a seismic refraction survey which identified a regional structure, particularly a small reef-like structure at the top of the underlying Dinantian, 1 kilometre below the tunnel. The Dinantian rocks in the area take the form of an alternating marine limestone-shale sequence. They thus provide a potential source rock and potential reservoir conditions in the reef, but it has been estimated that the reef could not, from evidence of its size, provide a sufficient methane outflow. The argument would be that the tunnel construction caused dewatering of the overlying rocks, so reducing water pressures sufficiently for the methane in solution to out-gas and bubble upwards. (The later BS 6164: 1990 (British Standards Institution, 1990a) formally recognises the point that methane gas can enter excavations and tunnel works, dissolved in water, as well as from the decay of organic deposits, and stresses the point that gas monitoring equipment should always be available on site.) A small proportion of the methane was of biogenic origin, and could be explained by the mixing of deep-seated water containing methane with a near-surface groundwater draining into a depressed water-take along the tunnel line. The conclusion that the events which led to the methane inflows to the tunnel could therefore have been triggered by the actual construction work can be said to have important implications not only for the design and construction of future tunnels but also in particular for the planning of site investigations.

8

POST-CONSTRUCTION PHASE

8.1   TECHNICAL AUDIT

A formal post-contract audit should be carried out by or under the direction of the Engineer in order to assess the relevance and accuracy of the original site investigation information with respect to the ground conditions actually revealed at the tunnel face and to highlight any conditions, omissions and procedures within the contract that were unhelpful to successful completion of the construction.   These audits draw on the careful recording of all relevant information derived during the course of the tunnelling contract   for   comparison   with   the   site investigation and other data. A site investigation audit should at least include notes on such matters as those given below.

• The results of any in-tunnel investigations undertaken   to   support   design   changes   for improving the ground conditions, and so increase progress, or for providing evidence in the case of, or in anticipation of, contractual claims. Typically these operations would include
   * Retrieval of soil and/or rock and/or water samples for laboratory testing, or in some cases testing on site.
   * Performing other *in situ* strength tests, the Schmidt hammer on rock, for example.
   * Logging of discontinuity features such as dips and dip directions in rock.
   * Recording the locations of strata changes, faults, igneous intrusions, and so on.
   * Measuring discrete inflows of water.

• The incidence of any probing ahead of the tunnel face for the definition of the ground and the incidence of water and water pressures.

• Graphs of time-related tunnel progress against a longitudinal geological section of the ground actually tunnelled.

• A record of any ground movement measurements that were taken in relation to the position of the tunnel face and any incidences of damage to buildings or other structures, above and below ground.

A full set of site investigation reports, plans, and sections should be kept for reference purposes together with a set of the original contract documents and the site records which would include   any   correspondence   relating   to contractual claims.   Although there will be a natural reluctance to commit post-contract resources to this work, such reluctance should be resisted by the employer and the audit completed within a pre-specified time frame.

It is also of obvious benefit to the industry that the results of such 'post-mortems' be published, again, however reluctantly and in whatever form, in order to reveal what went wrong, what and how mistakes were made, the manner in which those mistakes were rectified, or at least overcome, and what lessons were learned for the future.

8.2   DATA BANK

Whenever   large   tunnelling   projects   are undertaken, recent examples being the major sewerage schemes already completed and those under   construction   throughout   the   United Kingdom in response to the EU Bathing Waters Directive, a detailed data bank should ideally be assembled of the soil and rock properties identified in both the ground investigation and tunnel construction phases of the work.   This data bank would provide valuable information for future schemes of work undertaken in those areas. Digitisation of factual ground data and its entry on to disk 'for computer processing and transfer from one organisation to another' in order to 'facilitate assessment by geotechnical specialists both for the designer of the works and tenderers' is suggested by the Ground Board of the Institution of Civil Engineers (Institution of Civil Engineers, 1991c, p17).   Moves in this direction are under way under the auspices of the

UK Association of Ground Investigation Specialists, and as a further development in this direction the acceptance of knowledge based ('expert') systems in site investigation seems to be acquiring some momentum (*refer*, for example, to Toll, 1994).

As an alternative strategy, a body such as the British Geological Survey (BGS) at Keyworth near Nottingham could be further encouraged to collate and store this type of information in the form of a computer data base for rapid retrieval in the interests of the industry. A problem could arise in this respect because under most contract conditions the client retains ownership of the information for which he has paid (*refer to Section 4.9.3*). Early release of the information for it then to become freely available could put competitors at an advantage in, say, a situation where there is competitive bidding for land for the purposes of development or mineral extraction. The BGS is seen by some in the ground investigation industry as a competitor, and this complicates the position. On the other hand, the situation that has tended to exist in the past of the BGS providing factual information without interpretation, recommendation, or indeed payment, would be placed on a formal industrial basis. This has already happened with the initiatives pursued by the BGS - for example, a sub-surface data base for Central London (LOCUS) and the provision of more up-to-date geological maps, including digital maps. As a further constraint, however, there could be a specified delay between the provision of borehole information to the BGS and the time of its availability for release to the general public. Even at times of relative buoyancy in the tunnelling industry there would almost certainly be inertia and an unwillingness to provide the set-up support funding necessary for such a scheme to flourish, even with realistic payment for services, and so such a proposal is never likely to come to widespread fruition.

8.3   POST-CONSTRUCTION RESPONSIBILITIES OF CONSULTANTS AND CONTRACTORS

The trial of the Abbeystead case highlighted a continuing duty of consultants and contractors to warn former employers of advances in knowledge that could affect the proper functioning of the works. At the trial, the judge, referring to the responsibility of Binnie and Partners for the design of the valve house that exploded as a result of methane ignition in the system, issued a statement which implied that professionals offering a service to clients have a duty to keep abreast of technological and other developments, and report back to their former clients irrespective of the time that has elapsed since project completion. The judge said

> 'They (the Consultants) were to some degree negligent in not keeping abreast with, passing on to the Third Defendants (North West Water Authority, who owned the plant) and considering, in relation to design, developing knowledge about methane between handover and 1984'.

This same duty to inform would then seem to apply to a contractor actually constructing civils works, and it would certainly apply to a contractor who has engaged in any aspect of engineering design for those works. At the Appeal Court it was said that the statement of this trial judge, Mr Justice Rose, would be a novel and burdensome obligation for engineers, and that rather than the Abbeystead case being used to decide this continuing duty of care/duty to warn issue, the courts in future cases should determine the scope and limits of such duty on a case-by-case basis.

In a more general duty to warn setting, a situation may arise during the currency of a contract when a contractor is constructing the works strictly according to the design but realises that deficiencies in the design will lead to a sub-standard product. In the case of *Equitable Debenture Assets Corporation Limited* v *William Moss and Others* in 1984 the trial judge concluded that it was an implied term of the main contract (construction of a new office block in Ashford, Kent) that the main contractor would warn the architect or employer of design defects as soon as he discovered them. He also found the contractor to be liable in tort to the employer.

This requirement then of course raises the question of contractor skill and competency, specifically the ability of a contractor to actually detect defects in a design and thereby to warn the employer. Although there is an *a priori* assumption of contractor competence (for example, by being placed on a select list of tenderers for the civils work) there is still a relation between competence for the particular work and the design complexity of the work. The contractor may be competent to complete the civils work, but his overall expertise may be insufficient for him to identify shortcomings in the design for the particular work in hand. There might then be some difficulty in imposing liability for any problems that might arise because it would then be argued that the contractor could not reasonably have been aware of the particular defect, never having acquired the knowledge necessary to give the warning, and would therefore never be in breach of his duty to warn. As noted by Winter (1993), this really means that a more competent contractor would be treated less favourably than a less competent contractor.

REFERENCES

Abrahamson, M.W. (1979) *Engineering Law and the I.C.E. Contracts*, 4th Edition, Applied Science Publishers, London.

Ackers, G.L. (1989) Clause 12: the employer pays, *New Civil Engineer*, 7 December 1989, 49.

American Society of Civil Engineers (1991) *Avoiding and Resolving Disputes during Construction: Successful Practices and Guidelines*, Technical Committee on Contracting Practices of the Underground Technology Research Council, Sponsored by the American Society of Civil Engineers and the American Institute of Mining, Metallurgical and Petroleum Engineers (An updated and revised edition of *Avoiding and Resolving Disputes in Underground Construction* (1989)), 7, 82pp. (ISBN 0 87262 833 7.)

Amos, E.M., Blakeway, D. and Warren, C.D. (1986) Remote sensing techniques in civil engineering surveys. In: *Site Investigation Practice: Assessing BS 5930, Geol. Soc. Eng. Geol. Special Publication No. 2*, A.B. Hawkins (ed), 119-124.

Anon (1970) The logging of rock cores for engineering purposes, Engineering Group Working Party Report, *Quart. J. Eng. Geol.*, 3, 1-24.

Anon (1981) Basic geotechnical description of rock masses (ISRM Commission on Classification of Rocks and Rock Masses). *Int. J. Rock Mech. Min. Sci & Geomech. Abstr.*, 18, 85-110.

Anon (1987) *Guidelines for the Provision of Geotechnical Information in Construction Contracts*, Report prepared by a Construction Industry Committee convened by the Institution of Engineers, Australia, Published by the Institution of Engineers, Australia, National Headquarters, Canberra.

Anon (1990a) Shallow reflection seismic - where are we now? *Minerals Industry International (Bull. IMM)*, Number 992, 21 January 1990.

Anon (1990b) Fast track on trial in Worship St case, *New Civil Engineer*, 11 October 1990, 9.

Armstrong, W.E.I. (1991) *Contractual Claims under the 6th Edition of the ICE Conditions of Contract*, Studies in Contractual Claims 13, The Chartered Institute of Building, Ascot, Berkshire, England, 66pp. (ISBN 1 85380 036 8.)

Athorn, M.L. (1982) Geotechnical factors affecting selection of tunnelling systems, MSc thesis, University of Newcastle upon Tyne, England.

Atkin, R. (1991) Mechanised curved pipe jacking by giro compass, *Proc. 1st Int. Conf. Pipe Jacking and Microtunnelling, 23-24 October, 1991*, London, Pipe Jacking Association, 7.1-7.11.

Atkinson, D. (1992) Tort a lesson, *New Civil Engineer*, 31 December 1992/7 January 1993, 17.

Atkinson, D. (1995) Two cheers for ICE design + construct contract, *Tunnels and Tunnelling*, Jan. 1995, 23-26.

Atkinson, J.H. and Mair, R.J. (1981) Soil mechanics aspects of soft ground tunnelling, *Ground Engrg*, 14(5), 20-24, 26, 38.

Attewell, P.B. (1978) Ground movements caused by tunnelling in soil, *Proc. Conf. Large Ground Movements and Structures*, Cardiff, Pentech Press, 812-948.

Attewell, P.B. (1986) Noise and vibration in civil engineering, *Mun. Eng.*, 3, 139-158.

Attewell, P.B. (1988a) Aspects of risk assessment in engineering geology and geotechnics, *Proc. 1st Sino-British Geological Conf. on Geotechnical Engineering and Hazard Assessment in Neotectonic Terraines, Taiwan, April, 1987, Memoir of the Geological Soc. of China, No. 9*, 335-365.

Attewell, P.B. (1988b) An overview of site investigation and long-term tunnelling-induced settlement in soil. In: *Engineering Geology of Underground Movements (Proc. 23rd Annual Conf. of the Engineering Group of the Geological Society, Nottingham, Sept. 1987, Lead Lecture Session 2: Ground Movements due to Tunnelling)*, Geol. Soc. Eng. Group Special Publication No. 5, F.G. Bell, M.G.

Culshaw, J.C. Cripps and M.A. Lovell (eds), 55-61.

Attewell, P.B. (1993a) *Ground Pollution*, E. & F.N. Spon, London, 251pp. (ISBN 0 419 18320 5.)

Attewell, P.B. (1993b) The role of engineering geology in the design of surface and underground structures. In: *Comprehensive Rock Engineering, Principles, Practice and Projects*, J.A. Hudson, E.T. Brown, C. Fairhurst and E. Hoek (eds), 1(5), 111-154, Pergamon, London.

Attewell, P.B. and Boden, J.B. (1971) Development of stability ratios for tunnels driven in clay, *Tunnels and Tunnelling*, **3**, 195-198.

Attewell, P.B. and Farmer, I.W. (1976) *Principles of Engineering Geology*, Chapman and Hall, London, 1045pp. (ISBN 0 412 11400 3.)

Attewell, P.B. and Norgrove, W.B. (1984a) A flow chart guide to site investigations for tunnelling, *Mun. Eng.*, 1(4), 297-302.

Attewell, P.B. and Norgrove, W.B. (1984b) *Survey of United Kingdom Tunnel Contract and Site Investigation Costs*, Construction Industry Research and Information Association, RP 324, 154pp.

Attewell, P.B. and Woodman, J.P. (1982) Predicting the dynamics of ground settlement and its derivatives caused by tunnelling in soil, *Ground Engineering*, **15**(8), 13-22, 36.

Attewell, P.B. and Yeates, J.B. (1984) Tunnelling in soil. In: *Ground Movements and their Effects on Structures*, P.B. Attewell and R.K. Taylor (eds), Surrey University Press, London, 132-215.

Attewell, P.B., Yeates, J.B. and Selby, A.R. (1986) *Soil Movements Induced by Tunnelling and their Effects on Pipelines and Structures*, Blackie, Glasgow, 325pp.

Aydan, Ö, Akagi, T. and Kawamoto, T. (1993) The squeezing potential of rocks around tunnels: theory and prediction, *Rock Mech. Rock Engrg*, **26**(2), 137-163.

Baguelin, E., Jezequel, J.F. and Shields, D.H. (1978) *The Pressuremeter and Foundation Engineering*, Transtech Publications, Switzerland.

Ball, T.G., Beswick, A.J. and Scarrow, J.A. (1993) Geotechnical investigations for a deep radioactive waste repository: drilling. In: The Engineering Geology of Waste Storage and Disposal, *Proc. 29th Annual Conf. of the Engineering Group of the Geological Society of London*, 6-9 September 1993, S. P. Bentley (ed), Cardiff, 194-203.

Barber, J.M. (1989) Clause for thought, *New Civil Engineer*, 3 August, 1989, 34.

Barnes, M. (1992a) *CESMM3 Handbook*, Thomas Telford Ltd, London, 256pp. (ISBN 0 7277 1658 1.)

Barnes, M. (1992b) *CESMM3 Examples*, Thomas Telford Ltd, London, 116pp. (ISBN 07277 1657 3.)

Barr, M.V. and Hocking, G. (1976) Borehole structural logging employing a pneumatically inflatable impression packer, *Proc. Symp. Exploration for Rock Engrg*, Johannesburg, Z.T. Bieniawski (ed), A.A. Balkema, Rotterdam, 1, 29-34.

Barton, N. (1976) Recent experiences with the Q-system of tunnel support design. In: *Exploration for Rock Engineering*, Z.T. Bieniawski (ed), A.A. Balkema, Johannesburg, 107-115.

Barton, N. (1983) Application of Q-system and index tests to estimate shear strength and deformability of rock masses, *Proc. Int. Symp. Engrg Geol. Underground Constr.*, A.A. Balkema, Boston, 51-70.

Barton, N. (1988) Rock mass classification and tunnel reinforcement selection using the Q-system, *Proc. Symp.Rock Class. Engrg Purp.*, ASTM Special Technical Publication 984, Philadelphia, 59-88.

Barton, N., Lien, R. and Lunde, J. (1974) Engineering classification of rock masses for the design of tunnel support, *Rock Mech.*, **6**, 183-236.

Barton, N., Loset, F., Lien, R. and Lunde, J. (1980) Application of the Q-system in design decisions. In: *Subsurface Space*, M. Bergman (ed), Pergamon, New York, 553-561.

Baver, L.D. (1956) *Soil Physics*, 3rd Edition, John Wiley and Sons, New York.

Bellotti, R., Ghionna, V., Jamilokowski, M., Robertson, P. and Peterson, R. (1989) Interpretation of moduli from self-boring pressuremeter tests in sand, *Géotechnique*, **39**(2), 269-292.

Berry, N.S.M. (1980) Contribution to the British Tunnelling Society Meeting on 'Ground

Classification: Continental and British Practice', *Tunnels and Tunnelling*, **12**(6), 59-62.

Bieniawski, Z.T. (1973) Engineering classification of jointed rock masses, *Trans. S. Afr. Inst. Civ. Eng.*, **15**, 335-344.

Bieniawski, Z.T. (1974) Geomechanics classification of rock masses and its application in tunnelling, *Proc. 3rd Int. Congress on Rock Mechanics, Int. Soc. for Rock Mechanics*, Denver, Colorado, USA, **IIA**, 27-32.

Bieniawski, Z.T. (1975) The point load test in geotechnical practice, *Eng. Geol.*, **9**, 1-11.

Bieniawski, Z.T. (1976) Rock mass classifications in rock engineering, *Exploration for Rock Engineering*, A.A. Balkema, Johannesburg, 97-106.

Bieniawski, Z.T. (1978) Determining rock mass deformability: experience from case histories, *Int. J. Rock. Mech. Min. Sci.*, **15**, 237-247.

Bieniawski, Z.T. (1979a) *Tunnel Design by Rock Mass Classifications*, Technical Report GL-79-19 for the Office, Chief of Engineers, US Army, Washington DC 20314, USA under Purchase Order No. DACW39-78-M-3114.

Bieniawski, Z.T. (1979b) The geomechanics classification in rock engineering application, *Proc. 4th Int. Congress on Rock Mechanics*, Montreux, Switzerland, **1**, 41-48.

Bieniawski, Z.T. (1983) The geomechanics classification (RMR system) in design applications, *Proc. Int. Symp. Engrg Geol.Underground Constr.* LNEC, Lisbon, **2**, II.33-II.47.

Bieniawski, Z.T. (1984) *Rock Mechanics Design in Mining and Tunneling*, A.A. Balkema, Rotterdam, 272pp.

Bieniawski, Z.T. (1988) Rock mass classification as a design aid in tunnelling, *Tunnels and Tunnelling*, **20**(7) (July) 19-22.

Bieniawski, Z.T. (1989) *Engineering Rock Mass Classification*, Wiley-Interscience, New York, 251pp.

Boden, J.B. (1969) Site Investigation and Subsequent Analysis for Shallow Tunnels, Dissertation: MSc Advanced Course in Engineering Geology, University of Durham, England.

Boden, J.B. (1983) There is a clear need for an accurate and job-specific site investigation for all underground works. In: *Cost Benefit of Site Investigation for Tunnelling*, Joint Meeting of the Institution of Municipal Engineers, the Institution of Civil Engineers, and the Royal Institution of Chartered Surveyors, Newcastle upon Tyne Polytechnic, 12 April 1983.

Boyd, J.L. and Stacey, J.R. (1979) Wyresdale tunnel: design, the contract and construction, *Proc. Tunnelling '79 Symp.*, IMM, London.

Bradshaw, J. (1994) ADR: an amicable settlement, *Tunnels and Tunnelling*, Oct. 1994, 35-37.

Braybrooke, J.C. (1988) The state of the art of rock cuttability and rippability prediction, *Proc. 5th ANZ Geomechanics Conf.*, 22-26 August, 1988, Sydney, Australia, 13-42.

Brierley, G. and Cavan, B. (1987) The risks associated with tunneling projects, *Tunneling Technology Newsletter*, No. 58 (June 1987), 1-9; *see also* Brierley, G. (1998) Discussion on 'The risks associated with tunneling projects', *Tunneling Technology Newsletter*, No. 64 (December 1988), 1-5.

British Drilling Association (1992) *Guidance Notes for the Safe Drilling of Landfills and Contaminated Land*, British Drilling Association (Operations) Ltd, Essex, 28pp.

British Standards Institution (1974) *Core Drilling Equipment, Part 1: Basic Equipment*, BS 4019:1091, HMSO, London.

British Standards Institution (1975) *Methods of Sampling and Testing of Mineral Aggregates, Sands and Fillers*, BS 812:1975, HMSO, London.

British Standards Institution (1979) *Quality Systems*, BS 5750:1979 (*see also* BS 5750:1981 *Quality Systems, Part 6: Guide to the Use of BS 5750:Part 3 Specification for Final Inspection and Test*), HMSO, London.

British Standards Institution (1981) *Code of Practice for Site Investigations*, BS 5930:1981, HMSO, London, 149pp. (This Standard is in the process of being updated by BSI Sub-Committee B/526/1.)

British Standards Institution (1984) *British Standard Guide to Evaluation of Human Exposure to Vibration in Buildings (1Hz to 80Hz)*, BS 6472:1984, HMSO, London, 12pp.

British Standards Institution (1986) *British Standard Code of Practice for Foundations*, BS 8004:1986, HMSO, London, 149pp.

British Standards Institution (1988) *Code of Practice for the Identification of Provisionally*

Contaminated Land and its Investigation, Draft for Development DD 175, HMSO, London, March 1988.

British Standards Institution (1989) *Precast Concrete Pipes, Fittings and Ancillary Products - Part 120: Specification for Reinforced Jacking Pipes with Flexible Joints*, BS 5911: 1989, HMSO, London.

British Standards Institution (1990a) *Code of Practice for Safety in Tunnelling in the Construction Industry*, BS 6164:1990, HMSO, London, 96pp.

British Standards Institution (1990b) *Methods of Test for Soils for Civil Engineering Purposes*, BS 1377:1990, HMSO, London.
Part 1: *General Requirements and Sample Preparation.*
Part 2: *Classification Tests.*
Part 3: *Chemical and Electro-chemical Tests.*
Part 4: *Compaction Related Tests.*
Part 5: *Compressibility, Permeability and Durability Tests.*
Part 6: *Consolidation and Permeability Tests in Hydraulic Cells and with Pore Pressure Measurement.*
Part 7: *Shear Strength Tests (Total Stress).*
Part 8: *Shear Strength Tests (Effective Stress).*
Part 9: *In Situ Tests.*

British Standards Institution (1992a) *Noise Control on Construction and Open Sites: Part 4. Code of Practice for Noise and Vibration Control Applicable to Piling Operations*, BS 5228: 1992, HMSO, London 64pp. (ISBN 0 580 20381 6.)

British Standards Institution (1992b) *Guide to Evaluation of Human Exposure to Vibration in Buildings (1Hz to 80Hz)*, BS 6472:1992, HMSO, London, 18pp. (ISBN 0 580 19963 0.)

British Standards Institution (1992c) *Total Quality Management*, BS 7850: 1992, HMSO, London. *Part 1: Guide to Management Principles,* 9pp (ISBN 0 580 21156 8); *Part 2: Guide to Quality Improvement Methods*, 21pp. (ISBN 0 580 211576.)

British Standards Institution (1994a) *Quality Assessment Schedule to BS 5750: Part 1: ISO 9001-1994/EN 29001-1994 Relating to Ground investigation, Foundation design and Foundation construction*, HMSO, London.

British Standards Institution (1994b) *Specification for Environmental Management Systems*, BS 7750:1994, BSI, 2 Park Street, London W1A 2BS, 21pp. (ISBN 0 580 22829 0).

Broch, E. and Franklin, J.A. (1972) The point load strength test, *Int. J.Rock Mech. and Mining Sci.*, **9**, 662-697.

Broms, B.B. and Bennermark, H. (1967) Stability of clay at vertical openings, *Proc. A.S.C.E., Soil Mech. Found. Div.*, SM1, 71-94.

Brown, E.T. (1981a) Putting the NATM into perspective, *Tunnels and Tunnelling*, November, 1981, 13.

Brown, E.T. (ed) (1981b) *Rock Characterisation, Testing and Monitoring, ISRM Suggested Methods*, Pergamon Press, Oxford, 211pp.

Building Research Establishment (1981) *Concrete in Sulphate-bearing Soils and Groundwaters*, Digest Number 250, Garston, Watford WD2 7JR.

Building Research Establishment (1991) *Sulphate and Acid Resistance of Concrete in the Ground*, Digest Number 363, Garston, Watford WD2 7JR, England.

Building Research Establishment (1993) *Site Investigation for Low-rise Buildings: Soil Description*, Digest 383, Garston, Watford WD2 7JR, England.

Building Research Station (1975) *Protection of Buried Concrete against Attack by Acid Solutions*, Digest Number 174, London.

Bunni, N. G. (1991) *The FIDIC Form of Contract: The Fourth Edition of the Red Book*, Blackwell Scientific Publications, Oxford, 464pp. (ISBN 0 632 02514X.)

Burland, J.B. and Burbridge, M.C. (1985) Settlement of foundation on sand and gravel, *Proc. Inst. Civ. Eng.*, Part 1, **78**, 1325-1381.

Caldwell, R.A. (1986) Recent techniques in geophysics with special applications to engineering geology. In: *Site Investigation Practice: Assessing BS 5930, Geol. Soc. Eng. Group Special Publication No. 2*, A.B. Hawkins (ed), 157-162.

Cargill, J.S. and Shakoor, A. (1990) Evaluation of empirical methods for measuring the uniaxial compressive strength of rock, *Int. J Rock Mech. Min. Sci. & Geomech. Abstr.*, **27**(6), 495-503.

Chen, J.F. and Vogler, U.W. (1992) Rock cuttability/boreability, assessment of research at CSIR, *Proc. TUNCON '92, Design and Construction of Tunnels*, Maseru, South

African National Council on Tunnelling, 91-97.

Christiansson, R., Scarrow, J.A., Whittlestone, A.P. and Wilkman, A. (1993) Geotechnical investigations for a deep radioactive waste repository: in situ stress measurements. In: The Engineering Geology of Waste Storage and Disposal, *Proc. 29th Annual Conf. of the Engineering Group of the Geological Society of London*, 6-9 September 1993, S. P. Bentley (ed), Cardiff, 220-229.

Clarke, B.G., Newman, R.L. and Allan, P.G. (1989) Experience with a new high pressure self-boring pressuremeter in weak rock, *Ground Engrg*, **22**(5), 36-39; **22**(6), 45-51.

Clarke, T.M. (1993) A matter of contracts, *New Civil Engineer*, 25 November 1993, 11.

Clayton, C.R.I. (1986) Sample disturbance and BS 5930. In: *Site Investigation Practice: Assessing BS 5930, Geol. Soc. Eng. Geol. Special Publication No. 2*, A.B. Hawkins (ed), 33-40.

Coats, D.J., Carter, P.G. and Smith, I.M. (1977) Inclined drilling for the Kielder tunnels, *Quart. Jl of Eng. Geol.*, **10**, 195-205.

Commission of the European Communities (1989) *Draft Eurocode No. 7, Geotechnics, Design*, Nov. 1989.

Construction Industry Research and Information Association (1978) *Tunnelling - Improved Contract Practices*, CIRIA Report 79, London.

Construction Industry Research and Information Association (1982) *A Medical Code of Practice for Work in Compressed Air*, CIRIA Report 44, 3rd Edition, London.

Cooke, J.C. (1975) Radar transparencies of mine and tunnel rocks, *Geophysics*, **40**, 865-885.

Cording, E.J., Hansmire, W.H., MacPherson, H.H., Lenzini, P.A. and Vonderohe, A.D. (1976) *Displacements around Tunnels in Soil*, Final Report by the University of Illinois on Contract No. DOT FR 30022 to the Office of the Secretary and Federal Railroad Administration, Department of Transportation, Washington DC 20590, USA.

Cornes, D.L. (1994) *Design Liability in the Construction Industry*, Fourth Edition, Blackwell Scientific Publications, Oxford, 300pp. (ISBN 0 632 03261 8.)

Cottam, G. (1989) Clause 12 - not so much a nightmare, more a haze, *New Civil Engineer*, 17 August, 1989, 15.

Cottington, J. and Akenhead, R. (1984) *Site Investigation and the Law*, Thomas Telford Ltd, London, 184pp.

County Surveyors' Society (1987) *Coping with Landfill Gas*, County Surveyors' Society Committee Number 4 Report Number 4/4, 27pp.

Cox, D.W., Dawson, A.R. and Hall, J.W. (1986) Techniques for site investigation using trial pits. In: *Site Investigation Practice: Assessing BS 5930, Geol. Soc. Eng. Geol. Special Publication No. 2*, A.B. Hawkins (ed), 185-192.

Craig, R.N. (1983) *Pipe Jacking: a State-of-the-Art Review*, CIRIA Technical Note 112, London, 62pp. (ISBN 0 86017 205 8.)

Culverwell, D.R. (1989) *Professional Liability*, The Institution of Civil Engineers, Thomas Telford Ltd, London, 54pp.

Czurda, K.A. (1983) Freezing effects on soils: comprehensive summary of the ISGF 82, *Proc. Conf. Cold Regions Science and Technology*, **8**, 93-107.

Darling, P. (1994) Jacking under greenfield sites, *Tunnels and Tunnelling*, July 1994, 33-36.

Darracott, B.W. and McCann, D.M. (1986) Planning engineering geophysical surveys. In: *Site Investigation Practice: Assessing BS 5930, Geol. Soc. Eng. Geol. Special Publication No. 2*, A.B. Hawkins (ed), 119-124.

Davies, V.J. and Tomasin, K. (1990) *Construction Safety Handbook*, Thomas Telford Ltd, London, 264pp. (ISBN 0 7277 1385 X.)

Dawson, O. (1963) Compressed air and its applications. In: *Tunnels and Tunnelling*, C.A. Pequignot (ed), Hutchinson Scientific and Technical, London, 158-185.

Deaves, A.P. and Cripps, J.C. (1990) Engineering geological investigations for the Don Valley intercepting sewer (Sheffield/England), *Proc. 6th Int. Assoc. Engrg Geol. Congress*, Amsterdam, **4**, 2637-2644, A.A. Balkema, Rotterdam.

Deere, D.U. (1964) Technical description of cores for engineering purposes, *Rock Mech. Engrg Geol.*, **1**, 18-22.

Deere, D.U., Merritt, A.H. and Coon, R.F. (1969) *Engineering Classification of In-Situ Rock*, Technical Report No. AFWL-TR-67-144 under Contract AF 29(601)-6850 to the Air Force Weapons Laboratory, Air Force

Systems Command, Kirtland Air Force Base, New Mexico, USA, 272pp.

Department of the Environment Property Holdings (1990) *General Conditions of Contract for Building and Civil Engineering Works,* GC/Works/1 (Edition 3), December 1989, Revised 1990, HMSO, London.

Department of Transport (1987) *Specification and Method of Measurement for Ground Investigation,* HMSO, London.

Dumbleton, M.J. and West, G. (1974) *Guidance on Planning, Directing and Reporting Site Investigations,* TRRL Laboratory Report LR625, Crowthorne, Berkshire, England.

Dumbleton, M.J. and West, G. (1976) *A Guide to Site Investigation Procedures for Tunnels,* TRRL Laboratory Report LR740, Crowthorne, Berkshire, England.

Eggleston, B. (1993) *The ICE Conditions Sixth Edition: A User's Guide,* Blackwell Scientific Publications, Oxford, 304pp. (ISBN 0 632 03092 5.)

Eggleston, B. (1994a) *The ICE Design and Construct Contract: A Commentary,* Blackwell Scientific Publications, Oxford, 304pp. (ISBN 0 632 03697 4.)

Eggleston, B. (1994b) *Civil Engineering Contracts,* Blackwell Scientific Publications, Oxford, 464pp. (ISBN 0 632 03483 1.)

Einstein, H.H. and Schwartz, A.M. (1979) Simplified analysis for tunnel supports, *J. Geotech. Eng. Div. ASCE,* 105(GT4), April 1979, 499-518.

Eisenbud, M. (1987) *Environmental Radioactivity,* 3rd Edition, Academic Press Inc., New York.

Engelbreth, K. (1961) Discussion in Morgan, H.D. 'A contribution to the analysis of stress in a circular tunnel', *Géotechnique,* 11, 246-248.

Environmental Protection Act 1990, HMSO, London. (ISBN 0 10 544390 5.)

Evans, R.D., Harley, J.H., Jacobi, W., McClean, A.S., Mills, W.A. and Stewart, C.G. (1981) Estimate of risk from environmental exposure to radon-222 and its decay products, *Nature,* 290, 98-100.

Ewan, V.J. and West, G. (1983) *Appraising the Moisture Condition Test for Obtaining the Casagrande Classification of Soils,* Transport and Road Research Laboratory Sup-plementary Report 786, Crowthorne, Berkshire, England.

Fahey, M. and Jewell, R. (1990) Effects of pressuremeter compliance on measurement of shear modulus, *Proc. 3rd Int. Symp. on Pressuremeters,* Thomas Telford, London, 115-124.

Farmer, I.W. (1987) *Some Considerations Affecting the Selection of Tunnelling Methods,* Contractor's Report 51, Transport and Road Research Laboratory, Crowthorne, Berkshire, England.

Federation of Civil Engineering Contractors (1988) *Quality Assurance for Constructors,* London.

FIDIC (1989) *Conditions of Contract for Works of Civil Engineering Construction* (Fourth Edition).

Flanagan, R.F. (1993) Ground vibration from TBMs and shields, *Tunnels and Tunnelling,* Oct. '93, 30-33.

Fookes, P.G. and Hawkins, A.B. (1988) Limestone weathering: its engineering significance and a proposed classification, *Quart. J. Eng. Geol.,* 21, 7-31.

Fountain, F. (1991) *The Pocket-Guide to BS5750/ISO9000,* Cal-Tec, Hordern Road, Wolverhampton, Staffs WV6 0HS, England, 35pp.

Franklin, J.A. and Chandra, R. (1972) The slake durability test, *Int. J. Rock Mech. Mining Sci.,* 9, 325-341.

Fullalove, S. Dence, L., Clapham, H. and Hughes, F. (1983) New conditions seek ground survey pledge, Balance kept as contract evolved, Document tailored to upgrade industry, *New Civil Engineer,* 13 October, 1983, 20-24.

Gamon, T.I. (1986) A comparison between the core orienter and the borehole impression device. In: *Site Investigation Practice: Assessing BS 5930, Geol. Soc. Eng. Geol. Special Publication No. 2,* A.B. Hawkins (ed), 247-251.

Gates, A.E. and Gundersen, L.C.S. (eds) (1992) Geologic controls on radon, *Geol Soc. Amer. Special Paper 271,* 88pp.

Geological Society of London (1970) The logging of rock cores for engineering purposes, *Quart. Jl Eng. Geol.,* 3, 1-24.

Geological Society of London (1977) The

description of rock masses for engineering purposes, *Quart. Jl Eng. Geol.*, **10**, 355-388.

Glossop, N.H. (1982) The influence of geotechnical factors on tunnelling systems, PhD thesis, University of Newcastle upon Tyne, England.

Goodman, R.E. (1989) *Introduction to Rock Mechanics*, Wiley, New York, 562pp.

Goodman, R.E. and Shi, Gen-hua (1985) *Block Theory and its Application to Rock Engineering*, Prentice-Hall, Englewood Cliffs, New Jersey.

Great Britain Factory Inspectorate (1985) *The Abbeystead Explosion*, Health and Safety Executive, HMSO, London.

Greenwood, J.R., Cobbe, M.I. and Skinner, R.W. (1995) Development of the specification for ground investigation, *Proc. Inst. Civ. Engrs, Geotechnical Engrg*, **13**(1), 19-24.

Halleux, L (1994) Some aspects of modern engineering geophysics, *Proc. 7th Int. Congress IAEG*, Lisbon, 5-9 September 1994, R. Oliveira, L.F. Rodrigues, A.G. Coelho and A.P. Cunha (eds), LXXV-LXXXVII.

Hammond, R. (1963) Historical development. In: *Tunnels and Tunnelling*, C.A. Pequignot (ed), Hutchinson Scientific and Technical, London, 1-29.

Harper, T.R. and Hinds, D.V. (1978) The impression packer: a tool for recovery of rock mass fracture geometry, *Rockstore 77: Storage in Excavated Rock Caverns*, Pergamon Press, London, **2**, 259-266.

Harrison, R. (1992) *Assessing the Risk of Sulphate Attack on Concrete in the Ground*, BRE Information Paper 15/92, August 1992.

Hassani, F.P., Scoble, M.J. and Whittaker, B.N. (1980) Application of point load index test to strength determination of rock and proposals for new size correction chart, *Proc. 21st Symp. Rock Mech.*, Rolla, 543-564.

Haswell, C.K. (1986) The contract element in tunnelling, *Tunnels and Tunnelling*, **18**(4), 81-82.

Haswell, C.K. (1989) Clause discredited by contractors, *New Civil Engineer*, 10 August, 1989, 36.

Haswell, C.K. and de Silva, D.S. (1989) *Civil Engineering Contracts Practice and Procedure*, 2nd Edition, Butterworths, London, 234pp.

Hawker, G., Uff, J. and Timms, C. (1986) *The Institution of Civil Engineers' Arbitration Practice*, Thomas Telford Ltd, London, 244pp.

Hawkins, A.B. (1986) Rock descriptions. In: *Site Investigation Practice: Assessing BS 5930, Geol. Soc. Eng. Geol. Special Publication No. 2*, A.B. Hawkins (ed), 59-66.

Hawkins, A.B. and Pinches, G.M. (1986) Timing and correct chemical testing of soils/weak rocks. In: *Site Investigation Practice: Assessing BS 5930, Geol. Soc. Eng. Geol. Special Publication No. 2*, A.B. Hawkins (ed), 273-277.

Health and Safety Commission (1989) *The Control of Substances Hazardous to Health in the Construction Industry*, Construction Industry Advisory Committee, HMSO, London. (ISBN 0 11 885432 1.)

Health and Safety Commission (1992) *Management of Health and Safety at Work. Approved Code of Practice. Management of Health and Safety at Work Regulations*, HMSO, London.

Health and Safety Executive (1988a) *Control of Substances Hazardous to Health Regulations 1988, COSHH Assessments; a Step-by-Step Guide to Assessment and the Skills Needed for It*, HMSO, London.

Health and Safety Executive (1988b) *The Control of Substances Hazardous to Health Regulations*, Statutory Instrument 1988 No. 1657, HMSO, London. *See also* Health and Safety Commission (1988) Control of substances hazardous to health. Control of carcinogenic substances. Control of Substances Hazardous to Health Regulations 1988. Approved Codes of Practice, HMSO, London, 54pp. *See also* Guidance Note EH40 from the Health and Safety Executive (1990). Occupational exposure limits, HMSO, London.

Health and Safety Executive (1990) *Occupational Exposure Limits 1990;* Guidance Note Environmental Hygiene (EH) 40/90 (January 1990), HMSO, London.

Health and Safety Executive (1992) *Management of Health and Safety at Work - Approved Code of Practice*, HSE, London. (ISBN 0 7176 0412 8.)

Health and Safety Executive (1993) *Control of Substances Hazardous to Health Regulations - Approved Codes of Practice* (4th Edition), HSE, London. (ISBN 0 11 882085 0.)

Hencher, S.R. (1986) Discussion In: *Site

Investigation Practice: Assessing BS 5930, Geol. Soc. Eng. Geol. Special Publication No. 2, A.B. Hawkins (ed), 72.

Her Majesty's Inspectorate of Pollution (1989) *Waste Management Paper No. 27: The Control of Landfill Gas* (A technical memorandum on the monitoring and control of landfill gas), HMSO, London, 56pp.

Herrenknecht, M. (1994) EPB or slurry machine: the choice, *Tunnels and Tunnelling,* June 1994, 35-36.

Hewett, B.H.M. and Johannesson (1922) *Shield and Compressed Air Tunnelling,* McGraw-Hill, New York.

Hibberd, P.R. and Newman, P. (1994) *Alternative Dispute Resolution in Construction Contracts,* Blackwell Scientific Publications, 240pp. (ISBN 0 632 03817 9.)

Hinds, D.V. (1974) *A Method of Taking an Impression of a Borehole Wall,* Imperial College Rock Mechanics Research Report No. 32, November 1974

Hoek, E. and Bray, J.W. (1977) *Rock Slope Engineering,* Institution of Mining and Metallurgy, London, 402pp.

Hoek, E. and Bray, J.W. (1981) *Rock Slope Engineering,* 3rd Edition, Institution of Mining and Metallurgy, London.

Hoek, E. and Brown, E.T. (1980) *Underground Excavations in Rock,* Institution of Mining and Metallurgy, London, 527pp.

Hsieh, T.K. (1961) Discussion in Morgan, H.D. 'A contribution to the analysis of stress in a circular tunnel', *Géotechnique,* **11**, 364.

Hughes, J.S., Shaw, K.B. and O'Riordan, M.C. (1989) *Exposure of the UK Population - 1988 Review,* National Radiation Protection Board Report No. NRPB-R227.

Hunter, R.L.C. (1987) *The Law of Arbitration in Scotland,* T. and T. Clark, 440pp.

Huse, J. (1993) Preparation and presentation of construction claims, *Tunnels and Tunnelling,* **25**(1), 29-32 (January 1993).

!an Farmer Associates (1986) *Mechanical Excavation of Rock,* Report prepared for the Construction Industry Research and Information Association, London, 46pp.

Institution of Chemical Engineers (1992) *Model Form of Conditions of Contract for Process Plant Suitable for Reimbursable Contracts,* Second Edition, available from IChemE,

Davis Building, 165-171 Railway Terrace, Rugby, Warwickshire CV21 3HQ, 69pp.

Institution of Civil Engineers, Association of Consulting Engineers, Federation of Civil Engineering Contractors (1973) *ICE Conditions of Contract and Forms of Tender, Agreement and Bond for Use in Connection with Works of Civil Engineering Construction,* 5th Edition.

Institution of Civil Engineers (1983) *ICE Conditions of Contract for Ground Investigation,* Thomas Telford Ltd., London, 38pp.

Institution of Civil Engineers (1988a) *ICE Conditions of Contract for Minor Works.*

Institution of Civil Engineers (1988b) *Conciliation Procedure,* Thomas Telford, London.

Institution of Civil Engineers (1989) *Specification for Ground Investigation with Bill of Quantities,* Thomas Telford, London.

Institution of Civil Engineers, Association of Consulting Engineers, Federation of Civil Engineering Contractors (1991a) *ICE Conditions of Contract and Forms of Tender, Agreement and Bond for Use in Connection with Works of Civil Engineering Construction,* 6th Edition (January 1991), Thomas Telford Ltd, Thomas Telford House, 1 Heron Quay, London E14 4JD, 54pp. (ISBN 0 7277 16174.)

Institution of Civil Engineers (1991b) *Civil Engineering Standard Method of Measurement,* 3rd Edition (CESMM3), Thomas Telford Ltd, London, 109pp. (ISBN 0 7277 1561 5.)

Institution of Civil Engineers (1991c) *Inadequate Site Investigation,* Thomas Telford Ltd, London, 26pp. (ISBN 0 7277 1645 X.)

Institution of Civil Engineers (1992) *Design and Construct Conditions of Contract and Forms of Tender, Agreement and Bond for Use in Connection with Works of Civil Engineering Construction,* 52pp.

Institution of Civil Engineers (1993) *The New Engineering Contract,* 1st Edition, Thomas Telford Ltd, London. (ISBN (series) 0 7277-1616 6):

The Need for and Features of the NEC, 21pp.

The New Engineering Contract, 51pp.

A: Conventional Contract with Activity Schedule, 39pp.

B: Conventional Contract with Bill of Quantities, 39pp.

C: Target Contract with Activity Schedule, 39pp.
D: Target Contract with Bill of Quantities, 39pp.
E: Cost Reimbursable Contract, 38pp.
F: Management Contract, 33pp.
The New Engineering Sub-contract, 51pp.
Guidance Notes, 72pp.
Flow Charts.

Institution of Civil Engineers (1994) *The Adjudicator's Contract*, 5pp. (ISBN 0 7277 2337 5);
The Adjudicator's Contract Guidance Notes, 4pp;
Professional Services Contract, 25pp. (ISBN 0 7277 2338 3);
Professional Services Contract Guidance Notes, 22pp. (ISBN 0 7277 2339 1);
1st Editions, Thomas Telford, London.

Jardine, R.J. (1991) Discussion on 'Strain dependent moduli and pressuremeter tests', *Géotechnique*, **41**(4), 621-626.

Jefferis, S.A. and Wood, A.M. (1990) Ground conditions at Abbeystead, *Proc. Inst. Civ. Eng., Part 1, Design and Construction*, **88**, 721-725.

Jennings, J.N. (1985) *Karst Geomorphology*, Blackwell, Oxford,

Jerram, D. (1990) Seals come unstuck, *Civil and Structural Weekly*, 29 November 1990, 19.

Jing, C., Lu, G., Wei, Q., Grasso, P., Mahtab, A. and Xu, S. (1994) A reversible dilatometer for measuring swelling and deformation characteristics of weak rock, *Proc. 7th Int. Congress IAEG*, Lisbon, 5-9 September 1994, R. Oliveira, L.F. Rodrigues, A.G. Coelho and A.P. Cunha (eds), 339-345.

Joyce, M.D. (1982) *Site Investigation Practice*, Spon, London, 369pp.

Kathren, R.L. (1984) *Radioactivity in the Environment*, Harwood Academic Publishers, New York.

Kirby, C.E. and Morris, W. (1990) A response to COSHH, *Trans. Inst. Min. Metall.*, **99**, A138-A146.

Kirsch, G. (1898) Die Theorie der Elastizität und die Bedürfnisse der Festigkeitslehre, *Zeit. Ver. Deut. Ing.*, **42**(29), 797-807.

Knights, M.C. (1974) Exploratory shafts for in-town surveys, *Ground Engineering*, **17**(1), 43-45.

Kovári, K. (1994a) Gibt es eine NÖT? Fehlkonzepte der Neuen Österreichischen Tunnelbauweise (On the existence of the NATM: erroneous concepts behind the New Austrian Tunnelling Method), *Tunnel (Internationale Fachzeitschrift für unterirdisches Bauen)*, **1**, 16-25.

Kovári, K. (1994b) Erroneous concepts behind the New Austrian Tunnelling Method, *Tunnels and Tunnelling*, Nov. 1994, 38-42.

Krakowski, T.E. (1979) A universal tunnel design method for hand calculation, *Proc. Rapid Excav. and Tunneling Conf.*, A.C. Maevis and W.A. Hustrulid (eds), 1593-1611.

Ladanyi, B. (1974) Use of long-term strength concept in the determination of ground pressure on tunnel linings. In: *Advances in Rock Mechanics*, **II**, Part B, *Proc. 3rd Congress Int. Soc. for Rock Mech.*, Denver, 1974, Nat. Acad. Sci., Washington, DC, 1150-1156.

Lloyd's Register Quality Assurance Ltd (1989a) *Guidance Document to Quality Systems for Use in the Construction Industry,* Quality System Supplement Number 5000, Document ID: 0039E, dated 1988, Revision 0 Issue Date 28.06.89.

Lloyd's Register Quality Assurance Ltd (1989b) *Guidance Document to Quality Systems, Consulting Engineering Services,* Quality System Supplement Number: 8372, Document ID: 0039z, dated 1988, Revision 1 Issue Date 09.03.89.

Lyons, A.C. (1978) The design and development of segmental tunnel linings in the UK, *Proc. Symp. on Tunnelling under Difficult Conditions* Tokyo, I. Kitamura (ed), Pergamon, Oxford, 55-60.

McFeat-Smith, I. and Fowell, R.J. (1977) Correlation of rock properties and the cutting performance of tunnelling machines, *Proc. Conf. on Rock Engineering*, Newcastle upon Tyne, England, P.B. Attewell (ed), British Geotech. Soc., London, 581-602.

McFeat-Smith, I. and Fowell, R.J. (1979) The selection and application of roadheaders for rock tunnelling, *Proc. Rapid Excav. and Tunneling Conf.* A.C. Maevis and W.A. Hustralid (eds), **1**(Ch 16), 261-279, ASME, New York.

MacGregor, F., Fell, R., Mostyn, G.R., Hocking, G. and McNally, G. (1994) The estimation of rock rippability, *Quart. J. Engrg Geol.*, **27**(2), 123-144.

Manby, C.N.D. and Wakeling, T.R.M. (1990)

Developments in soft-ground drilling, sampling and *in-situ* testing, *Trans. Inst. Min. Metall.*, **99**, A91-A97.

Manson, K. (1993) *Law for Civil Engineers: an Introduction*, Longman, London. 336pp.

Megaw, T.M. and Bartlett, K. (1981) *Tunnels: Planning, Design and Construction*, **1** and **2**, Ellis Horwood Ltd, Chichester, England.

Mesri, G., Pakbaz, M.C. and Cepeda-Diaz, A.F. (1994) Meaning, measurement and field application of swelling pressure of clay shales, *Géotechnique*, **44**(1), 129-145.

Morgan, H.D. (1961) A contribution to the analysis of stress in a circular tunnel, *Géotechnique*, **11**, 37-46.

Morgan, J.M., Barratt, D.A. and Hudson, J.A. (1979) *Tunnel Boring Machine Performance and Ground Properties*, Transport and Road Research Laboratory Report SR 464, Crowthorne, Berkshire, England.

Moses, H., Stehney, A.F. and Lucas, H.F. Jr (1960) The effect of meteorological variables upon vertical and temporal distributions of atmospheric radon, *J. Geophys. Res.*, **65**(4), 1223-1238.

Muller, L. (1990) Removing misconceptions on the New Austrian Tunnelling Method, *Tunnels and Tunnelling*, Summer, 1990, 15-18.

Mustill, M.J. and Boyd, S.C. (1989) *The Law and Practice of Commercial Arbitration in England*, 2nd Edition, Butterworths, London, 835pp.

NEDC (1964) *National Economic Development Council Working Party (Sir Harold Banwell) Report on the Placing and Management of Contracts for Building and Civil Engineering Works*, Ministry of Public Buildings and Works, HMSO, London.

NEDO and EDC for Civil Engineering (1968) (Harris) *Report on Contracting in Civil Engineering since Banwell*, HMSO, London.

New, B.M. and O'Reilly, M.P. (1991) Tunnelling induced ground movements, predicting their magnitude. *Proc. 4th Int. Conf. Ground Movements and Structures*, Cardiff, Pentech Press, 671-697.

Norbury, D.R. (1986) The point load test. In: *Site Investigation Practice: Assessing BS 5930, Geol. Soc. Eng. Geol. Special Publication No. 2*, A.B. Hawkins (ed), 325-329.

Norbury, D.R., Child, G.H. and Spink, T.W. (1986) A critical review of Section 8 (BS 5930) - soil and rock description. In: *Site Investigation Practice: Assessing BS 5930, Geol. Soc. Eng. Geol. Special Publication No. 2*, A.B. Hawkins (ed), 331-342.

Norgrove, W.B. and Attewell, P.B. (1984) Assessing the benefits of site investigation for tunnelling, *Mun. Engr*, **1**(2), 99-106.

Norgrove, W.B., Cooper, I. and Attewell, P.B. (1979) Site investigation procedures adopted for the Northumbrian Water Authority's Tyneside Sewerage Scheme with special reference to settlement prediction when tunnelling through urban areas. In: *Tunnelling '79* (Proc. Int. Symp., 12-16 March 1979, London, M.J. Jones (ed)), IMM, London, 79-104.

Norris, P. (1992) The behaviour of jacked concrete pipes during site installation, DPhil thesis, University of Oxford, England.

Obert, L. and Duvall, W.I. (1967) *Rock Mechanics and the Design of Structures in Rock*, Wiley, New York, 650pp.

O'Reilly, M.P. (1993) *Principles of Construction Law*, Longman Scientific and Technical, Harlow, 300pp.

O'Rourke, T.D. (ed) (1984) *Guidelines for Tunnel Lining Design*, Prepared by the Technical Committee of Tunnel Lining Design of the Underground Technology Research Council, ASCE, New York, USA.

Ortigao, J.A.R. (1994) Dilatometer tests in Brasília porous clay, *Proc. 7th Int. Congress IAEG*, Lisbon, 5-9 September 1994, R. Oliveira, L.F. Rodrigues, A.G. Coelho and A.P. Cunha (eds), 359-365.

Panek, L.A. and Fannon, T.A. (1992) Size and shape effects in point load tests of irregular rock fragments, *Int. J. Rock Mech. Min. Sci. and Geomech. Abstr.*, **22**(2), 51-60.

Parker, D. (1994) Designs on Europe, *New Civil Engineer*, 24 February 1994, 31-35.

Peck, J.G. (1991) Simulation of blowouts in compressed air tunnels using a numerical method, MSc Thesis, University of Durham, England.

Peck, R.B. (1969) Advantages and limitations of the observational method in applied soil mechanics (Ninth Rankine Lecture), *Géotechnique*, **19**(2), 171-187.

Peck, R.B., Hendron, A.J. and Mohraz, B. (1972) State of the art of soft ground tunneling, *Proc. North Amer. Rapid Excav. and Tunneling Conf.*, Chicago, Ill, 5-7 June 1972, K.S. Lane and L.A. Garfield (eds), **1**, 259-286.

Pells, P.J.N. (1985) Engineering properties of the Hawkesbury Sandstone. In: *Engineering Geology of the Sydney Region*, P.J.N. Pells (ed), A.A. Balkema, Rotterdam, 179-197.

Pender, M.J. (1980) Elastic solutions for a deep circular tunnel, Technical Note,*Géotechnique*, **XXX**(2), 216-222.

Penner, E., Eden, W.J. and Gillot, J.E. (1973) Floor heave due to biochemical weathering of shale, *Proc. 8th Int. Conf. Soil Mech. Found. Engrg.*, Moscow, USSR, **2**(2), Session 4, 151-158.

Pettifer, G.S. and Fookes, P.G. (1994) A revision of the graphical method for assessing the excavatability of rock, *Quart. J. Engrg Geol.*, **27**(2), 145-164.

Powderham, A.J. (1994) An overview of the observational method: development in cut and cover and bored tunnelling projects, *Géotechnique*, **44**(4), 619-636.

Powell-Smith, V. (1990) *GC/Works/1 - Edition 3: The Government General Conditions of Contract for Building and Civil Engineering*, Blackwell Scientific Publications, Oxford, 216pp. (ISBN 0 632 02633 2.)

Powell-Smith, V. and Stephenson, D. (1989) *Civil Engineering Claims*, BSP Professional Books, Oxford, 181pp.

Powell-Smith, V. and Stephenson, D. (1993) *Civil Engineering Claims*, Second Edition, Blackwell Scientific Publications, Oxford, 240pp. (ISBN 0 632 03606 0.)

Power, R.D. (1985) *Quality Assurance in Civil Engineering*, Construction Industry Research and Information Association, Report 109, London, 70pp.

Priest, S.D. (1985) *Hemispherical Projection Methods in Rock Mechanics*, George Allen and Unwin, London, 124pp. (ISBN 0 04 622007 0.).

Priest S.D. (1993) *Discontinuity Analysis for Rock Engineering*, Chapman and Hall, London, 473pp.

Priest, S.D. and Hudson, J.A. (1976) Discontinuity spacings in rock, *Int. J. Rock Mech. Min. Sci. and Geomech. Abstr.*, **13**, 135-148.

Priest, S.D. and Hudson, J.A. (1981) Estimation of discontinuity spacing and trace length using scanline surveys, *Int. J. Rock Mech. Min. Sci. and Geomech. Abstr.*, **18**, 183-197.

Reynolds, P. (1990) Further treatment, *New Civil Engineer Water Supplement*, September, 68-72.

Roxborough, F.F. (1987) The role of some basic rock properties in assessing cuttability, *Seminar on Tunnels, Wholly Engineered Structures*, I.E. Aust./A.F.C.C., Sydney, Australia, April 1987.

Russell, H. (1994) Echo chamber, *New Civil Engineer*, 27 Jan. 1994, 31-32.

Russell, M.J. (1992) Mining, metallurgy and the origin of life, *Minerals Industry International (Bull. Inst. Mining Metall.)*, November 1992, 4-8.

Savin, G.N. (1961) *Stress Concentration around Holes*, Pergamon, Oxford, 430pp.

Schenck, W. and Wagner, H. (1963) Luftverbrauch v. Uberdeckung beim Tunnelvortrieb mit Druckluft, *Bautechnik*, **2**.

Selmer-Olsen, R. and Palmström, A. (1989) Tunnel collapses in swelling clay zones, *Tunnels and Tunnelling*, **21** (11), 49-51.

Serafim, J.L. and Pereira, J.P. (1983) Considerations of the geomechanics classification of Bieniawski, *Proc. Int. Symp. Eng. Geol. Underground. Construct.*, LNEC, Lisbon, **1**, II.33 - II.42.

Singh, S.P. (1989) A simple criterion for the machinability of hard rocks, *Int. J. Mining and Geol. Engrg*, **7**, 257-266.

Singh, V.K. and Singh, D.P. (1993) Correlation between point load index and compressive strength for quartzite rocks, *Geotech. and Geol. Engrg*, **11**, 269-272.

Site Investigation Steering Group (1993) *Site Investigation in Construction: 1. Without Site Investigation Ground is a Hazard*, 45pp. (ISBN 0 7277 1982 3); *2. Planning, Procurement and Quality Management*, 30pp. (ISBN 0 7277 1983 1); *3. Specification for Ground Investigation*, 103pp. (ISBN 0- 277 1984 X); *4. Guidelines for the Safe Investigation by Drilling of Landfills and Contaminated Land*, 34pp. (ISBN 0 7277 1985 8); Thomas Telford, London.

Skipp, B.O. (1984) Dynamic ground movements. In: *Ground Movements and their Effects on Structures*, P.B. Attewell and R.K. Taylor (eds), Surrey Univ. Press, London, 381-434.

Spangler, M.G. (1948) Underground conduits - an appraisal of modern research, *Trans. Amer. Soc. Civ. Engrs*, **113**, 316-374.

Steeples, D.W. and Mileer, R. (1990) Seismic

reflection methods applied to engineering, environmental and groundwater problems, *Soc. Explor. Geophys., Geotech. and Environmental Geophys.*, **1**, 1-30.

Stephenson, D.A. (1993) Disputing the problem, *New Civil Engineer*, 14 October 1993, 16.

Stroud, M.A. (1974) The standard penetration test in insensitive clays and soft rocks, *Proc. European Symp. on Penetration Testing*, Stockholm, **2.2**, 367-375.

Stroud, M.A. and Butler, F.G. (1975) The standard penetration test and the engineering properties of glacial materials, *Proc. Symp. on Engineering Behaviour of Glacial Materials*, Birmingham.

Széchy, K. (1973) *The Art of Tunnelling*, 2nd Edition, Akadémiai Kiadó, Budapest, Hungary, 1097pp.

Tarkoy, P.J. (1973) Pedicting tunnel boring machine penetration rates and cutter costs in selected rock types, *Proc. Ninth Canadian Symp. on Rock Mechanics*, Montreal, 263-274.

Temple, K. and Delchamps, E. (1953) Autotrophic bacteria and the formation of acid in bituminous coal mines, *Applied Microbiol.*, **1**(5).

Terzaghi, K. (1946) Rock defects and loads on tunnel supports. In: *Rock Tunneling with Steel Supports* by R.V. Proctor and T.L. White, Commercial Shearing and Stamping Co., Youngstown, Ohio, USA, 17-99.

Terzaghi, K. and Pack, R.B. (1967) *Soil Mechancis and Engineering Practice*, Second Edition, Wiley International Edition, New York, 729pp.

Thorburn, S. (1986) Field testing: the standard penetration test. In: *Site Investigation Practice: Assessing BS 5930, Geol. Soc. Eng. Geol. Special Publication No. 2*, A.B. Hawkins (ed), 21-30.

Toll, D.G. (1994) Interpreting site investigation data using a knowledge based system, *Proc. 13th Int. Conf. Soil Mech. and Found. Engrg*, New Delhi, 5-10 January 1994, 1437-1440.

Totterdill, B. (1994) On post-Latham dispute resolution, *New Civil Engineer*, 15 September 1994, 11.

Uff, J. (1983) Ground conditions: who is responsible, *New Civil Engineer*, 24 November 1983, 24-25.

Uff, J. (1989) The Clause 12 nightmare, *New Civil Engineer*, 6 July, 1989, 19-20.

Uff, J. and Clayton, C.R.I. (1986) *Recommendation for the Procurement of Ground Investigation*, CIRIA Special Publication 45, London, 44pp.

Uff, J.F., Hawker, G.F., Hacking, Lord, Bishop, D., Blackmore, J., Ashley, D.B. and Abrahamson, M.W. (1989) The professional engineer at risk, *Proc. Inst. Civ. Eng.*, **83**, 553-566.

Vallally, C.O. (1986) The valvate system related to percussive boring and sampling techniques. In: *Site Investigation Practice: Assessing BS 5930, Geol. Soc. Eng. Geol. Special Publication No. 2*, A.B. Hawkins (ed), 413-417.

Varley, P.M. (1990) Susceptibility of Coal Measures mudstone to slurrying during tunnelling, *Quart. J. Engrg Geol.*, **23**, 147-160.

Vickers, J.A., Francis, J.R.D. and Grant, A.W. (1968) *Erosion of Sewers and Drains*, Construction Industry Research and Information Association, Report 14, London.

Wallis, S. (1993) Better times for Storebælt, *Tunnel Construction, Internationale Fachzeitschrift für unterirdisches Bauen* (Offizielles Organ der STUVA, Köln), 4/93, 186-194.

Wallis, S. (1994) UK EPB machine conquers Bordeaux's karstic limestone, *Tunnels and Tunnelling*, July 1994, 43-46.

Ward, W.H. (1970) *Yielding of the Ground and Structural Behaviour of Linings of Different Flexibility in a Tunnel in London Clay*, Building Research Station Current Paper 34/70, Garston, Watford, England.

Ward, W.H., Burland, J.B. and Gallois, R.W. (1968) Geotechnical assessment of a site at Mundford, Norfolk, for a large proton accelerator, *Géotechnique*, **18**, 399-431.

Ward, W.H., Coats, D.J. and Tedd, P. (1976) Performance of tunnel support systems in the Four Fathom Mudstone. In: *Tunnelling '76* (Proc. Int. Symp., London, M.J. Jones (ed)), IMM, London, 329-340.

Ward, W.H. and Thomas, H.S.H. (1965) The development of earth loading and deformation in tunnel linings in London Clay, *Proc. 6th Int. Conf. on Soil Mech. and Found. Engrg*, **II**, Toronto, 8-15 Sept. 1965, Univ. of Toronto Press, 432-436.

Washbourne, J. (1993) Development and use of microtunnellers in the UK, *Tunnels and Tunnelling*, **25**(3), 69-71.

Weaver, J.M. (1975) Geological factors significant in the assessment of rippability, *The Civil Engineer in South Africa*, **17**, 313-316.

Weltman, A.J. and Head, J.M. (1983) *Site Investigation Manual*, CIRIA Special Publication 25, London, 144pp.

West, G. (1986) Air photograph interpretation and reconnaissance for site investigation. In: *Site Investigation Practice: Assessing BS 5930, Geol. Soc. Eng. Geol. Special Publication No. 2*, A.B. Hawkins (ed), 9-13.

West, G. (1989) Rock abrasiveness testing for tunnelling, *Int. J. Rock Mech. Min. Sci. and Geomech. Abstr.*, **26**(2), 151-160.

West, G., Carter, P.G., Dumbleton, M.J. and Lake, L.M. (1981) Rock mechanics review: site investigation for tunnels, *Int. J. Rock Mech. Min. Sci. and Geomech. Abstr.* **18**, 345-367.

Whitfield, J. (1994) *Conflicts in Construction: Avoiding, Managing, Resolving*, Macmillan, London, 158pp. (ISBN 0 333 60671 X.)

Wilbur, L.D. (1982) Rock tunnels. In: *Tunnel Engineering Handbook*, J.O. Bickel and T.R. Kuesal (eds), Van Nostrand, New York, Ch. 7, 123-207.

Wilson, R. (1992) Project management and the promoter, *Proc. Inst. Civ. Eng.* **92**(2), 54-55.

Windle, D. and Wroth, C.P. (1977) The use of a self-boring pressuremeter to determine the undrained properties of clays, *Ground Engrg*, **10**(6), 37-46.

Winter, J.B. (1993) The duty to warn, *Proc. Inst. Civ. Engrs*, Aug. 1993, 102-103.

Wood, A.M. Muir (1975) The circular tunnel in elastic ground, *Géotechnique*, **25**(1), 115-127.

Wood, A.M. Muir (1979) *Ground Behaviour and Support for Mining and Tunnelling*, 14th Sir Julius Wernher Memorial Lecture of the IMM, 12 March 1979, Institution of Mining and Metallurgy, London, 12pp.

Wood, A.M. Muir (1994) Can the newcomer stand up? *Tunnels and Tunnelling*, Sept. 1994, 30-32.

Wood, A.M.Muir and Kirkland, C.J. (1987) Tunnels and underground chambers. In: *Ground Engineer's Reference Book*, F.G. Bell (ed), Butterworths, London, 41/1-41/19.

UK STATUTES

Arbitration Act 1950.

Arbitration Act 1979.

Arbitration Act (Northern Ireland) 1937.

Arbitration (Scotland) Act 1894.

Building Regulations 1991.

Companies Act 1989.

Consumer Protection Act 1987.

Control of Pollution Act (1974) Chapter 40, HMSO, London, 136pp.

Courts and Legal Services Act 1990.

Employment Medical Advisory Service Act 1972.

Environmental Protection Act 1990.

Explosives Act 1875.

Explosives Act 1923.

Health and Safety at Work etc Act 1974, HMSO, London.

Latent Damage Act 1986, Chapter 37, HMSO, London, 6pp.

Law of Property (Miscellaneous Provisions) Act 1989.

Limitation Act 1980.

Local Government Act 1972.

Local Government Act 1988.

Misrepresentation Act 1967

Public Health Act 1961.

Public Utilities and Street Works Act 1950,

Roads and Street Works Act 1991.

Sale of Goods Act 1893.

Construction (Design and Management) Regulations 1994.

Construction (General Provisions) Regulations 1961.

Control of Substances Hazardous to Health Regulations, amended by SI 1990/2026, SI 1991/2431, and SI 1992/2382. (ISBN 0 11 087657 1.)

Diseases and Dangerous Occurrences Regulations 1986.

Health and Safety (First Aid) Regulations 1981. (ISBN 0 11 016917 4.)

Lifting Plant and Equipment (Records of Test and Examination etc) Regulations 1992.

Management of Health and Safety at Work Regulations 1992, SI 2051, HMSO, London. (ISBN 0 11 02051 6.)

Manual Handling Operations Regulations 1992.

Noise at Work Regulations 1989, SI 1790, HMSO, London.

Personal Protective Equipment at Work

Regulations 1992, SI 2966, HMSO, London. (ISBN 0 11 025832 0.)

Provision and Use of Work Equipment Regulations 1992, SI 2966, HMSO, London. (ISBN 0 11 025832 0.)

Reporting of Injuries, Diseases and Dangerous Occurrences Regulations 1985, SI 1457, HMSO, London.

Utilities Supply and Works Contract Regulations 1992.

Work in Compressed Air Regulations 1958.

Workplace (Health, Safety and Welfare) Regulations 1992, SI 3004, HMSO, London. (ISBN 0 11 025804 5.)

EU STATUTES

76/160/EEC, Bathing Waters Directive.
90/531/EEC, Utilities Directive.

2

HAZARDOUS GASES WITH PARTICULAR REFERENCE TO THOSE THAT MIGHT BE ENCOUNTERED WHEN TUNNELLING IN COAL MEASURES STRATA

There is a range of flammable, toxic and asphyxiating gases that might be encountered in the environment of a tunnel, particularly one driven through Coal Measures strata.

Flammable gases are, in the main, methane, carbon monoxide and hydrogen, but the actual presence depends on whether the tunnel is in the proximity of coal or ore. Carbon monoxide and hydrogen produced by the thermal decomposition of organic materials can be explosive and lead to dust explosions under generally dry conditions.

Toxic gases comprise carbon monoxide, nitrogen dioxide, nitric oxide and hydrogen sulphide.

Asphyxiant gases are those gases which displace oxygen.

There is no absolute safe action level in the UK for a hazardous gas. However, to give any satisfactory guidance, practical action levels must be set lower than maximum exposure limits for toxic and asphyxiant gases, and at concentrations below lower explosive limits for combustible gases and gas mixtures. Table SI 2.1 notes some gases for which exposure limits are available and Table SI 2.2 shows the range of inflammability for certain gases.

Table SI 2.1 Some occupational exposure limits for a range of hazardous gases

Gas	Limit (8 hours) ppm	Limit (10 hours) ppm
Carbon dioxide	5000	15000
Hydrogen sulphide	10	15
Carbon monoxide	50	400
Nitrogen monoxide	25	35
Nitrogen dioxide	3	5
Sulphur dioxide	2	5
Ammonia	25	35
Hydrogen cyanide	-	10
Phosgene	0.1	-
Chlorine	1	3
Hydrogen chloride	5	5

Table SI 2.2 Limits of inflammability of a range of gases at normal temperature and pressure

Gas	Lower explosive limit (%)	Upper explosive limit (%)
Methane	5	15
Ethane	3	12.5
Propane	2.1	9.4
Hydrogen sulphide	4.3	45.5
Hydrogen	4	74.2
Carbon monoxide	12.5	74.2

Table SI 2.3 Typical compositions of gases containing methane (after Card, 1993)

Source	Gas composition percentage volume in air							
	CH_4	C_2H_6	C_3H_8	CO_2	CO	H_2S	N_2	O_2
Landfill gas	20-65			16-57	$<1x10^{-4}$	$2x10^{-5}$	0.5-37	<0.3
Mine gas								
seam	80-95	8	4	0.2-6			2.9	
pumped drainage[1]	22-95	3	1	0.5-6	0-10		1-61	
Wetlands/peatlands								
freshwater muds[2]	3-86			0.3-13			16-94	
saltwater muds	55-79			2-13				
marsh gas	11-88						3-69	
buried peats and organics soils	45-97						1.6-54	
mains/natural gas[3]	94	3.2	0.6	0.5			1.2	

Notes on Table SI 2.3:

[1] Gas mixed with air   [2] Composition varies with depth   [3] Also 0.2% C_4H_{10}

Methane (CH_4) is a colourless, odourless and tasteless gas approximately 0.6-times the density of air. When mixed with air, methane is flammable at concentrations of between 5% and 15%. If a mixture between these limits is confined, it is explosive. The presence of other gases may affect these limits. It is usual to express concentrations of methane as a percentage of the lower explosive limit (LEL) and, as a guide, alarm triggers for the evacuation of underground personnel are usually set for concentrations of 20% or 25% of LEL. A second alarm set at 40% of LEL may sometimes be used for additional safety.

Methane, the composition of which is shown in Table SI 2.3, can be found naturally anywhere where organic decay occurs in anoxic conditions. The decay may be related to organic-rich silts, to the biogenic anaerobic decay of organic wastes in landfill sites, or in natural marshes or swamps. It is also found at depth in association with both coal and oil deposits where its formation is both pressure- and temperature-controlled. The gas may thus be recent or geologically ancient, and may be biogenic or thermogenic in nature. It follows that methane migration may occur both from geological strata at depth and from shallower surface deposits. Migration is also encouraged by differential pressures and/or differential concentrations between source and release points. It is just the problem in site

investigation to anticipate its presence, its flows and its releases, not only in the context of pressure and concentration but also more immediately in the local and regional geological setting. In this respect special reference should be made to the discussion on the Abbeystead case in Part One of the book, Section 7.10.5, and to the Health and Safety Executive (1985).

It must be understood that methane (and other hydrocarbon gases) are soluble in water and so are readily amenable to transportation by this means from quite remote locations - a point made in BS 6164:1990 (British Standards Institution, 1990).

UK coal seams are usually of very low permeability (much less than 1 millidarcy), and so in such strata gas is only usually released in substantial quantities near to geological disturbances, or to disturbances caused by old mine workings. Methane quantities in UK coal seams seem to vary quite significantly, from a trace up to about $20m^3$/tonne. Monitoring for gas in closed atmospheres is usually done by hand-held instruments (methanometers for methane), or by sampling in tubes for later analysis by gas chromatography or infra-red adsorption, fixed remote sampling, and remote transducers wired to a logger.

There must be an awareness, both at the site investigation stage and during the tunnel construction, of the possible presence of gas if a tunnel is driven in the vicinity of a landfill site comprising controlled (domestic and trade) waste (*see* Her Majesty's Inspectorate of Pollution, 1989). Landfill gas may be heavier or lighter than air and can permeate not only through the ground but also through service ducts and cracks in foundations. The landfill gas problem could be particularly acute in the case of old, unsealed waste disposal sites operating on the 'dilute and disperse' principle and which are still active in the sense of anaerobic decomposition. If not already installed prior to tunnelling, provision would need to be made for suitable borehole gas monitors at the perimeter of the landfill in addition to continuous monitoring in the tunnel environment.

A special gas hazard can arise during the construction of connections with and entrance to existing operational sewers. In circumstances such as these a system of forced ventilation may be required together with the use of appropriate gas monitoring equipment.

Groundwater can be contaminated with hydrocarbons and so create a safety hazard. A £5.6m outfall sewer, originally scheduled for completion in March 1995, was constructed by Kier Construction using a Dosco roadheader in Mansfield, Nottinghamshire, to the instructions of the District Council acting as agent for Severn Trent Water. The purpose of the sewer was to stop pollution of the River Maun. Groundwater contaminated by hydrocarbons comprising petrol of different grades was encountered during the construction. At 117m of a 760m drive, gas monitors indicated gas at 20% of the lower flammable limit and so the tunnel was evacuated. So that driving could proceed several measures were undertaken: water sprays were directed onto the picks; there was a 50% increase in the ventilation; air movers were fitted in the shield; there was air purging of the switchgear; flame and gas detectors were fixed in the shield and there were interlocks to ensure automatic shutdown in the event of an alarm. To inhibit ingress of the contaminated water, high pressure cement/bentonite grout injected from the road surface was used to fill larger voids in exposed rock around, in front of, and behind the shield, and MQ5, a water soluble resin powder, was used to seal finer fissures.

Under the ICE Conditions of Contract, responsibility for safety during construction rests with the contractor. It therefore follows that information on any possible methane hazard should be made available to the bidding contractors at the tender stage so that they can fully assess the risks and price for the necessary ventilation and monitoring equipment, and for the automatic isolations of electrical plant. The contractor must also develop his working plan to take account of the need to minimise accumulations of the gas; for example, the ventilation system should be able to maintain a 0.5m/s velocity of fresh air at the tunnel face, with particular attention paid to clearing accumulations from the crown of the tunnel. Methane gas sampling systems comprise simple pressure pumps and cylinders used in conjunction with gas chromatographic or infra-red spectrographic analysis. A level of 70 parts per million of the gas indicates, in practice, a possible methane gas hazard.

If tunnelling construction must take place under methane risk conditions then the equipment and working practices must be the same as or similar to those used in United Kingdom coal mines, as defined by the Regulations in the Mines and Quarries Act.

CO_2 is a toxic gas and an asphyxiant. The Health and Safety Executive set 10-minute and 8-hour occupational exposure limits (OELs) of 1.5% and 0.5% by volume, respectively. The lower limit of detection of O_2 is 1% by volume, concentrations of less than 18% by volume being likely to induce asphyxiation.

Toxic hydrogen sulphide gas tends to be associated mainly with sewers but its presence is not exclusive to them. Construction on a 2.3 mile, $490 million stretch of the Los Angeles Metro Red Line was delayed for six months beyond its scheduled September 1994 start when borings revealed high concentrations of H_2S gas between 40 and 45 feet below ground surface. Options for avoiding the gas or mitigating the problem included treatment technologies, re-alignment and raising the tunnel levels and considering a cut-and-cover solution. Discovery that the primary lining did not meet specification for dry-pack concrete in the expansion gaps - reports of plywood having been used to fill some gaps, with the plywood then having been plastered over - could, without correction, have exacerbated this problem.

REFERENCES

British Standards Institution (1990) *Code of Practice for Safety in Tunnelling in the Construction Industry*, BS 6164:1990, HMSO, London, 96pp.

Card, G.B. (1993) *Protecting Development from Methane*, A report in CIRIA's research programme 'Methane and associated hazards to construction', Funders Report/CP/8, Construction Industry Research and Information Association, London, March 1993, 185pp.

Health and Safety Executive (1985) *Report of the Investigation into the Abbeystead Explosion*, HMSO, London, 22pp.

Her Majesty's Inspectorate of Pollution (1989) *The Control of Landfill Gas* (A technical memorandum on the monitoring and control of landfill gas), Waste Management Paper No. 27, HMSO, London, 56pp.

3

APPLICATION OF BAYESIAN STATISTICS TO SITE INVESTIGATION

Site investigations for building foundations and tunnels are rarely supported by any theoretical or statistical considerations. There is the possibility, however, of applying, with due caution, some simple Bayesian theory to the exercise. The discussion below takes the form of a re-statement and extension of the material published in Attewell and Farmer (1976). Further reference may be made to the paper by Attewell (1987).

Bayes' equation is first expressed in the form of equation 1.

$$P(A_i \mid B) = \frac{P(B \mid A_i)\,P(A_i)}{\sum\limits_i P(B \mid A_i)\,P(A_i)} \quad \text{................(1)}$$

where A_i are mutually exclusive events. If, for example, A_i represents the value of some physical property of the ground which varies with location,

then

$P(A_i)$ describes the form of the variation,

and

$P(B \mid A_i)$ is the conditional probability of B relative to the hypothesis that A_i has occurred.

When $P(B) \neq 0$,

then

$P(A_i \mid B)$ is the conditional probability of A_i relative to the hypothesis that B has occurred (i.e. it is the probability that A_i is the true value of A, given that B is the result of a test);

$P(A_i)$ is the prior probability;

$P(B \mid A_i)$ is the pre-posterior probability(a modification of the prior);

and

$P(A_i \mid B)$ is the posterior probability.

SITE INVESTIGATION FOR A TUNNEL

Suppose that, according to borehole and map evidence, the ground in question should be a *firm laminated clay*, but it could change to a *saturated silt*. Thus there is uncertainty at the site investigation stage as to the type of material that will be tunnelled, the manner in which it will be tunnelled and the plant that will be required, and the ground movements that could be caused by the tunnelling operation. This uncertainty can also presents problems for the billing and pricing of the work and raises the expectation of contractual claims.

Initial Inspection by a Geologist [P(A)]

As a result of his initial inspection, the geologist assigns his perceptions of the tunnel face geology:

Laminated clay:	80% probability
Saturated silt:	20% probability

These values define the <u>prior probabilities</u>.

Later Inspection of Ground and Buildings by the Geologist and an Engineer [P(B|A)]

On a *granular soil* the foundations of a building would tend to <u>hog</u>. In simple terms this would tend to cause tension cracking at an upper level in the building.

On a *clay soil* the foundations of a building would tend to <u>sag</u>. Again, in simple terms this would tend to cause tension cracking at the centre of a foundation slab or basement level. There could therefore be some evidence for re-interpretation of the probabilities.

As a result of this inspection of structural damage to nearby buildings, it is now decided that the following probabilities apply:

Laminated clay:	10% probability
Saturated silt:	90% probability

These are the <u>pre-posterior probabilities</u>.

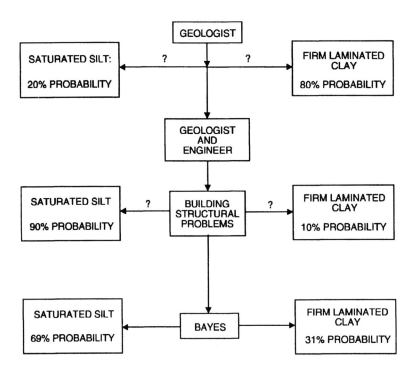

Figure SI 3.1 Application of Bayes' equation to tunnelling

Posterior Probabilities

It is now possible, from equation 1, to calculate the posterior probability of encountering *saturated silt*. The necessary assumption here, of course, is that the soil conditions expected to be encountered at the tunnel face will be the same as those inferred from the evidence at the ground surface. Additional evidence would clearly be required in order to support this assumption, and since that evidence could only be provided by an in-ground investigation the point of the exercise is to some degree questionable.

$$P(A_i \mid B) = \frac{P(B \mid A_i) P(A_i)}{\sum_i P(B \mid A_i) P(A_i)}$$

$$= \frac{0.90 \times 0.2}{(0.90 \times 0.2) + (0.10 \times 0.8)} = 69\%$$

The effect of the joint inspection by the geologist and engineer has thus been to raise the probability of encountering silt from its prior value of 20% to

its posterior value of 69% as a result of the high pre-posterior probability value of 90%. This is a quite significant increase. By similar calculation it will be seen that the probability of encountering clay has been downgraded from its prior value of 80% to a posterior value of 31%.

There would be some logic in engaging in an iteration procedure by placing the posterior probability value as a new prior, and then seeking more evidence for a pre-posterior probability value. The rather simple exercise outlined above and the iterations can be most readily presented in flow-chart form, as shown in Figures SI 3.1 and SI 3.2 below. However, it will be readily understood that this line of analysis, if undertaken at all, would never be conducted in isolation. As part of the 'site walk' and subsequent appraisal of the ground conditions it would merely serve as an adjunct to a detailed desk study followed by a full drilling and/or trial pitting and sampling ground investigation.

DISCUSSION ON THE INTERPRETATION OF PRIOR PROBABILITIES

It is necessary to begin with a clear idea of the population of events over which the prior distribution $P_1(\bullet)$ is defined. Otherwise, when it comes to interpreting the posterior distribution

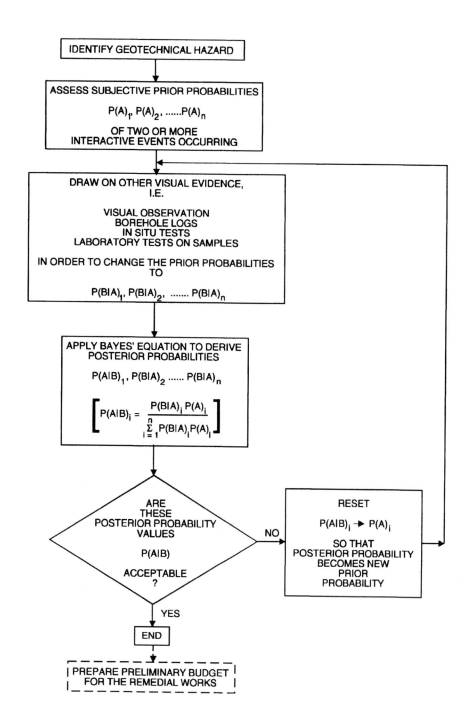

Figure SI 3.2 Flow chart to demonstrate the implementation of Bayes' equation in site investigation

$P_2(\bullet \,|\, B)$ consequent to an observation B one will be 'swimming around aimlessly'.

The posterior is, in principle, similar in nature to the prior. Logically, $P_1(\bullet) = P_1(\bullet\,|$ prior data) is no different in form from $P_2(\bullet \,|\, B) = P_2(\bullet \,|\, B,$ prior data). It is possible to construct a sequence of distributions

$$P_n(\bullet\ ) = P_n(\bullet \,|\, \text{data } n,\, ..., \text{data } 1)$$

in which successive members are, respectively, prior and posterior to one another. In this equation 'data' can be numerical data or knowledge expressed by language, and they are read from right to left. The prior data restricts the population of events from the infinity of possibilities. The allowable events are those having properties consistent with prior data, and the population appropriate to P_2 is that of P_1 restricted by the observation B.

Developing from the earlier example, let a large area of land, say in County X, be one of the two types, clay or silt, and the population of events be the set of soil type instances {clay, clay, silt, clay,..........} (one element/grid reference). Let this prior state of knowledge, with the same probabilities as before, be summarised by

$$P_1(clay) = \sum_{\substack{all\ clay \\ events}} P(clay\ event) = 0.8 \qquad \ldots\ldots\ldots\ldots (2)$$

$$P_1(silt) = 0.2$$

where the number of clay/silt material elements have simply been summed and the results normalised:

$$P_1(clay) + P_1(silt) = 1 \qquad \ldots\ldots\ldots\ldots\ldots\ldots\ldots (3)$$

The engineer (with the geologist) can observe two outcomes, failure f and non-failure $\bar{f}$, and we are given, say,

$$P(f\,|\,clay) = 0.1;\ P(\bar{f}\,|\,clay) = 0.$$
$$\text{(i.e. sum = 1)} \qquad \ldots\ldots\ldots\ldots\ldots\ldots\ldots (4)$$

$$P(f\,|\,silt) = 0.4;\ P(\bar{f}\,|\,silt) = 0.6$$
$$\text{(i.e. sum = 1)} \qquad \ldots\ldots\ldots\ldots\ldots\ldots\ldots (5)$$

There is, already, an implicit assumption that we will not come across buildings straddling both soil types. We may know the values for

$$P(f\,|\,x\%\ clay,\ [100\text{-}x]\ \%\ silt),\ \text{and}$$

$$P(\bar{f}\,|\,x\%\ clay,\ [100\text{-}x]\ \%\ silt),$$

but the prior given by equation 2 is inadequate to utilise this. We would need to know something about the spatial variation of clay and silt. The effective assumption is that structures are small relative to soil spatial variation.

Re-writing Bayes theorem:

$$P_2(A_i\,|\,B) = \frac{P(B\,|\,A_i)P_1(A_i)}{\sum_j P(B\,|\,A_j)P_1(A_j)} \qquad \ldots\ldots\ldots\ldots (6)$$

$$P_2(clay\,|\,f) = \frac{0.1\times0.8}{(0.1\times0.8)+(0.4\times0.2)} = \frac{1}{2} \qquad \ldots\ldots\ldots\ldots\ldots\ldots (7)$$

$$P_2(silt\,|\,f) = \frac{0.4\times0.2}{(0.1\times0.8)+(0.4\times0.2)} = \frac{1}{2} \qquad \ldots\ldots\ldots\ldots\ldots\ldots (8)$$

$$P_2(clay\,|\,f) + P_2(silt\,|\,f) = 1 \qquad \ldots\ldots\ldots\ldots (9)$$

$$P_2(clay\,|\,\bar{f}) = \frac{0.9\times0.8}{(0.9\times0.8)+(0.6\times0.2)} = \frac{6}{7} \qquad \ldots\ldots\ldots\ldots\ldots\ldots (10)$$

$$P_2(silt\,|\,\bar{f}) = \frac{0.6\times0.2}{(0.9\times0.8)+(0.6\times0.2)} = \frac{1}{7} \qquad \ldots\ldots\ldots\ldots\ldots\ldots (11)$$

$$P_2(clay\,|\,\bar{f}) = P_2(silt\,|\,\bar{f}) = 1 \qquad \ldots\ldots\ldots\ldots (12)$$

Assuming that failure has been observed, then from equation 7 there is a 50:50 chance that the soil under the building is clay. However, this does not mean that the whole site has a 50% chance of comprising clay. In order to come to that conclusion we require a further assumption that the variation of soil type is small with respect to the size of the area of interest. This is a quite fundamental assumption and needs to be justified by prior data or knowledge.

In the above example, the model is well-defined and unique. What has gone before applies equally to what follows, but one can turn to the realistic case when there is ambiguity about the probability model.

In the first place one can sometimes still use a geographical model. It is accepted that a county proportioned into two areas has been somewhat extreme, but if we consider each item of prior data successively restricting the allowable area, we finish up with a subset of County X corresponding to the probability model.

(There is an established betting model in which the population of events is the set of bets. The fact that only one bet is normally made is a problem. However, it is possible to get round this by considering a lifetime of similar situations and corresponding bets, or the bookmaker's viewpoint. Alternatively, there are frequency models where the number of times right/wrong provides a measure of probability.)

Notwithstanding what is said in the last paragraph, it should not be inferred that geologists, engineers and the like are gamblers, nor that spinning coins is a suitable substitute for observation and experience!

Finally, when all else fails, measures of probability represent nothing but themselves - real numbers - but real numbers treated in a logical and consistent way with respect to data acquisition.

REFERENCES

Attewell, P.B. (1987) Aspects of risk assessment in engineering geology and geotechnics, *Mem. Geol. Soc. China*, No. 9, 335-365.

Attewell, P.B. and Farmer, I.W. (1976) *Principles of Engineering Geology*, Chapman and Hall, London, 1045pp.

4

WEAK ROCK

Weak rock has an unconfined compressive strength (UCS) of between 1.25 and 5MPa. The general characteristics of weak rock mean that it is often unsuitable for engineering construction where it is required to offer a foundation support capability or a self-support capability in the long term when exposed not only to deforming pressures but also to groundwater and atmospheric effects. On the other hand, it can be excavated with relatively little effort. It represents a large percentage of exposed rock material on the earth's crust and has been the cause of several notable civil engineering failures.

Hatheway (1990) has described weak rock as 'a consolidated earth material possessing an unusual degree of bedding or foliation separation, fissility, fracturing, weathering, and/or alteration products, and a significant content of clay minerals, altogether having the appearance of a rock, yet behaving partially as a soil, and often exhibiting a potential to swell or slake, with the addition of water; some weak rocks are also subject to time-dependent release of stored tectonically-induced stress'. He notes that when weak for reasons other than weathering or alteration, weak rock is generally Cretaceous or younger in age. Weak rocks are often thought of as being synonymous with 'shales' and 'mudstones' ('mudrocks - *see* Supplementary Information 6), although other non-clay mineral rocks, such as chalk, are also weak. Meigh and Wolski (1979) addressed the subject on which there was also a major conference in 1980 (Akai *et al.*, 1980). The following associations, attributable to Hatheway, relate weak rock to its geological environment.

GEOLOGICAL ASSOCIATIONS OF WEAK ROCK

Age and lithologic associations

1) Cretaceous or younger age: insufficient time for lithification or favourable diagenesis.

2) Permian and younger marl, chalk, or other dirty (clay-rich) carbonate rock types.

3) Most rocks classified as 'claystone' (preferred term) or 'mudstone' (non-preferred term: Underwood, 1967: mudrocks of the Commonwealth countries). Those rocks which have lithified only as a result of consolidation ('compaction' of classical geology) and which have little or no cementation (Mead, 1937).

4) Rock classified as 'flysch', the interbedded sequence of shale and mudstone of a marine turbidite origin.

5) Sedimentary rock with an organic depositional origin resulting in the presence of sulphate minerals, calcium, iron, or magnesium, now prone to swelling and mineral bond disintegration.

6) Sedimentary rock with a lack of cementing agent(s) or dissolution of, or cation exchange removal of, pore and void cement.

7) Clay-rich sedimentary rock having been within the near proximity (say hundreds of metres) to a dyke or sill; the result is thermally induced alteration short of low-grade metamorphism. The damage is probably greater if the host rock was wet and poorly lithified at the time of the intrusion. This situation is doubly difficult to anticipate or to detect in site exploration.

8) Rocks of greenish colours, denoting the presence of chlorite as an undesirable mineral or precursor to swelling clay minerals.

9) High clay or silt content; when un-metamorphosed these are the 'shaly' rocks which are also thinly bedded and subject to considerable rock mass strength-reducing jointing.

10) Volcaniclastic rock in general. Unfavourable combination of clay, as primary minerals or as products of alteration, along with pyroclastic glass shards, or their alteration products, and ferromagnesian minerals

subject to degradation on oxidation.

11) Tropical marine limestone; generally within 20 degrees of the Equator.

Structural-tectonic-geographic associations

12) Plate margins, former subduction zones and zones of transform or transcurrent faulting; common provenance of volcaniclastic and/or alteration-degraded ultramafic rock.

13) Zones of tectonic deformation or shear; mainly physiographic boundaries of a structural nature, or plate-margin, terrestrial faults or major intraplate fault zones.

14) Volcanic and volcaniclastic rocks found in a presently tropical climate and within the zone of weathering.

DETERIORATION SEQUENCE FOR WEAK ROCK

Breakdown of weak rock, once exposed by excavation or as a long-term phenomenon in some embankments, has been described by Hatheway (1990) as follows:

1) *Mechanical breakdown* by stress relief of newly excavated rock

2) *Hydration* of any expansive clay minerals present in the rock.

3) *Dispersion* (re-adjustment) of the more active clay minerals.

4) *Alteration* of layer silicate minerals.

5) *Leaching* of cementing agents.

6) *Concentration of stresses* as a result of clay mineral (crystal lattice) and bound-water (adsorption) expansion.

7) *Softening* of the rock; loss of strength and durability.

SUSCEPTIBILITY TO BREAKDOWN

Hatheway notes the tests on weak rocks in Table SI 4.1 as being useful for identifying their susceptibility to breakdown.

Since weak rocks are often fine-grained, it will not be possible, without optical microscopic facilities, to determine the particle shape. Determination of clay mineralogy by X-ray diffraction tests and the use of ethylene glycol for resolving crystal lattice swelling potential are outlined in Supplementary Information 7. Although plasticity of a soil is readily determined by an Atterberg limits test, plasticity of a weak rock is perhaps best easily resolved by an unconfined compression test and inspection of the shape of the loading stress (ordinate)-strain (abscissa) curve (convexity upwards), and also in the unloading (concave upwards) mode if that is possible, although it is not immediately obvious how that particular parameter would be indicative of a potential for disintegration.

Table SI 4.1 Some tests for indicating a potential for disintegration of weak rocks

Test	Use/Indication
Size gradation	Indicates presence of loose or friable material; can be used before or after other forms of physical or chemical stressing
Particle size	Inspect after swelling or slaking occurred
Clay mineralogy	Presence of swelling minerals
Plasticity	Indication of presence and release of swelling minerals
Ethylene glycol absorption	Qualitative test for swelling or slaking potential
Methylene blue absorption	Qualitative test for swelling or slaking potential
Cation exchange adsorption of lime	Suggestive of presence of swelling clay minerals
Ultrasonic disaggregation forces	Models a variety of field environments; wetting and agents; breakdown
Relative compression	Indication of degree of lithification strength of excavation-run fragments

REFERENCES

Akai, K., Hayashi, M. and Nishimatsu, Y. (eds) (1980) Weak rock - soft, fractured and weathered rock, *Proc. Int. Symp. on Weak Rock*, Tokyo (3 volumes).

Hatheway, A.W. (1990) Perspectives No. 5: Weak rock, poorly lithified cockroaches and snakes, *AEG News*, **33**(3), 33-36.

Mead, W.J. (1937) Geology of dam sites in shale and earth, *Civil Engrg*, **7**(6), 392.

Meigh, A.C. and Wolski, W. (1979) Design parameters for weak rocks, *Proc. 7th European Conf. on Soil Mech. and Found. Engrg.*, Brighton, England, **V**, 59-79.

Underwood, L.B. (1967) Classification and identification of shales, *Proc. Amer. Soc. Civ. Engrs, J. Soil Mech. and Found. Div.*, **93**(SM6), 97-116.

5

STANDARD PENETRATION TEST AND CONE PENETRATION TEST

STANDARD PENETRATION TEST (SPT)

The standard penetration test is the most widely used field test world-wide. It involves driving a 50mm outside diameter, 35mm internal diameter split spoon (barrel) sampler, length 675mm, into the ground at the bottom of a cased borehole by means of a 93.6kg sliding hammer falling 762mm on to the top of the boring rods to the end of which the sampler is connected. A trip-release mechanism and guide assembly are usually adopted to control the drop of the hammer, and an anvil at the lower end of the assembly serves to transmit the blow to the boring rods. Before testing, the hole must be cleaned out to the necessary depth, but with care taken to ensure that the ground at the bottom of the hole is not disturbed. Jetting should not be used for location, and the casing must not be driven below the level in the ground at which the test is to be conducted.

Operationally, the spoon is driven an initial 150mm both for seating purposes and to clear any disturbed ground. During this period the blow count recorded for each of two 75mm increments. If the blow count N is greater than 50, the test is completed. The blow count for the next 300mm is recorded for each 75mm increment, since this allows for the better assessment of the depth of any disturbance. In the case of granular materials, the tests are carried out at 1m intervals and in gravels a 60° angle solid cone is used to avoid damage to the spoon. Slightly higher results seem to be obtained in the same material when the 60° cone replaces the normal driving shoe.

Different methods of applying the test in various countries lead to different ratios between the energy actually used for penetrating the ground and the free fall energy of the hammer. This rod energy ratio seems to vary from 45% to 78%. In the UK, for rod lengths greater than 10m, a ratio of 60% seems to be appropriate, and this should be adopted as standard. Recorded N-values should then be normalised to this standard and be denoted N_{60}. A further correction has also been proposed if the total rod length is less than 10m (for example, a correction factor of 0.75 if the length is 3-4m). The N-values are also somewhat sensitive to borehole diameter, with power values being associated with larger diameters (for example, correction factors of 1.05 and 1.15, respectively, being applied to 150mm and 200mm diameter boreholes compared with test results from a 115mm borehole).

Numerous empirical correlations exist between SPT N-values and the soil properties, as well as direct correlations between N-values and performance. In the case of buildings, it is estimated that 80% to 90% of routine foundation design is carried out using SPT N-values.

Most correlations apply to sands and gravel. Fine and silty sands and silts below the water table can give abnormally high penetration resistances because the excess pore pressures set up during driving do not dissipate. Terzaghi and Peck (1967) suggest that if N-values exceed 15 in these soils a corrected value N' should be calculated from the expression:

$$N' = 15 + 0.5(N - 15).$$

Empirical relations between N-values and c_u (undrained shear strength), m_v (compressibility) and E (the deformation modulus) for cohesive soils have been derived by Stroud and Butler (1975). These are quite crude and only relevant to stiff clays. There are relations between N-values and friction angle (Peck *et al.* 1974), density (Terzaghi and Peck, 1967; Skempton, 1986) and compressibility of sand (Burland and Burbridge, 1985). N-values have also been used for classifying soft rocks (*see*, for example, Meigh and Wolski, 1979).

There may be occasions, when evaluating the effects of tunnelling-imposed settlements on buildings (*see* Supplementary Information 16 for information on calculating ground loss and consolidation movements in soil caused by tunnelling), to take into account the effects of allowable bearing pressures and also, perhaps, of pile design if a piled foundation has been adopted. Reference may be made to the widely used design chart by Terzaghi and Peck (1967) which relates allowable bearing pressure to foundation width and *N*-values, and also to Peck *et al.* (1974).

There are many correlations using *N*-values for settlement of foundations on sands and gravels, but they do tend to give widely differing results. The pseudo-elastic methods used by D'Appolonia *et al.* (1968) and Parry (1971) seem to have been most consistently reliable. However, the method put forward by Burland and Burbridge (1985) based on a large data set is probably to be preferred. Again, it must be emphasised that the SPT blow count can never be anything more than a crude indicator of compressibility.

Meyerhof (1976) provides charts of end bearing resistance and skin friction for bored and driven piles in sands. Here skin friction correlations are poor, and the relations take no account of variations in pile type or the hammer used to install them.

Table SI 5.1 relates the SPT blow count to the relative density of a soil, and is the one most frequently adopted at the site investigation stage. These values, however, take no real account of the fact that SPT is also a function of the effective stresses at the measurement depth, an omission that could be of particular significance in tunnelling ground investigation. Several proposals have been made for corrections (N_1) of the measured *N*-values (N), a linear relation between the two being expressed as

$$N_1 = C_N N.$$

Correction factor C can be related to effective overburden pressure σ'_0 (kN/m^2), ranging from about 1.75 at 25kN/m^2 through 1 at 100kN/m^2 and 0.7 at 200kN/m^2 to 0.45 at 500kN/m^2.

Values of N_1, together with values of density index (I_D) after Gibbs and Holtz (1957), are also given in Table SI 5.1.

Meyerhof also related the SPT *N*-value, density index I_D, and the effective overburden pressure σ'_0 in kN/m^2:

$$(N/I^2_D) = a + b(\sigma'_0/100)$$

where values of parameters a and b have been given by Skempton (1986). The characteristics of a granular soil can be represented by $(N_1)_{60}$ and $(N_1)_{60}/I^2_D$, where $(N_1)_{60}$ is the standard penetration resistance normalised to a UK rod energy ratio of 60% and an effective overburden pressure of 100kN/m^2. Values of $(N_1)_{60}$ after Skempton, appropriate to normally consolidated sands, are included in Table SI 5.1.

In addition to the technical factors, noted above, which affect standard penetration resistance, SPT is also influenced by material properties such as the shape and grading of the granular soil particles, the amount of overconsolidation and the amount of time during which the ground has experienced over-consolidation. The evidence generally suggests that SPT increases with increasing particle size, increasing overconsolidation ratio, and this time factor.

Table SI 5.1 Standard penetration test values

Relative density classification of cohesionless soils	SPT *N*-values (blows per 300mm penetration)	Density index $I_D\%$	Correction value $(N_1)_{60}$
Very loose	0-4	0-15	0-3
Loose	4-10	15-35	3-8
Medium dense	10-30	35-65	8-25
Dense	30-50	65-85	25-42
Very dense	>50	85-100	42-58

Problems can arise in loosely packed sands and in sands which contain appreciable quantities of highly compressible material such as mica, shell fragments, organic matter, or particles composed of soft materials such as carbonates. In such materials, where $N < 10$, it is suggested that other test procedures should be used.

SPTs give only very rough estimates of deformations likely to occur in stiff clays and soft rocks. Results from the test are also highly equipment- and procedure-sensitive.

CONE PENETRATION TEST (CPT)

In the cone penetration test, which was originally developed in the Netherlands for pile design, a steel cone of standard dimensions is thrust into the ground at a constant rate, and measurements are taken of the resistance to its penetration. The CPT is a useful tool for soil profiling, for interpolation of ground conditions between boreholes, information from which is used to calibrate the CPT, and for obtaining geotechnical parameters. It is the only *in situ* test able to provide a continuous stratification profile with depth, and can recognise thin layers of ground of only a few centimetres thickness which would usually remain unresolved during a conventional boring and sampling programme. The review of the test by Meigh (1987) is particularly appropriate, and the paper by Erwig (1993) may be consulted. British Standard 5930: 1981 describes only the basic principles of the test and the two types of equipment that are in widespread use: the mechanical and electrical systems. An international reference test procedure, published in 1988 (International Society for Soil Mechanics and Foundation Engineering, 1988) was based on earlier European recommended practice, but it also included further developments, in particular the piezo-cone which is discussed below.

When the CPT is used in granular soils, the most common correlations between geotechnical parameters that can be derived relate cone end resistance to relative density Dr, soil friction angle ϕ, and the deformation moduli E and G. Estimates of undrained shear strength c_u, stress history in terms of the overconsolidation ratio (OCR), and the deformation moduli may be made from the test in cohesive soils.

The cone comprises a $60°$ apex angle of area 1000mm^2 which bears on the ground, and usually a friction sleeve of area $15\,000\text{mm}^2$. Early friction cones had a thin tubular load cell in the annulus between the upper section of the friction sleeve and the cylindrical core of the instrument. As a result, the core was too thin and broke, and the load cell was also prone to damage. In current models the friction sleeve is attached to the cone at mid-height. Below this attachment point a load cell measures cone tip force and above it a second load cell measures the combined force of the tip and the sleeve. The cone is pushed into the ground at a rate of 20mm/s by means of hydraulic rams operating at a force of 2 tonnes (standard) or 10 (or 17) tonnes (stiff clays and dense gravels). Screw pickets or kentledge are used for reaction. In the case of a *mechanical cone*, the force to advance the cone over a 70mm deep section of a borehole is measured to give the cone end resistance q_c. After this 70mm thrust, the friction sleeve operates, and the force to move both the cone and the sleeve is measured. The difference in force gives the frictional resistance f_s. This test is repeated at 200mm intervals. In the case of an *electrical cone*, with a load cell behind the cone and strain gauges on the friction sleeve, these together provide a continuous output of q_c and f_s. The *friction ratio* (FR) is a measure of the respective contributions:

$$FR = 100(q_c/f_s)\%$$

Unfortunately, mechanical and electrical cones do not give identical results.

Angles of friction can be assessed for sands (Durgunoglu and Mitchell, 1975; Meyerhof, 1976). A cone factor N_k is used for correlating CPT with undrained shear strength c_u:

$$c_u = (q_c - \sigma_v)/N_k.$$

There are also correlations between CPT and shear vane strengths which depend on the plasticity

Table SI 5.2 Coefficient of constrained modulus for normally consolidated and lightly overconsolidated clays and silts (mantle cone) (after Sanglerat, 1979)

Soil	Classification	α_M ($=M/q_c$)	
		Mantle cone (M)	Reference tip (R)
Highly plastic clays and silts	CH,MH	2 to 6	2.5 to 7.5
Clays of intermediate or low plasticity	CI, CL		
$q_c < 0.7 MN/m^2$		3 to 8	3.7 to 10
$q_c > 0.7 MN/m^2$		2 to 5	2.5 to 6.3
Silts of intermediate or low plasticity	MI, ML	3 to 6	3.5 to 7.5
Organic silts	OL	2 to 8	2.5 to 10
Peat and organic clay	Pt, OH		
$50 < w < 100\%$		1.5 to 4.0	1.9 to 5.0
$100 < w < 200\%$		1.0 to 1.5	1.25 to 1.9
$w > 200\%$		0.4 to 1.0	0.5 to 1.25

Table SI 5.3 Coefficient of constrained modulus for overconsolidated clays and silts (mantle cone) (after Sanglerat, 1979)

Soil	Classification	α_M ($=M/q_c$)	
		$1.2 < q_c < 2.0$ (MN/m^2)	$q_c > 2.0$ (MN/m^2)
Highly plastic silts and clays	MH, CH	2 to 6	-
Clays of intermediate of low plasticity	CI, CL	2 to 5	1 to 2.5
Silts of low or intermediate plasticity	MI, ML	3 to 6	1 to 3

index (Lunne *et al.*, 1976). However, these authors found that a modified cone factor N_k^* could be derived that incorporated Bjerrum's correction to vane strength such that

$$c_{u(field)} = \mu c_{u(vane)} = (q_c - \sigma_v)/N_k^*$$

and did not depend on the plasticity index.

Sanglerat (1979) has tabulated values of α_M ($=M/q_c$), where M is the constrained modulus (m_v^{-1}). These values are shown in Table SI 5.2 (for normally consolidated and lightly over-consolidated clays and silts) and Table SI 5.3 (for overconsolidated clays and silts). As is the case with SPT values, q_c derived from a cone penetration test is unlikely to give a good indication of compressibility.

In weak soils there tends to be a loss of accuracy where readings are taken over the lower part of the measurement range, but accuracy can be improved by the use of amplified cones that are calibrated over different measurement ranges.

In chalk, CPTs are frequently used to check on weathering profiles and to probe for solution features or man-made cavities (Power, 1982; Bracegirdle *et al.*, 1989). The presence of flints can create a problem, causing considerable scatter in the results and sometimes excessive wear to the cones (Manby and Wakeling, 1990). Power (1982) gives a chalk grade classification based on CPT results, as well as relationships between SPT and Young's modulus with q_c, but both correlations are somewhat tenuous.

Refusal to penetrate weak rock occurs when either the maximum load capacity (100-150kN) of the cone or the total available thrust capacity (200kN) of the equipment is reached.

The piezo-cone has developed from the standard electrical cone test. It comprises a cone into which, or in the immediate vicinity of which, a porous filter has been inserted to measure, by means of a pore pressure sensor, the porewater pressure induced at the interface between the penetrometer tip and the soil during penetration. The measured pore pressure can increase or decrease in response to compression or dilation of the saturated soil as a result of shearing in the locality of the cone. This added information on pore pressure, together with a knowledge of the friction ratio, can assist in the interpretation of soil type, and equilibrium groundwater conditions (piezometric profile) can also be determined during a stop in penetration. As would be expected, piezo-cone results are affected by the stress conditions that pertain in the ground at the tip of the cone. At Oxford University, Houlsby and Hitchman (1988) have shown that for a sand compacted at a specified density there is a firm relation between cone end resistance q_c and the *in situ* horizontal stress, with the angle of soil friction being capable of determination from q_c when the effective horizontal stress $\sigma_h{'}$ is estimated. In the case of normally consolidated and overconsolidated clays, piezo-cone results seem to be affected by the overconsolidation ratio and there is some evidence of relations between pore pressure ratio, resistance at the tip and the overconsolidation ratio.

There are several other types of cone - seismic

cone, density-measuring probe, lateral stress cone, nuclear density probe, electrical density probe, electrical conductivity cone, and thermal conductivity cone. These have been reviewed by Mitchell (1988). The seismic cone is particularly significant in that shear wave velocities c_s in the ground can be obtained during the profiling, and from c_s can be derived the low-strain dynamic shear modulus G which seems to correlate well with G obtained from other low-strain laboratory tests. The shear wave velocity is a useful parameter in its own right to derive, since its value for a particular type of soil is a function of moisture content.

Using equipment capable of providing a thrust of 20 tonnes, penetrations of up to 40m or 50m in weak soils and 15m to 25m in stiff to very stiff clay soils or in medium to dense sands can be achieved. There are certain ground conditions, such as sands below the water table, when the SPT values may be incorrect due to possible ground disturbance, in which case the CPT tends to come into its own since the results from this test are not affected by this problem. In homogeneous ground, repeatability of CPT results exceeds those from the SPT. CPT test penetrations of up to 150m/day or even 200m/day can be achieved under good ground conditions.

REFERENCES

Bracegirdle, A., Mair, R.J. and Daynes, R.J. (1989) Construction problems associated with an excavation in chalk at Costessey, Norfolk, *Int. Chalk Symp.*, Brighton, Thomas Telford, London, 385-391.

Burland, J.B. and Burbridge, M.C. (1985) Settlement of foundation on sand and gravel, *Proc. Inst. Civ. Eng.*, Part 1, **78**, 1325-1381.

D'Appolonia, D.J., D'Appolonia, E.E. and Brissette, R.F. (1968) Settlement of spread footings on sand, *Amer. Soc. Civ. Eng.*, **94**, SM3, 735-760.

Durgunoglu, H.T. and Mitchell, J.K. (1975) Static penetration resistance of soils, *Am. Soc. Civ. Engrs*, Raleigh (N. Carolina), June 1975, **1**, 151-188.

Erwig, H. (1993) Development of electric static cone penetration testing in Great Britain from 1972-1992, *Ground Engrg*, **26**(2), 30-34.

Gibbs, H.J. and Holtz, W.G. (1957) Research on determining the density of sands by spoon penetration testing, *Proc. 4th Int. Conf., Soil Mech and Found. Engrg.*, London, **1**, Butterworths, London, 35-39.

Houlsby, G.T. and Hitchman, R. (1988) Calibration chamber tests of a cone penetrometer in sand, *Géotechnique*, **38**(1), 39-44. Also Discussion: *Géotechnique* **39**(4), 727-731.

International Society for Soil Mechanics and Foundation Engineering (1988) Technical Committee on Penetration Testing, CPT Working Party. In *Proc. 1st Int. Symp. on Penetration Testing*, Orlando, A.A. Balkema, Rotterdam, **1**, 27-51.

Lunne, T., Eide, O. and de Ruiter, J. (1976) Correlations between cone resistance and vane shear strength in some Scandinavian soft to medium clays, *Canad. Geotech. J.*, **13**(4), 430-441.

Manby, C.N.D. and Wakeling, T.R.M. (1990) Developments in soft-ground drilling, sampling and *in situ* testing, *Trans. Inst. Min. Metall.*

(Sect.A: Min. industry), **99**, A91-A97.

Meigh, A.C. (1987) *Cone Penetration Testing, Methods and Interpretation*, CIRIA Ground Engineering Report: In-situ Testing, Butterworths, London, 141pp.

Meigh, A.C, and Wolski, W. (1979) Design parameters for weak rock, *Proc 7th European Conf, Soil Mechanics and Foundation Engineering*, Brighton, **5**, 59-79.

Meyerhof, G.G. (1976) Bearing capacity and settlement of pile foundations, *Amer. Soc. Civ. Eng.*, **102**, GT3, 195-228.

Mitchell, J.K. (1988) New developments in penetration tests and equipment, *Proc 1st Int. Symp. on Penetration Testing*, Orlando, A.A. Balkema, Rotterdam, **1**, 245-261.

Parry, R.H. (1971) A direct method for estimating settlements in sand from SPT values, *Proc. Symp. on Interaction of Structures and Foundations*, University of Birmingham, 29-37.

Peck, R.B., Hanson, W.E. and Thorburn, T.H. (1974) *Foundation Engineering*, Wiley, New York.

Power, P.T. (1982) The use of the electric static cone penetrometer in the determination of the engineering properties of chalk, *Proc. 2nd European Symp. Penetration Testing*, Amsterdam (Netherlands National Society for Soil Mechanics and Foundation Engineering), **2**, 769-774.

Sanglerat, G. (1979) *The Penetrometer and Soil Exploration*, Second Edition, Elsevier, Amsterdam.

Skempton, A.W. (1986) Standard penetration test procedures and the effects in sands of overburden pressure, relative density, particle size, ageing and overconsolidation, *Géotechnique*, **36**(3), 425-447.

Stroud, M.A. and Butler, F.G. (1975) The standard penetration test and the engineering properties of glacial materials, *Proc. Symp. on Engineering Behaviour of Glacial Materials*, University of Birmingham, England.

Terzaghi, K and Peck, R.B. (1967) *Soil Mechanics in Engineering Practice*, Wiley, New York.

BIBLIOGRAPHY

Campanella, R.G., Robertson, P.K. and Gillespie, D. (1986) *Seismic Cone Penetration Test. Use of In-situ Tests in Geotechnical Engineering*, ASCE Geotech. Spec. Publ. No. 6, 116-130.

Kulhawy, F.H. and Mayne, P.W. (1994) On interrelationships among mechanical, electric, and piezoelectric cone penetrometer test results, *Proc. 7th Int. Congress IAEG*, Lisbon, 5-9 September 1994, R. Oliveira, L.F. Rodrigues, A.G. Coelho and A.P. Cunha (eds), 229-235.

Livneh, M. and Livneh, N.A. (1994) Subgrade strength evaluation with the extended dynamic cone penetrometer, *Proc. 7th Int. Congress IAEG*, Lisbon, 5-9 September 1994, R. Oliveira, L.F. Rodrigues, A.G. Coelho and A.P. Cunha

(eds), 219-227.

Nixon, I.K. (1982) Standard penetration test, state of the art report, *Proc. Europ. Symp. on Penetration Testing II*, Amsterdam, **1**, 3-24.

Sully, J.P. and Echezuria, H.J. (1988) In-situ density measurement with nuclear cone penetrometer, *Proc. 1st Int. Symp. on Penetration Testing*, Orlando, A.A. Balkema, Rotterdam, **2**, 1001-1005.

Tanaka, Y., Karube, D. and Tanimoto, K. (1994) Engineering properties of soft marine clay as studied by seismic cone testing, *Proc. 7th Int. Congress IAEG*, Lisbon, 5-9 September 1994, R. Oliveira, L.F. Rodrigues, A.G. Coelho and A.P. Cunha (eds), 323-329.

6

MUDROCKS

Because of the widespread presence of mudrocks within the temperate regions of the world and the fact that many tunnels are driven through these rocks it is useful to note rather briefly some of their compositional, structural and strength characteristics which can contribute to an engineering classification.

• The disruptive processes most commonly reported to affect mudrocks used in construction in the UK are dissolution of calcite and oxidation of pyrite. Mudrocks formed under marine conditions are much more likely to contain these minerals than their non-marine counterparts. Because of this, and the increased incidence of fine laminae build-up under slow rates of sedimentation in deep marine environments, marine mudrocks are much more susceptible to both chemical and physical weathering than are non-marine mudrocks.

• Textural features in a mudrock (fabric and microcracking, mainly as a result of air-drying) may affect in a much more significant way the longer-term weathering and disintegration of mudrocks than will the mineral composition (Olivier, 1990). The rock may thus shrink anisotropically, and to some extent irreversibly, and there may be similar partial irreversible expansion when subjected to capillary action stemming from moisture absorption, the moisture coming from atmospheric humidity or from direct wetting. Excavation operations - mechanical pick/disk action or blasting - can help to open up small fractures. After excavation the rock should be given an impermeable spray coating as soon as possible in order to inhibit slaking action.

• By definition, mudrocks contain substantial quantities of clay minerals. These minerals can be checked by X-ray diffraction methods (*see* Supplementary Information 7). An illite-rich mudrock will have originated from warm, upland regions, in contrast to a kaolinite-rich rock which will have had its source in tropical lowland swamps. A greater presence of vermiculite and smectite indicates increased weathering in the source area. The presence of kaolinite at the expense of chlorite reflects an overall increase in humidity (sub-tropical weathering) in the formational environment. Marls should be considered as mudrocks. For example, the Mercia Mudstone (previously known as the Keuper Marl) comprises aeolian silts, dominantly illitic (70-80%), but with some chlorite, expandable mixed layer clays and authigenic clays such as sepiolite, palygorskite, corrensite and attapulgite.

• The boundary between mudrocks and stiff clays is not readily discernible. From work on Mesozoic and Tertiary rocks which are not heavily indurated, Morgenstern and Eigenbrod (1974) suggested defining mudrock by a procedure of allowing the material to soften in water and then testing its strength. If the cohesive strength is then $>1.72 \text{ MN/m}^2$ (UCS $> 3.45 \text{MN/m}^2$), and this strength also does not drop more than 40%, the material is deemed to be a mudrock. Applied to British clays and mudrocks, this criterion indicates that only rocks of Lower Lias age and older are likely to be durable and to be mudrocks. Younger beds would at best be stiff and hard clays. Grainger (1984) placed the limit at a UCS value of 3.6MN/m^2 together with a slake durability of at least 90%. Grainger (1984) also placed an upper strength (UCS) limit of 100MN/m^2 ('strong' to 'very strong' boundary) on mudrocks. Above this strength it is implied that they should not be called mudrocks; they could be perhaps an argillite (a non-fissile and slightly meta-morphosed mudrock), or a slate (fissile and metamorphosed mudrock), or a meta-argillite (a non-fissile and metamorphosed mudrock).

(Following from the quantitative description of a mudrock, a clay (soil) will have an initial UCS greater than 3.6MN/m^2 and would lose more than 60% of this strength on soaking. A further classification sub-division is based on the time of softening during which more than 50% of the initial UCS is lost.)

• An average unconfined compressive strength (UCS) for a range of mudstones and shales is $10.2MN/m^2 \pm 7.10$ and for siltstones is $19.27MN/m^2 \pm 7.92$.

• A reasonable relation between UCS and point load strength (I_s) is:
  – UK Coal Measures shales and mudstones (excluding siltstones):
    UCS $(MN/m^2) = 4.70I_s \pm 1.68$ at natural moisture content.
  - Overconsolidated clays and shales, of Oligocene and Devonian age:
    UCS $(MN/m^2) = 5.6I_s \pm 1.64$ at natural moisture content.

• Mead (1936) differentiated between mudrocks and overconsolidated clays on the basis of interparticle cementation:
  - Compacted or soil-like shales
      little cementation.
  - Cemented or rock-like shales
      significant cementation.
He suggested using the slaking test to differentiate between the two.

• Any leaching by groundwater removes salts in the pore solution. This decrease reduces the activity, liquid limit, and the undisturbed and remoulded shear strength of the material. Also, at any given temperature the shear strength decreases as the pH of the permeating fluid decreases. Any change in porewater chemistry that removes $CaCO_3$ causes a decrease in the cementation strength, resulting in increased compressibility and reduced shear resistance in the material.

Table SI 6.1 Mudrock terms (after Stow, 1981)

Basic Terms

Unlithified	Lithified/non-fissile	Lithified/fissile	Approximate proportions (grain size)
Silt	Siltstone	Silty-shale	>2/3 silt-size (2-63μm)
Mud	Mudstone	Shale	Silt & clay mix (<63μm)
Clay	Claystone/mudstone	Clay-shale	>2/3 clay-size (<2μm)

Metamorphic Terms

Name	State	Composition
Argillite[+]	Slightly metamorphosed/non-fissile	Silt & clay mixture
Slate	Metamorphosed (heat & pressure)/fissile	Silt & clay mixture

Other Terms

Textural Descriptive Terms	Approximate Proportions
Silty	>10% silt-size
Muddy	>10% silt- or clay-size (applied to non-mudrock sediments)
Clayey	>10% clay-size
Sandy, pebbly, etc.	>10% sand-size, pebble-size, etc.
Composition Descriptive Terms	**Approximate Proportions**
Calcareous	>10% $CaCO_3$ (foraminiferal, nannafossil, etc.)
Siliceous	>10% SiO_2 (diatomaceous, radiolarian etc.)
Carbonaceous; pyritiferous	> 1% organic carbon
Ferruginous, micaceous, and others	commonly adopted for contents > about 1-5%

- *Cation exchange*: Substitution of sodium by potassium into the clay mineral lattice causes an increase in shear strength in illites and montmorillonites *in their undisturbed state*. As a result of the substitution, the *remoulded strength* is increased even more (a sensitivity decrease from 7.5 to 4.5). However, a remoulded potassium montmorillonite has a much lower shear strength than a corresponding sodium montmorillonite.

- There are several *classifications* other than those based on strength.

(a) Grain size: A mudrock does not disaggregate easily, but if it can satisfactorily be broken down then a mudrock would comprise more than 50% passing a British Standard 63μm sieve (or a 75μm US sieve).

(b) Size and composition: These values are given in Table SI 6.1. In this table,

- Definition of a mudrock:
  >50% siliciclastic* ; >50% less than 63μm.
- Siliciclastic means minerals having a silicate composition derived from a pre-existing landmass.
- An argillite (marked +) has been heated at depths greater than about 6000m.

(c) Average mineral composition: These percentages are given in Table SI 6.2.

Table SI 6.2 Average* mineral composition (%) of mudrocks

Minerals	References				
	Yaalon (1962)	Shaw & Weaver (1965)	Pettijohn (1975)	Smith (1978) UK	N. America
Clay minerals	59	66.9	58	60+	61
Quartz	20	36.8	28	29	36
Feldspar	8	4.5	6	1	3
Carbonates	7	3.6	5	4	3
Iron oxide	3	0.5	2	3	1
Organic carbon	-	1.0	-	1	1
Miscellaneous	-	0.2	-	-	-

Notes:
* >50% siliciclastic grains of size <63μm. In these averages:
siliciclastics clay + quartz + feldspar = 94% of which 60% are clay minerals and 26% is quartz.
+ Figure distorted by the presence of fuller's earth.

Table SI 6.3 Mudrock classification based on quartz percentage (after Spears, 1980)

Quartz percentage	Non-fissile rock	Fissile rock
>35	Siltstone	Flaggy siltstone
30-35	Very coarse mudstone	Very coarse shale
20-30	Coarse mudstone	Coarse shale
10-20	Fine mudstone	Fine shale
>10	Very fine mudstone	Very fine shale

(d) **Quartz content** (*see* Table SI 6.3). Quartz content is also a useful parameter for defining the material boundaries of the mudrock (Spears, 1980).

In respect of Table SI 6.3, >65% quartz may be suggested for the sandstone boundary (but there will be a notable feldspar presence also in sandstone). A figure of 35% quartz marks the siltstone/mudstone boundary.

As the silt/clay size ratio increases, then so does the quartz/clay minerals ratio. Since quartz content affects physical and engineering properties, Spears (1980) has recommended that quartz content, determined directly, be used as a means of classification.

(e) **Ratio of combined silica to alumina:** This information can only be derived by the use of X-ray diffraction techniques (*see* Supplementary Information 7). The combined silica comprises 'free' (detrital) quartz and the silica bound in the clay minerals. A ratio approaching 3.3 is indicative of the relatively inert clay mineral illite and a low ratio suggests the presence of kaolinite. Tests on numerous mudrocks show a well-defined decrease in the ratio with increasing quartz content.

(f) **Thickness of laminations** (*see* Table SI 6.4).

(g) **Fissility** (*see* Table SI 6.5).

Lundegard and Samuels (1980), on the other hand, claim that fissility is not a valid classification feature since it develops very late in the rock's history as a weakening phenomenon and does not really relate to primary depositional features.

Table SI 6.4 Classification of laminated shales on the basis of lamination thickness
(after Potter *et al.*, 1980)

Terminology	Thickness (mm)
Very thin	<0.5
Thin	0.5 - 1
Medium	1 - 5
Thick	5 - 10

Table SI 6.5 Classification of laminated shales on the basis of fissility
(after Potter *et al.*, 1980)

Terminology	Thickness (mm)
Paper	<0.5
Fissile	0.5 - 1
Platy	1 - 5
Flaggy	5 - 10
Slabby	-
Blocky	-
Massive	-

REFERENCES

Grainger, P. (1984) The classification of mudrocks for engineering purposes, *Quart. J. Engrg Geol.*, **17**, 381-387.

Lundegard, P.D. and Samuels, N.D. (1980) Field classification of fine-grained sedimentary rocks, *J. Sed. Pet.*, **50**, 781-786.

Mead, W.J. (1936) Engineering geology of dam sites, *Trans. 2nd Int. Congr. Large Dams*, Washington, D.C., **4**, 183-198.

Morgenstern, N.R. and Eigenbrod, K.D. (1974) Classification of argillaceous soils and rocks, *Proc. Geotech. Div., Amer. Soc. Civ. Engrs,*

100, 1137-1156.

Olivier, H.J. (1990) Some aspects of the engineering geological properties of swelling and slaking mudrocks, *Proc. 6th Int. IAEG Congress,* Rotterdam, **1**, 707-712, A.A. Balkema, Rotterdam.

Pettijohn, F.J. (1975) *Sedimentary Rocks,* Harper and Bros, New York, 618pp.

Potter, P.E., Maynard, J.B. and Pryor, W.A. (1980) *Sedimentology of Shale,* Springer-Verlag, N.Y., 306pp.

Shaw, D.B. and Weaver, C.E. (1965) The mineralogical composition of shales, *J. Sed. Pet.*, **35**, 213-222.

Smith, T.J. (1978) Consolidation and other geotechnical properties of shales with respect to age and composition, PhD thesis, University of Durham, England, 452pp.

Spears, D.A. (1980) Towards a classification of shales, *J. Geol. Soc.* London, **137**, 125-129.

Stow, D.A.V. (1981) Fine grained sediments: terminology, *Quart. J. Engrg Geol.*, **14**(4), 243-244.

Taylor, R.K. and Spears, D.A. (1981) Laboratory investigation of mudrocks, *Quart. J. Engrg Geol.*, **14**, 291-309.

Yaalon, D.H. (1962) Mineral composition of average shale, *Clay Minerals, Bull.*, **5**, 31-36.

7

MINERALOGICAL COMPOSITION OF ROCKS

Geological terminology will often give us a reasonable indication of the petrographic nature of a rock and a broad range of strength. A sandstone will contain a substantial content of free quartz which will abrade cutting picks and, if the rock is strong, render excavation difficult. When the grain sizes are within the capacity of a petrographic microscope it is important to estimate the amount (volume) and size of the quartz grains, and the nature of the cementing material which is often of an amorphous nature. A point count facility is used for estimating the relative volumes of the different minerals and a graticule or calibrated micrometer eyepiece is used for estimating grain size.

The analytical technique requires the preparation of thin sections of rock, stabilised with Canada balsam or Lakeside cement, and the mounting of the sections between cover-strips for observation through a transmitted light microscope. The method of preparation is described in Brown (1981).

Chemical composition, as distinct from mineralogical content, can be determined by atomic absorption spectrophotometry or by X-ray fluorescence analysis. Fine-grained rocks - shales, mudrocks, clays - can be identified and their composition determined by infra-red absorption spectrography, differential thermal analysis, or X-ray diffraction.

X-RAY DIFFRACTION ANALYSIS

The technique of X-ray diffraction analysis has particular application to the identification and quantification of clay minerals (see, for example, Klug and Alexander, 1954, for full details of the methods that can be used). Mineral analysis may be achieved by comparing peak positions on a diffractogram to data given by the Joint Committee on Powder Diffraction Standards (JCPDS). There are several important factors in the implementation of X-ray analysis and mineral determination. Both text books and appropriate learned journal papers need to be

consulted, but the processes include the following:

(a) **Choice of tube.** Although a Cu tube tends to be preferred for the analysis of clay minerals having low levels of iron in their composition, if high levels of Fe are present, then due to the diffraction height backgrounds caused by iron fluorescent radiation (FeK_α) an iron or cobalt tube is preferred.

(b) **Treatment of the material to be analysed.** It may be necessary chemically to pre-treat an X-ray sample of rock in order to remove cementing agents. Carbonates can be dissolved out by treatment with a weak organic acid (for example, by heat in a buffered solution of sodium acetate and glacial acetic acid), organic material by treatment with 35% hydrogen peroxide (H_2O_2), and iron compounds using sodium dithionite ($Na_2S_2O_4$) together with a citrate-chelating agent. Clay mineral powders should be washed several times in distilled water to remove soluble salts and then the $<2\mu m$ fraction from sedimentary analysis transferred to a beaker of distilled water having a glass slide on the bottom, the water then being allowed to evaporate at 60°C in a warm oven. Such a procedure produces an orientated sample due to the platy nature of the clay minerals, but the sample may be inhomogeneous as a result of the differential settling of particles of different size. This problem may be partially overcome by vacuum-settling a clay mineral suspension on to a porous substrate (such as a ceramic tile). Other methods of (partially) overcoming preferred orientation effects include adding a non-crystalline material to inhibit clay mineral orientation without complicating the diffraction pattern, or embedding the sample in a resin and then re-grinding to provide approximately equant particles which will not orientate.

(c) **Detection of expandable clay minerals.** For a qualitative identification of these minerals a diffraction pattern is produced from the original powder. This pattern comprises a series of peaks above background radiation on the chart recorder output, each peak denoting an integrated reflection from a single crystallographic plane in all the diffracting crystals exposed to the incident radiation or a combined integrated reflection from more than one plane. (Such composite peaks can sometimes be separated out by changing the X-ray tube to one having a different target anode.) In the case of smectites and certain mixed layer clays the spacing of the important basal (001) crystallographic planes varies, and so the diffraction peaks are often broad and asymmetric. These problems can often be alleviated by then placing the slide containing the powder in a desiccator in which the drying agent has been replaced by ethylene glycol or a similar organic liquid. After leaving for 4 to 6 hours in an oven at 60°C the expandable species will have absorbed the chemical at the expense of water and the basal spacing will have expanded to its maximum value of 17 angstrom units (Å), with an associated shift of the diffraction peak on re-analysis. By this means, the reflection will not interfere with that of chlorite 14Å. Further heating to a temperature of 375°C will then collapse the smectite (and illite-smectites) to a basal plane spacing of 10Å without affecting the other clay minerals. The effects of this should again be detectable on the X-ray diffraction chart output. Even further heating of the slide to a temperature of 550°C for a period of 2 to 4 hours will destroy any kaolinite present. This is an important procedure because in the untreated state, with radiation from a copper anode, there will be interference both between the kaolinite 001 peak at 7.1Å and the chlorite 002 peak at 7.15Å and between the kaolinite 002 peak at 3.58Å and the chlorite 004 peak at 3.55Å. Elimination of kaolinite means that any remaining diffraction peak can be attributed to chlorite. Mixed layer clays present many problems of interpretation; reference may be made to Reynolds and Hower (1970).

(d) **Quantitative mineral analysis.** Peak intensity is best represented by integrating the area under a peak and the estimated background. One quick method is to use the height of the peak multiplied by its width at half height. Sometimes the height of the peak can be used on its own. The relative intensities of the peaks may give some indication of the relative abundances of each mineral present. However, different mineral species show different responses to X-rays. Probably the most popular method (Griffin, 1954) involves adding internal standards first to a single mineral powder and then to the sample requiring analysis. The mineral boehmite is a most useful internal standard because it projects a diffraction peak close to the major basal plane peaks of the clay minerals. The boehmite peak does not interfere with the clay mineral peaks, it has similar absorption characteristics to them, and it is of relatively high intensity when it is used in low concentrations. Reference should be made to Griffin (1954) for details of the method, which is particularly appropriate for the identification of minerals which are capable of causing severe ground engineering problems.

REFERENCES

Brown, E.T. (ed) (1981) *Rock Characterization, Testing and Monitoring, ISRM Suggested Methods, Suggested Method for Petrographic Description of Rocks*, 75-77, Pergamon Press, Oxford, England.

Brown, G. and Brindley, G.W. (1980) X-ray diffraction procedures for clay mineral identification. In: *Crystal Structures of Clay Minerals and their X-ray Identification*, G.W. Brindley and G. Brown (eds), Mineralogical Society, London, 305-360.

Griffin, O.G. (1954) *A New Internal Standard for the Quantitative X-ray Analysis of Shales and Mine Dusts*, Res. Rep. No. 101, S. Afr. Mines Res. Est., 1-25, Min. of Fuel and Power, Sheffield.

Hardy, R. and Tucker, M.E. (1988) X-ray powder diffraction of sediments. In *Techniques in Sedimentology,* M. E. Tucker (ed), Blackwell Scientific Publications, 191-228.

Klug, H.P. and Alexander, L.E. (1954) *X-ray Diffraction Procedures for Polycrystalline and Amorphous Material,* Wiley, New York.

Reynolds, R.C. and Hower, J. (1970) The nature of interlayering in mixed-layer illite-montmorillonites, *Clays Clay Minerals,* **18,** 25-36.

8

GROUT INJECTION

GENERAL

Some grouts act mainly as water stops to reduce the porosity (and permeability) of the soil before the use of, or as substitute for, compressed air. These grouts are not intended to increase the strength of the ground to any significant extent, since excavation must proceed without difficulty. Other grouts greatly enhance the compressive strength of the ground, so causing it to require more excavation effort. As a general procedure, grout should be injected in small quantities at closely spaced intervals, and repeated.

COMPENSATION GROUTING

Stiff, high-viscosity grouts may be used to compact (densify) *granular soils* and may also be used to fill known voids - for example, around building foundations and pipes that may have settled. Low viscosity grouts may be injected at pressures sufficiently to actually split *cohesive soils* and so create fissures along which the grout flows. This latter process can be conducted in successive injection stages, with sufficient time being allowed for grout hardening between injection stages. An alternative injection process for *granular soils* is pressure filtration grouting, which involves the injection of a low-viscosity grout having a high solids content. The grout is designed to bleed soon after injection so that the solid particles are deposited in the soil near to the point of injection.

Grout injected under pressure above a tunnel lining can be used to compensate for the effects of relatively large (say, greater than 12mm) tunnel-induced settlements in built-up areas. Typically, the aim is, after each shove of a shield, to place a low slump cement, having about 13:1 pulverised fuel ash/cement mix with 4% bentonite, through holes drilled from ground surface and thereby replace the volume of ground lost around the tunnel during advance. At suitable pressures the grout tends to densify the soil surrounding the grout bulb, heaving the overlying soils and pushing down on the tunnel lining. The cost of an operation such as this, and the possible difficulties of access for drilling, have to be balanced against the projected savings in building reinstatement costs. Also to be taken into account is the effect of the grouting on pipes and cables in the ground and the fact that in some instances it has been claimed that the drilling for the grout tubes has induced more settlement than that attributable to the tunnelling alone. It must be notes that common law claims for attributable damage can be made up to six years after completion of construction (*see* Section 7.10.2 in Part One of the book on the subject of latent damage).

Compaction grouting was successfully used in the late 1970s in the USA for building protection on the Baltimore subway construction. The technique, employing careful monitoring (electrolevels), has more recently been used in London on the Crossrail tunnels and at Waterloo station (Harris *et al.*, 1994), and is planned to be used on the Jubilee underground line extension. Its adoption has also been considered for the Los Angeles Metro Red Line in front of Mann's Chinese Theater on Hollywood Boulevard, the well-known tourist attraction. Pressure filtration grouting used at Waterloo Station, undertaken through *tubes-à-manchette*, was directed mainly at the Thames Gravel/London Clay interface and was used to prevent a London Underground 8m diameter tunnel from affecting the Victory Arch there and the Waterloo and City Line tunnels a few metres above. (*Tube-à-manchette* grouting will characteristically provide uniformity of treatment and is more or less essential where the layers to be treated are only a few metres thick.) It is claimed that by this means the settlement was limited to 10 to 15mm, with no consequential damage to the superadjacent structures, whereas settlements in the absence of compensation grouting had been estimated at 50 to 100mm with considerable damage.

Sometimes, as an alternative to the use of slurry walls for isolating tunnels from adjacent properties, it may be necessary to grout soil beneath a building in order to provide added 'underpinned' support before tunnelling. Grouting may be from the ground surface ahead of the tunnel or, more expensively, from a pilot tunnel or from (and ahead of) the main tunnel.

The types of injection grouts range from particulate (suspension), usually cementitious, grouts through emulsions, such as bitumens and foams, to solutions such as silicate grouts, to organic polymer grouts, the choice being determined by cost and by the particle size of the soil into which the injection is to be made.

Suspension (particulate) grouts

For a *cement grout*, ordinary Portland cement is mixed in slurry with a water-cement ratio of 0.6 or less. This grout is both inexpensive and strong. It is used in particular for grouting fractured rocks in order to reduce their permeability and increase their strength. Additives can be used to accelerate setting (eg. calcium chloride), to achieve expansion (eg. Ferrogrout), to improve lubrication (eg. bentonite), and to lower the viscosity (eg. Lignasol). Fine sand can be added in order to achieve bulk when grouting open fissures and cavities in rocks. Fluid cement grouts have a relatively weak structured framework and, although behaving generally in a 'plastic' manner rheologically, they function as Newtonian fluids when very dilute.

In the case of a particulate grout, the maximum size d_p of particle capable of passing in suspension through the minimum pore cross-section of a soil will have a diameter in the region of $0.15d_0$, where d_0 is the diameter of the soil particles, assumed to be of spherical shape. In practice, the maximum diameter, d_p, of injection particle may be taken to be $0.1d_0$, where d_0 is equated to the D_{10} soil grain size in a mass of irregular-size particles. D_{10} is the size below which 10% of the particles will be finer, and is determined at the site investigation stage by sieve analysis. Ordinary portland cements normally contain particles up to $100\mu m$ in size, and this restricts the use of these cement grouts to medium sands and coarser. Flow through holes (pore space or fissures) having an opening smaller than about three times this diameter can

lead to a separation of cement and water, leading to bleeding and a reduction in grout strength. High early strength cements having a maximum grain size of 20 microns ($m \times 10^{-6}$) could be used to grout all sands having a permeability as low as $10^{-4}m/s$. Prediction of groutability from *in situ* permeability tests, or vice versa, may be made on the basis of the Hazen formula which was developed for single-size filter sands and gives an approximate value for hydraulic conductivity:

$$\text{permeability } k(\text{m/s}) = C(D_{10})^2/10^4.$$

In this equation D_{10}, in millimetres, is the size at which 10% of the particles are finer, and C is a constant which varies from about 70 to 170, but which for single-size material and for a first approximation of permeability would be taken as equal to 100. Thus, taking C equal to 100 the permeability can be written as

$$k \text{ (m/s)} = (D_{10})^2/100.$$

For the operation of this equation the sand is graded by particle size distribution in accordance with BS 1377 (British Standards Institution, 1990).

Natural pozzolans consist of volcanic glass, clay minerals and hydrated aluminium oxides. Pulverised fuel ash (pfa), an artificial pozzolan, is often used for grouting on its own or mixed with cement. When cement grouts hydrate, lime is produced and this can have a weakening effect. If pfa is present it reacts with the lime and water to produce a stable cementitious material. A pfa/cement grout tends, therefore, to be stronger and longer lasting than a cement grout. The pfa particle size approximates to that of a Portland cement and so it can be used as a filler in a cement grout without affecting its injectability. In its bound state as a grout, pfa poses few or no environmental problems. Because of its low cost and wide availability it also tends to be adopted as an bulk infilling medium for larger cavities, a free-flowing 50%-50% pfa-water mixture being used. However, because of potential pollution problems with pfa in this unbound form there are environmental objections to its use for this purpose and it is likely that it will be banned by the European Union. Alternatives as fillers in cement grout for the treatment of large openings in the ground are fly-ash or pea gravel (preferred).

Since the particle size of *clays* is generally less than 2μm, these natural materials can be used to grout soils having permeabilities as low as 10^{-4} to 10^{-5}m/s. However, clay minerals such as montmorillonite, illite and kaolinite are often preferred in clay grouts since these minerals possess good water absorption and gelling characteristics. The clay must be substantially free from silt or sand particles because, as discussed above, these can inhibit grout movement through a soil structure. Clay can also be used as a filler in cement grouts for the improvement of soils.

Dilute clay suspensions behave somewhat in a Newtonian manner. Clay grouts will behave approximately as Bingham materials and often exhibit thixotropy. A pH of 8 to about 10.5 (slightly alkaline) avoids flocculation of the clay particles.

Bentonites are montmorillonite-rich clays derived from weathered volcanic ash. Bentonite suspensions containing between about 5% and 25% solids behave rheologically as Bingham bodies in which the shear strength and the viscosity increase almost exponentially with the clay content. Here the suspensions behave thixotropically, flowing freely if disturbed but forming an elastic gel if remaining undisturbed. This thixotropic behaviour can be changed by the addition of chemicals such as sodium silicate and sodium polyphosphate. *Cement-bentonite grouts* can be used both to reduce the permeability of, and increase the strength of fissured rocks and granular soils, with the bentonite helping to prevent sedimentation of the cement particles in the mix.

Emulsions

Bitumens (coal tars and petroleum asphalts) resist attacks by water, acids and alkalis, and so can be applied where other grouts would be washed away or dissolved by aggressive ground and groundwater. *Bitumen emulsions*, having a low viscosity approximating to that of water, usually contain 55%-65% bitumen, 45%-55% water and 1% emulsifier, usually a form of soap. Cold bitumen solutions with cement may be sprayed onto the surfaces of newly exposed soil and rock surfaces to improve their strength,

reduce degradation, and resist water ingress. Hot bitumen solutions can be used to grout medium sands but they offer only low cohesive strength. Bitumen is also suitable for grouting fissured rocks in cases where high water flows would have removed other grouts, since the bitumen strength increases quickly when it meets the cool water and so seals the fissures securely.

Foam grouts are gaseous emulsions of ordinary (usually cement-based) grouts and have a stiffness which resists flow when the foam leaves the injection pipe. The gas-liquid volume ratio can be up to 3:1 for cement-based foams or 50:1 for organic foams. Surface tension of the liquid surrounding the foam bubbles gives rigidity to the foam. Preparation of the foam is by adding soap solution and agitating or by adding suitable materials which will react together and produce gas. The actual method of preparation controls the size and shape of the bubbles in the foam, and these determine the stability, flow and expansion features of the foam. Also, organic materials containing polyurethane, an oil-based polymer having a relatively high viscosity, can be designed to foam in water and the foam, although weak, assist in preventing water circulation in open fissures during early grouting operations.

Solutions

Chemical grouts, which are more expensive than other forms of grout, behave as Newtonian fluids with a lower viscosity than chemical grouts. Because they are non-particulate they can be used to treat fine-grained soils and rocks having very small discontinuities.

Silicate grouts are based on an aqueous solution of sodium silicate, with chemical additives giving an irreversible time-dependent gelation of the solution. Initial viscosity and final gel strength increase with increasing sodium silicate in the solution, so the mix design is essentially a compromise between achieving optimum flow characteristics while still obtaining a good gel strength. Silicate grouts having a low viscosity of about 2 to 5 centipoise can still be injected at rates up to ten times higher than those achieved with cement grouts. They do not shrink much on setting and resist attack by salt water and sulphate-rich groundwater.

The chemical compositions of silicate grouts vary depending upon the use to which the grout is put. The well-known Joosten I process involves a double injection, or two-shot technique, in which the injection of a concentrated sodium silicate solution is followed by a strong salt solution, the reaction between the two components leading to gelation. The Joosten II method uses a first shot of alkaline sodium silicate solution followed by calcium chloride. The Joosten III method is a one-shot process involving the injection of sodium silicate in a solution containing a heavy metal salt and ammoniacal colloid. This one-shot process is less expensive but produces a lower strength than a two-shot process. The two-shot process requires a closely spaced grid of boreholes for the grouting so that the two components are suitably mixed. Set retarders can be used in one-shot grouts to lower viscosity and thereby grout finer-grained soils using hole spacings as high as 3m and quite low grout injection pressures. In the Guttman process, viscosity of the silicate is reduced by dilution with sodium carbonate to improve penetration but at the expense of lower gel strength. One-shot grout gelling occurs at a rate which can be controlled by changing the design of the mix, and materials such as lime or sodium bicarbonate can be used to assist the gelling action.

As an example of a silicate grout, Siroc grout acts in a one-shot manner and consists of four components: a modified silicate (20%-60%), an amide gelling agent (2%-14%), water (20%-70%), and a chloride or aluminate accelerator. The initial set is controlled by the accelerator and the gelling agent controls the process to the final strength. Depending on the conditions of the ground to be grouted, the concentrations of these two components can be changed to adjust the initial set and gelling rates. An increase in the water content reduces the viscosity and so allows the treatment of finer-grained soils. Siroc-cement, or Siroc-bentonite mixes can be used to control substantial groundwater flow velocities, the Siroc acting to lubricate the cement particles and effect gelling within a period of 10 seconds to 10 minutes.

Silicate grouts are mainly used to treat fine-grained soils, increasing their compressive strengths up to as high as $10MN/m^2$. They are also used for the high-pressure grouting of low-permeability rocks.

Lignins, which are low-viscosity grouts for penetrating fine sands, can be developed from lignosulphite (a by-product of the wood pulp industry) in combination with bichromates. The firm gelatinous grouted mass has a setting time of from 10 minutes to 10 hours depending on the bichromate concentration. A drawback with this grout is its toxicity and its need for careful handling.

Another chemical grout is formed by the polymerisation of an *aqueous solution of acrylamide and methylene-bis-acrylamide* (MBA). Setting times, as in the case of AM-9 (American Cyanamid Co.), are affected by salts and pH. Polymerisation can be induced by catalysts - dimethylaminoproprionitrile (DMAPN) and ammonium persulphate. Acrylamide and MBA concentrations of up to 20% are used to give viscosities of 1.2 centipoise, which similar to that of water (1 centipoise). Ammonium persulphate is an accelerator, and in concentrations of 0.5% to 1% by weight produces setting times of 6 seconds to 20 minutes, depending on the temperature and pH. The grout takes the form of a water-soluble, water-stopping elastic resin, its low viscosity and resistance to chemical attack making it suitable for controlling groundwater flow in fine-grained soils and fine-fissured rocks having permeabilities as low as 10^{-7} m/s but not as a medium for increasing the strength of soils or rocks in the mass.

Other chemical grouts are of a resin-based nature and are produced by the polymerisation of organic chemicals. *Resorcinol-formaldehyde grout* is injected as an aqueous solution of resorcinol and formaldehyde that is polymerised by changing the pH to produce a material having part elastic/part plastic properties. The resin is stronger than an acrylamide - methylene-bis-acrylamide grout. *Urea-formaldehyde* is a one-shot grout that has a higher strength than resorcinol-formaldehyde. *Cyanaloc* is a water-mixable resin grout which forms a stiff gel when catalysed by sodium bisulphite. It has a viscosity of about 8 centipoise to 14 centipoise, depending upon the mix concentrations, and is used for treating fractured rock. Injected as a particle suspension, its penetration is low and so it is less suitable as a permeation grout. *Geoseal* is the name for a range of low-viscosity grouts in the form of water-soluble resins. Borden (UK)'s Geoseal MQ-4, being of tannin/formaldehyde composition, is catalysed by caustic soda and is especially resistant to saline groundwaters. MQ-

14, of resorcinol/formaldehyde composition, is a low-viscosity alkaline grout which contains formaldehyde as the hardener and is designed for treating low-permeability soils and rocks. *Terranier grouts* are polyphenolic polymers in aqueous solution which form insoluble gels in the presence of suitable catalysts such as formaldehyde. They are mainly used in sands and gravels. Within the range are different concentrations of polymer and catalyst producing a variety of viscosities from 4 centipoise to 10 centipoise and rates of gel formation from a few minutes to several hours.

SAFETY

Many emulsions and chemical grouts are toxic and carcinogenic. They should be handled and used with care according to the requirements under Health and Safety legislation (Health and Safety at Work etc Act 1974) and COSHH Regulations. It must also be recognised that slow release of chemicals from a grouted soil could reach the groundwater table, so creating a possible pollution hazard. Chrome lignin-based products are a problem in this respect. Organic digesters in certain silicate grouts are of low toxicity, but other minor components could be toxic from their origin in caprolactam synthesis for nylon manufacture. Hardeners, partially miscible with silicate, can separate, so affecting grout setting and leading to environmental pollution. In addition, sodium silicate has a pH of 13, but most widely used grouts made from it have pH values of 10 or less. In Japan, all grouts having a pH above 8.6 are banned and classified as dangerous to human health and the environment. Acrylamides can be absorbed by inhalation, ingestion, or through skin contact. Acrylamide, of which AM9 manufactured by Cyanamid was a notable grout until banned by the World Health Organisation, is a suspected neurotoxin that affects the central nervous system and can cause long-term personality changes. Further, if the acrylamide percentage is above about 20 at room temperature, then an exothermic reaction can cause an explosion in the storage containers. Acrylamide replacement grouts are not, in fact, non-toxic but merely less toxic than the originals. Some other grouts in the acrylamide family include Rhône-Poulenc's Rocagil:BT, Rocagil:BT2 and Rocagil:AO6 (acrylate composition), and the Company's Siprogel (silicate and acrylate composition). TAM International Consultants also supply Gelacryl-2000 (an aqua-reactive acrylic) and 2006, and SNF supply a range of polyacrylamide grouts comprising acrylate resins. Formaldehyde and isocyanide grouts, which tend to be particularly poisonous, are used frequently in the industry but formaldehyde is used at modest concentrations in grouts and is usually degraded to innocuous products by microorganisms in the ground. However, formaldehyde is an irritant to the eyes and breathing passages and is a suspected carcinogen.

REFERENCES

British Standards Institution (1990) *Methods of Test for Soils for Civil Engineering Purposes*, BS 1377:1990, HMSO, London.
Part 1: General Requirements and Sample Preparation.
Part 2: Classification Tests.
Part 3: Chemical and Electro-chemical Tests.
Part 4: Compaction Related Tests.
Part 5: Compressibility, Permeability and Durability Tests.
Part 6: Consolidation and Permeability Tests in Hydraulic Cells and with Pore Pressure Measurement.
Part 7: Shear Strength Tests (Total Stress).
Part 8: Shear Strength Tests (Effective Stress).
Part 9: In situ Tests.
Harris, D.I., Mair, R.J., Love, J.P., Taylor, R.N. and Henderson, T.O. (1994) Observations of ground and structure movements for compensation grouting during tunnel construction at Waterloo Station, *Géotechnique*, **44**(4), 691-713.

BIBLIOGRAPHY

Nonveiller, E. (1989) *Grouting Theory and Practice*, Elsevier, Netherlands, 251pp.

9

TUNNELLING THROUGH CONTAMINATED GROUND

SOURCES OF GROUND CONTAMINATION

The industrial processes that need to be checked at the desk study stage of the site investigation include the following.

(Former) public utilities and transportation

These include old gasworks, coal carbonisation plants and ancillary by-product works, coal and metal mines, old sewage farms and works (where the concentrations of metals in the soil may be high), railways and sidings, dockyards, and any former hospital sites.

Gasworks sites have been contaminated by: indiscriminate dumping of chemical or process waste while the site was in operation; spillages/leakages from tanks, pipes and through any production process; residual material, such as tanks not being drained after closure and stockpiles of materials not being removed; more generally, contamination spread during demolition, for example, high silica bricks from the furnaces.

It is the practice of the now privatised British gas industry to clear former gasworks sites to ground level, filling cavities and removing all tars and residues 'where appropriate'. However, British Gas does not give guarantees as to the condition of the site after clean-up and the sale of a site would include a clause informing the buyer of its (past) contamination. Reference may be made to the document by Environmental Resources Ltd (1987) and Waste Management Paper No. 27 (Department of the Environment, 1991).

The basic process for coke, gas and tar production is the same, involving the combustion of coal in an oxygen deficient atmosphere to produce a complex mixture of gaseous, liquid and solid products the proportion of which depends on the coal composition and the temperature of combustion. Coal carbonisation commenced industrially in 1805. Coke production followed the growth of the iron and steel industries, first appearing in the mid-18th century and rising to a peak production of 30 million tonnes/year in 1956. There has been a steady contraction since then, falling to 8 million tonnes/year in 1984. The strike in the mining industry virtually eliminated the remaining production. Estimates for the total number of coal carbonisation sites in Britain are 3000 gasworks, 8600 coke ovens in 174 plants and 378 tar distilleries, these figures relating to the industries at their peak.

Processing of primary materials

Examples of this type of processing are asbestos works, smelters, foundries and metal finishing works.

Chemical process works

These works include, in particular, oil refining, storage and distribution, pharmaceuticals, ceramics, plastics, pesticides, paint, solvents, chlorinated sludges, materials containing polynucleararomatic hydrocarbons (PAHs), wood preserving, paper, tanning and munitions (explosives).

Miscellaneous categories

These include such industries as scrapyards, munitions factories, old processes such as gelatine works and creosote manufacture, and waste disposal/landfill sites. Scrapyards (see ICRCL Guidance Note 42/80, October 1983) are a notable source of contamination from oil-soaked soil to a substantial depth, spillage from batteries, and de-greasing fluids. Cemeteries also create ground contamination, not only on site but also off site and at depths sufficient to cause concern for tunnelling works when pollutants may have migrated through low-permeability soil. Biological risks can also arise from animal carcasses, with an increased risk if the buried animals had suffered from BSE. Anthrax spores also create a potential hazard.

Transfer of pollutants through soils is assisted by the growth of plant roots and the activity of invertebrates. Discontinuities in rocks provide the main contribution to water transmission capacity at depth, while intergranular porosity has an important aquifer storage function.

RANGE OF CONTAMINANTS

The general *range of contaminants* includes the following:

- *Methane* (*see* Department of the Environment, 1991) and other gases, particularly from existing and abandoned landfill sites, which are combustible, carcinogenic, toxic or asphyxiants (CO_2). Of the 80 million tonnes or so of mixed household, commercial and industrial waste discarded in Britain every year about 28 million tonnes is domestic and trade waste having a potential for gas (and toxic liquor) generation. *Hydrogen sulphide*, which can emanate from a contaminated land site, is also highly toxic and inflammable, exploding in a manner similar to methane and having explosive limits 4.3 to 43.5% in air. Its toxicity much precedes its ability to explode. Gases could be in solution under pressure, but could also come out of solution as pressure is reduced during construction.
- *Pathogens* and *carcinogens*.
- *Asbestos fibres* arising from the demolition of industrial premises, factories, offices and old power stations. Asbestos contamination is also associated with poorly operated and managed waste disposal sites and from illegal and indiscriminate dumping.
- High concentrations of *cadmium, mercury, lead* and *other heavy metals* which can constitute a hazard to health.
    Reference in respect of cadmium may be made to the following Council of the European Communities directive and resolution:
    83/513/EEC Council Directive of 26 September 1983 on limit values and quality objectives for cadmium discharges, OJ L 291 24.10.83 p.1; D by 390L 0656 (OJ L 353 17.12.90 p.59).
    Council resolution of 25 January 1988 on a Community action programme to combat environmental pollution by cadmium, OJ C 030 4.2.88 p.1.

Reference in respect of mercury may be made to the Council of the European Community Directive:
82/176/EEC Council Directive of 22 March 1982 on limit values and quality objectives for mercury discharges by the chlor-alkali electrolysis industry, OJ L 081 27.03.82 p.29; D by 390L0656 (OJ L 353 17.12.90 p.59).
Note is also made of Council of the European Community directives relating to titanium dioxide:
82/883/EEC Council Directive of 3 December 1982 on procedures for the surveillance and monitoring of environments concerned by waste from the titanium dioxide industry, OJ L 378 31.12.82 p.1; M by 185I.
89/428/EEC Council Directive of 21 June 1989 on procedures for harmonising the programmes for the reduction and eventual elimination of pollution caused by waste from the titanium dioxide industry, OJ L 201 14.07.89 p.56.

- *Tars, phenols, benzenes and other organic compounds* from old gasworks which affect not only health but can also cause problems for new construction. Typically at such a site, in addition to coal and coke, relatively harmless glassy clinker, flue dust and high silica bricks (used in the ovens), various liquors and tarry products associated with high levels of phenol and other toluene extractable matter, there are total and free cyanides due to waste product, highly acidic spent oxides ('blue billy'), containing ferric ferrocyanide ('Prussian Blue'), produced when iron oxide is used to purify the gas and remove hydrogen sulphide, hydrogen cyanide, ammonia, organic sulphur and tarry compounds. Sulphite, sulphate and also heavy metals are found. Ponded surface water is highly acid and contaminated with oil-tar material. Contaminants will be in the soil at depths up to 4 metres but it must not be assumed that they cannot have penetrated further into the ground.
    Coal tar is a highly complex combination of hydrocarbons which form the insoluble residue of the carbonisation process. It contains toxic compounds such as benzene, toluene, xylene, ethyl benzene, styrene, phenols, cresols,

xylenols, polyhydric phenols, naphtha, naphthalenes, acenaphthelene, fluorene, diphenyl oxide, anthracene, phenanthene, carbazole, tar bases and pitch (refer to ICRCL 18/79, 1986). Polycyclic aromatic hydrocarbons (PAHs) and polynuclear heterocyclic hydrocarbons (PHHs) commonly form between 30-40% or up to 50%, and up to 5% of the total material, respectively. Both are carcinogenic. Ammoniacal liquor, which is an important by-product of gas manufacture, may also be present (its most important constituent being ammonia) usually in the form of ammonium salts, chiefly the cyanide, sulphide and carbonate ('free ammonia') but also as the chloride, thiocyanate, thiosulphate, sulphate and ferrocyanide ('fixed ammonia'). Also, lighter hydrocarbons such as paraffins, benzene and toluene may be present, and they are skin irritants prone to cause dermatitis with repeated exposure. A 1g dose is considered fatal for a 20kg child, the concentration in the soil determining the quantity of soil that must be ingested for a fatality. Adults are also considered to be at risk from the chronic effects due to any repeated exposure to contamination. Exposure to soil containing more than 500mg/kg PAH is a significant hazard with respect to potential skin cancers. This is also the level at which direct ingestion is thought to be a hazard, with the potential to cause cancers through the body. Together with road tar, pitch and creosote are primary products of tar refining. Dust produced by pitch is a known carcinogen and has produced cases of lung cancer in gasworks site workers. Care must therefore be taken when the material is disturbed.

Phenols, the term applied to the family of phenolic compounds found in coal tar and ammoniacal liquors, are highly aromatic hydrocarbons which can penetrate high density polyethylene (HDPE) pipes and contaminate water supplies, imparting an antiseptic taint to chlorinated water at low concentrations. The average fatal dose is 10-30g, although as little as 1g has apparently been reported as being fatal; 1g can be considered to be a fatal dose for a 20kg child. (Data from former gasworks sites suggests a maximum phenol concentration of about 2g/kg, corresponding to a child ingesting 0.5kg of soil, which is most unlikely.)

Absorption through the skin is the most common form of exposure, usually causing dry skin and irritation, and at high concentrations blistering may occur. Skin tumours can be caused by repeated exposure to solutions containing more than 50g/litre of phenol. Concentrations over 15g/litre can cause whitening and blistering of the skin, and solutions with over 0.2% phenol can cause allergic dermatitis in some individuals. Concentrations between 1 and 10mg/litre in surface water or groundwater may cause toxic effects in fish and plant life, and a maximum acceptable concentration in drinking water is 0.5 micrograms/litre. The regional water companies are wary about laying mains in ground having phenol concentrations exceeding 5 parts per million (ppm). Phenol solutions can also affect concrete when at concentrations of 0.1% by volume.

A 5g mouthful of soil containing free cyanide at concentrations over 2500mg/kg would be dangerous for a child playing in a contaminated area. Groundwater pollution may also occur, although this is uncommon since cyanide is usually degraded rapidly in water. Free cyanide presents few problems to plant life, and poses little risk to buildings and services. Of most concern to site workers is the inhalation of hydrogen cyanide gas. Vapour build-up in a deep trench would present a hazard at concentrations of 300mg/kg free cyanide in the soil.

Spent oxide contains free sulphur, sulphates, cyanide in various forms, manganese and heavy metals. Of these, it is the cyanide compounds which are the most toxic, but due to the levels found in spent oxide it is not realistic to consider the material as being potentially lethal. At concentrations of about 10 000mg/kg spent oxide is visible as a blue coloration in the soil (the 'blue billy' referred to above). This is equivalent to 500mg/kg of complex cyanides. The spent oxide at concentrations of more than 500mg/kg is phytotoxic (harmful to plant life). However, spent oxides are relatively non-toxic and concentrations of 40 000mg/kg are needed to produce toxic effects by ingestion, which can cause vomiting and convulsions. Levels of 50 000mg/kg, equivalent to 2500mg/kg of complex cyanides, can cause skin irritation.

Generally, thiocyanates are not thought to pose a problem with toxicity unless encountered at concentrations much higher than those expected on a coal carbonisation site. Cyanide compounds may be converted by chlorinated water into cyanogen chloride. This has a pungent odour, is detectable at 1ppm and is irritating at levels of 1-4ppm. It is also particularly toxic to aquatic life at such levels. If not converted, discoloration of groundwater may occur; for example, thiocyanate can cause water to become stained red at concentrations of 5-10mg/litre. Free cyanide and hydrogen cyanide can be developed from complex cyanides, and sulphur dioxide is produced from sulphur when the oxide is heated. Due to the high sulphur content of the oxide, it may maintain combustion once it is ignited. The main hazard is the generation of sulphur dioxide - an irritant at low concentrations and toxic at high concentrations.

Chronic toxicity is not considered to pose a problem with sulphides. Skin contact with spent oxide concentrations which exceed 50 000mg/kg may cause skin irritation due to the sulphide content. When the soil is disturbed, sulphides can oxidise to sulphates which can then in turn attack buildings (Building Research Establishment, 1991) and services. Under acid conditions sulphides can produce hydrogen sulphide gas when their concentrations in the soil reach 200-1100mg/kg. This gas has a pungent, rotten egg smell at concentrations of 0.25% of those concentrations required to produce toxic effects. However, prolonged exposure does tend to dull sensitivity.

As an element, sulphur presents no toxicity problems but, with oxidation, sulphate levels in the soil can be increased and attack on buildings and services will begin. Through reduction of the chemical, sulphide and thence hydrogen sulphide can be produced. Sulphur is combustible and its presence will significantly increase the ability of a soil to burn. It is particularly common in spent oxides, and may constitute up to 50% of the oxide. Skin contact with sulphur dust will cause irritation and there will be irritation of the respiratory system if it is inhaled.

The hazards associated with sulphates are mainly related to the metal compounds of the salt rather than the sulphate itself. The most likely toxic soluble sulphate found on gasworks sites is ferrous sulphate. Gastro-intestinal irritation may occur at concentrations greater than 400ppm in drinking water, and the acceptable level in treated water is 250ppm. Concentrations as high as 4000ppm have been reported for disused gasworks sites - a toxic condition. The effect of sulphates on concrete is most severe - concrete 'cancer' is caused by the reaction of sulphates with aluminates in the concrete, the product having a higher volume than the original salt and so causing internal disruption/spalling.

- *Petroleum vapour* from leaking storage tanks. It is hazardous and flammable between 1% and 7% by volume in air.
- *Explosives* that will be in a state of decay.
- *Acidic sites* which present problems for new building, especially with respect to the effect of the acids and sulphates on concrete, as noted above. Acidic pH values as low as 2 can also be found in soils on these sites and will cause irritation if handled without protection. Acidic conditions promote a range of reactions including the production of hydrogen cyanide from sulphides in the presence of moisture, increases in the mobility of heavy metal ions, and attacks on building materials and services. Conversely, a high pH causes fewer problems and assists none of the above reactions. Concrete also tends to be insensitive to high pH unless exposed for long periods of time under very caustic conditions.
- *Colliery waste tips* containing burnt or partially burnt shale. There may be leachate migration together with carbon monoxide and sulphur dioxide emissions.

GROUND INVESTIGATION

Contaminated land sites will present ground investigation specialists with special problems. Reference on this subject should be made to the Site Investigation Steering Group (1993) while bearing in mind the provisions of BS 5930:1981 (British Standards Institution, 1981) which is being updated under BSI Sub-Committee B/526/1). The document DD 175 (British Standards Institution, 1988), although somewhat overtaken, should also be consulted. Suitable emergency procedures and contacts should be established well in advance of the work being started. There should also be some training programmes in first aid and in the recognition of diseases associated with working in contaminated land.

In the first instance there should be a carefully thought out drilling and pitting strategy (*see*, for example, Ferguson, 1992 and also Health and Safety Executive, 1989), but there will always be the possibility of missing highly contaminated 'hot spots'. Second, there will need to be full compliance with Regulations under the Health and Safety at Work etc Act 1974 and the COSHH assessments (Health and Safety Executive, 1988a) when working on these sites - drilling holes; taking samples of soil and water; digging and working in trial pits and/or trenches. The investigators must always wear clothing that is appropriate for the task - stout boots, gloves and full coverage of the skin, and face masks in some instances. No food or drink should be consumed on site, and there must be suitably designated areas for washing and disposal of dirty clothing. Any disturbance of asbestos should be left to a team that is suitably masked and trained for that type of work (*see* Health and Safety Executive, 1985, 1988b, 1990).

The second point concerns the question of induced contamination and cross-contamination of soil and groundwater during drilling operations. Some of the points that need to be addressed are noted below.

- Drill string joints should not be glued.
- Threads should not be lubricated.
- Attempts should be made not to store drilling materials on site.
- Temporary casings and well liners can introduce contamination even if they are, in effect, clean. Mild steel, for example, will corrode; it supports Fe bacteria and may add Fe to a sample; and it forms OH complexes. Galvanised steel can add Zn to samples and form organo-metal complexes. Stainless steel type 304, 316, which is expensive, will corrode in low pH or saline water.
- Polyvinylchloride (PVC) absorbs volatile organics, deteriorates in contact with ketones, esters and PAHs, and leaches plasticisers and fillers. High density polyethylene (HDPE) leaches less than PVC, but deteriorates in contact with aromatic and halogenated

hydrocarbons. Polypropylene deteriorates in contact with oxidising acids, aliphatic and aromatic hydrocarbons. Fluorocarbon resins are immune to chemical attack and corrosion, but in addition to a low structural strength do actually absorb some organics.

- Brass, chrome-plated or galvanised material should not be used for sampling if water is being tested for heavy metals. Iron or steel alloys are acceptable when testing for iron and manganese.
- Special equipment and care in procedures is needed when sampling volatile organic substances and highly toxic constituents in very low concentrations. Glass containers should generally be used when major ion analyses are proposed.
- Even when a borehole is cased, the perforation between the extrados of the casing and the ground can admit water-borne contaminants from ground surface and can transmit contaminants up and down the perforation under local water pressures. When sampling from several water-bearing horizons, a sampling borehole should ideally be used for each aquifer to avoid any cross-contamination from vertical leakage. If this is not possible, then a multi-packer system can be used with control on the pumping rate in particular packer zones. Water samples can be obtained by bailing from a borehole or from continuous pumping. Single bailing tubes allow rapid sampling at a relatively low degree of accuracy for relatively high ionic concentrations. If any pump sampling is undertaken, the pumping should also be maintained until the pH, the electrical conductivity and/or temperature have attained constant values. Suitable equipment comprises suction pumps, plunger pumps, submersible electrical pumps and gas-driven diaphragm pumps. But suction pumps are unsuitable for determining dissolved gas concentrations because of the high degree of induced aeration. Petrol-driven pumps should not be used for sampling low concentrations of hydrocarbons.

TESTING

The contaminants for which testing should be undertaken are given in Tables SI 9.1 to SI 9.4.

Table SI 9.1 Summary of possible <u>primary contaminants</u> for which testing should be undertaken

Arsenic	Cadmium	Chromium	Hexavalent Chromium (undertaken if total chromium content is >25mg/kg)
Lead (total)	Mercury (total)	Selenium (total)	Boron (water soluble)
Copper (total)	Nickel (total)	Zinc (total)	Cyanide (alkali extraction methods)
Cyanide complex*	Cyanide free*	Thiocyanate*	Phenols total
Sulphide	Sulphate	Sulphur free	pH value
Toluene Extractable Matter	Coal tar/Polyaromatic Hydrocarbons (undertaken if toluene extractable matter is >2000mg/kg dry mass of soil)	Asbestos (asbestos content determination shall be carried out by visual examination and polarised light microscopy)	

* Tests for cyanide complex, cyanide free and thiocyanate should be carried out if the total cyanide exceeds 25mg/kg dry mass, the methods to follow alkali extraction.

Table SI 9.2 Summary of possible <u>secondary contaminants</u> for which testing should be undertaken

Antimony	Barium	Beryllium	Vanadium
Cyclohexane Extractable Matter	Freon Extractable Matter	Mineral oils	Chloride

Table SI 9.3 Summary of <u>contaminants in water</u> for which testing should be undertaken

Arsenic	Cadmium	Chromium	Hexavalent Chromium
Lead (total)	Mercury total	Selenium total	Boron - water soluble
Copper (total)	Nickel total	Zinc total	Cyanide total
Cyanide complex	Cyanide free	Thiocyanate	Phenols total
Sulphide	Sulphate	Sulphur free	pH value
Polyaromatic Hydrocarbons	Antimony	Barium	Beryllium
Vanadium	Chloride	Ammoniacal Nitrogen	Nitrate Nitrogen
Chemical Oxygen Demand	Biochemical Oxygen Demand	Total Organic Carbon	Volatile Fatty Acids
Iron	Manganese	Calcium	Sodium
Magnesium	Potassium		

Table SI 9.4 Summary of <u>gases</u> for which samples should be taken for analysis

Carbon Dioxide	Hydrogen	Hydrogen Sulphide	Methane
Nitrogen	Oxygen	Ethane	Propane
Carbon Monoxide			

Unfortunately, results of tests are not always absolute but rather seem to be dependent on the equipment and the laboratory used for the testing. The concentrations of chemical contaminants are checked against guideline levels, sometimes referred to as 'standards'. The best-known of these in the UK are the ICRCL 'standards' which are expressed in terms of 'trigger levels' and 'action levels'. These guidelines are, in fact, now obsolete and it is expected that 1995 will see the publication of standard methods of analysing various groups of contaminants, the report prepared under the auspices of the Laboratory of the Government Chemist serving to update and expand the ICRCL guidelines. Nevertheless, when reporting on degrees of contamination reference is usually made to the ICRCL guidelines, together with those attributable to the former Greater London Council (GLC)/Kelly (Kelly, 1979a, 1979b) and The Netherlands (Anon, 1987; *see* also van Ommen, 1994).

REFERENCES

Anon (1987) *Soil Clean-up Guidelines (Leidraad Bodemsaniering)*, Dutch Ministry of Housing, Physical Planning and the Environment, The Hague (SDU). (Revised guidelines are expected to be published in 1994/95.)

British Standards Institution (1981) *Code of Practice for Site Investigations*, BS 5930:1981, London.

British Standards Institution (1988) *Draft for Development Code of Practice on the Identification of Potentially Contaminated Land and its Investigation*, BS DD 175, BSI, London.

Building Research Establishment (1991) *Sulphate and Acid Resistance of Concrete in the Ground*, Digest Number 363, Garston, Watford WD2 7JR, England.

Department of the Environment (1987) *Problems Arising from the Redevelopment of Gasworks and Similar Sites*, 2nd Edition, HMSO, London.

Department of the Environment (1991) *Landfill Gas*, Waste Management Paper No. 27, 2nd Edition, HMSO, London.

Environmental Resources Ltd (1987) *Problems Arising from Redevelopment of Gas Works and Similar Sites*, 2nd Edition, DoE, HMSO, London.

Ferguson, C.C. (1992) The statistical basis for spatial sampling of contaminated land, *Ground Engineering*, June 1992, 34-38.

Health and Safety Executive (1985) *Respiratory Protective Equipment for Use against Asbestos*, Guidance Note EH 41, HMSO, London.

Health and Safety Executive (1988a) *COSHH Assessments: a Step by Step Guide to Assessments and the Skills Needed for It*, Control of Substances Hazardous to Health Regulations, HMSO, London.

Health and Safety Executive (1988b) *Work with Asbestos Insulation and Asbestos Coating and Asbestos Insulating Board: Approved Code of Practice*, HMSO, London.

Health and Safety Executive (1989) *Monitoring Strategies for Toxic Substances*, Guidance Note EH 42, HMSO, London.

Health and Safety Executive (1990) *Asbestos - Exposure Limits and Measurement of Airborne Dust Concentrations*, Guidance Note EH 10, HMSO, London.

Kelly, R.T. (1979a) Site investigation and materials problems, *Proc. Conf. on Reclamation of Contaminated Land, Eastbourne, 1979*, Society of Chemical Industry, London, B2/1-14.

Kelly, R.T. (1979b) *GLC Investigations into Problems Created by Chemicals in the Environment*, Lorch Foundation, High Wycombe, England, 9-10 May 1979, Harwell (Harwell Environmental Seminars), Oxfordshire, England, 83-101.

Site Investigation Steering Group (1993) Guidelines for the safe investigation by drilling of landfills and contaminated land, *Site Investigation in Construction*, 4, Thomas Telford, London. (ISBN 0 7277 1985 8.)

van Ommen, H.C. (1994) Contaminated land, the Dutch experience, *Proc Symp. on Contaminated Land, from Liability to Asset*, Birmingham, 7-8 February 1994, The Institution of Water and Environmental Management, London, 1-14.

BIBLIOGRAPHY

Attewell, P.B. (1993)   *Ground Pollution: Environment, Geology, Engineering, Law*, E. & F.N. Spon, London, 251pp.   (ISBN 0 419 18320 5.)

Harris, M. and Herbert, S. (1994) *ICE Design and Practice Guide, Contaminated Land: Investigation, Assessment and Remediation*, Thomas Telford Services Ltd, London, 77pp. (ISBN 0 7277 2016 3.)

Health and Safety Executive (1991) *Protection of Workers and the General Public during the Development of Contaminated Land*, HS(G) 66, HMSO, London.

Health and Safety Executive (1993) *Occupational Exposure Limits*, EH 40, HMSO, London.

Interdepartmental Committee on the Redevelopment of Contaminated Land (1983) *Notes on the Redevelopment of Scrap Yards and Similar Sites*, Guidance Note 42/80, 2nd Edition, October 1983, 12pp, Department of the Environment, London.

Interdepartmental Committee on the Redevelopment of Contaminated Land (1983) *Notes on the Redevelopment of Sewage Works and Farms*, Guidance Note 23/79, 2nd Edition, November 1983, 15pp, Department of the Environment, London.

Interdepartmental Committee on the Redevelopment of Contaminated Land (1986) *Notes on the Redevelopment of Gasworks Sites*, Guidance Note 18/79, 5th Edition, April 1986, 12pp, Department of the Environment, London.

Interdepartmental Committee on the Redevelopment of Contaminated Land (1986) *Notes on the Fire Hazards of Contaminated Land*, Guidance Note 61/84, 2nd Edition, July 1986, 7pp, Department of the Environment, London.

Interdepartmental Committee on the Redevelopment of Contaminated Land (1987) *Guidance on the Assessment and Redevelopment of Contaminated Land*, Guidance Note 59/83, 2nd Edition, July 1987, 19pp, Department of the Environment, London.

Interdepartmental Committee on the Redevelopment of Contaminated Land (1990) *Notes on Development and After-use of Landfill sites*, Guidance Note 17/78, 8th Edition, December 1990, 27pp, Department of the Environment, London.

10

ROCK QUALITY DESIGNATION AND OTHER ROCK MASS DISCONTINUITY DESCRIPTORS

Table SI 10.1 includes several of the parameters that would be measured for the purposes of design in rock. RQD and fracture frequency are routinely derived from measurements on rock cores taken during the ground investigation. Mass factor j requires deformation modulus values to be taken in the field at ground surface or at underground openings, as appropriate for the future construction design, and values of Young's modulus to be derived from laboratory unconfined compressive strength tests on carefully prepared, visibly intact rock specimens. The velocity index requires field measurements to be taken of the speed of wave transmission (celerity) following the input of an impulsive disturbance, usually at ground surface, for comparison with the wave velocity obtained from ultrasonic tests on small specimens of the same visibly intact rock.

RQD

Developed by Deere (1963), Rock Quality Designation is a simple observational index of rock engineering quality, to be quantified at the time that the rock core is removed from the borehole. It expresses the degree of natural intactness of a rock mass, which is indicative of mass strength. For reliability, an RQD index

number must be calculated in a standard manner.

RQD is defined as the ratio of the summed length of all single pieces of naturally unfragmented rock core lengths equal to or greater than 100mm (4 inches) to the total length of rock core attempted in the particular core run being assessed. Thus, for the numerator of the ratio, when the core is laid out in the core box all the stick lengths of rock equal to or greater than 100mm are added together. The total length drilled should be entered as the numerator term. (The total length of core in the box is sometimes used as the denominator term, but this is incorrect.)

It is important to be aware that Deere defines RQD as applying universally to NX rock core size (54.7mm, or 2.16 inches, diameter). Larger diameter cores may promote artificially higher RQD values for the same rock formation(s) drilled than would be derived from an NX core. Subsequent work reported by Deere and Deere (1989) has indicated that smaller diameter cores (such as NQ, 27.6mm or 1.88 inches diameter) as well as larger diameter wireline core are also valid for the calculation of RQD. As good practice, it is recommended that when core sizes differ from NX, the RQD values be qualified with the core size so that the reader of the report can infer the effects of size for himself.

Table SI 10.1 Rock mass parameters useful for the purposes of design

Term	RQD%	Fracture frequency (m^{-1})	Mass factor j E_f/E_{lab}	Velocity index $(c_{Pf}/c_{Plab})^2$
Very poor	0 to 25	>15	0.2	0 0 - 0.2
Poor	25 to 50	15 - 8	0.2	0.2 - 0.4
Fair	50 to 75	8 - 5	0.2 - 0.5	0.4 - 0.6
Good	75 to 90	5 - 1	0.5 - 0.8	0.6 - 0.8
Excellent	90 to 100	<1	0.8 - 1.0	0.8 - 1.0

Assessment of the nature and presence of the rock fracturing that controls RQD should take account of all discontinuities, persistent and non-persistent, such as non-systematic microfractures, bedding planes, foliation planes, shear planes and faults. It will also be necessary to assume that each of these natural breaks in the continuity of the rock mass is equal in its effect on RQD. Lithological boundaries impose their own fractures, so when assessing a rock core particular attention should also be paid to the distribution pattern of the fractures between such boundaries. Fractures imposed on the core by the drilling and core retrieval process should be discounted in the RQD assessment, although the presence and density of such fractures are themselves indicative of rock weakness. It is usually stated that such artificially created fractures can be identified by clean break surfaces, but this is not always the case. Sluicing with clean water would clear drilling mud away, but this operation cannot be recommended, particularly if the core comprises weaker rock, such as mudrock, prone to fail in the presence of water. Although there will often be peripheral evidence to assist, identification of imposed fractures may be inexact. Thus, too much engineering credence should not be placed on the RQD operation nor on the values that it produces.

GENERAL USES OF RQD (GENERALLY AFTER HATHEWAY, 1990)

Strength: An indicator of the overall compressive strength of rock in the mass. A stronger, more brittle rock will resist closely spaced fracturing by tectonic influences. Zones of low RQD rock in a core probably indicate shear zones or fault zones.

Presence of discontinuities: An RQD below about 70% denotes rock of concern when designing for stability in that ground.

Structural domains: The distribution pattern in a core of RQD values below about 70% may suggest changes in lithological strengths between bedding boundaries, as noted above.

Rock mass character for empirical design: RQD is an important element in the three most-

used rock mass characterisation systems - Rock Mass Rating (RMR of Bieniawski, 1974, 1979a, b, 1984, 1989), Rock Mass Quality ('Q' of Barton *et al.*, 1974 and Barton, 1976), and Rock Structure Rating (RSR of Wickham *et al.*, 1972, 1974).

Stand-up time and ground support: For unsupported underground openings RQD is an indication of the extent to which an opening can advance before temporary or permanent support has to be installed. There is a direct relation between RQD and the need to provide artificial support for underground openings. Lower RQD values generally indicate a need for some form of permanent ground support to be in place within hours to days of the excavation being formed. The absolute and relative amplitudes, and the directions, of the *in situ* ground stresses apply at least as great an influence on the self-support capacity of the ground as does the fracture frequency.

Ease of excavation: As a general indication, rocks having RQD values less than about 50 to 60% can be cut without significant problems. This observation is based on an indirect indication of intrinsic strength and the rock actually being cut. Of course, the lower the RQD value the greater is the *direct* contribution of the fracture density to the excavation process, since there is then the possibility of greater volumes of rock being *displaced* for removal rather than being cut.

Groundwater flow: Greater volumes will tend to flow through zones of rock having lower RQD values.

Contaminant transport: As is the case with water, contaminants will usually follow zones of low RQD value.

CAUSES OF LOW RQD (GENERALLY AFTER HATHEWAY, 1990)

Pervasive bedding plane and foliation discontinuities, reflecting the actual origin of the rock.

Metamorphic influences caused by temperatures and pressures of igneous intrusions having sizes which range from batholiths to dykes. Such influences would apply both to the metamorphic rock itself and to contiguous sedimentary rock(s).

Tectonic forces that have taken the rock to and beyond its elastic limit in a brittle state. In compression the deviatoric components of stress must be sufficient to induce shear. Uplift accompanied by dominant horizontal components of stress can induce shear and tensile fractures.

Mineral alteration from temperature, pressure and by cation exchange as a result of diagenesis and/or groundwater movement.

Groundwater dissolution of soluble minerals and/or interstitial cements.

Near-surface weathering effects.

CONTROL GUIDELINES FOR THE DETERMINATION OF RQD (GENERALLY AFTER HATHEWAY, 1990)

(a) Never allow the drillers to handle the core beyond removal from the core barrel. This will avoid unnecessary breakage.

(b) The engineering geologist/geotechnical engineer doing the logging should be responsible for placing the core into the core box.

(c) Except when the core is water-degradable, or when there are other compelling site characterisation reasons, the core should be carefully washed on first inspection as it is placed in the core box.

(d) Place a strike mark, in indelible ink, perpendicularly across all fractures that are judged on first inspection to be 'natural', and therefore denoting true discontinuities.

(e) Align the core in each core box trough so that the length of core from each core run fits together as closely as possible. Then measure the individual segments of core, resolving which are greater than or less than 100mm. Always divide the total length of rock core fragments, each greater than

100mm, by the total length of the attempted core run.

(f) When an interval of poor quality rock denoting 'bad ground' is encountered as a successive string, consider re-calculating the RQD value for only their combined lengths. This additional RQD determination is then written in the log with a special marker denoting a realistic determination of the actual presence and location of poorer quality rock.

FRACTURE FREQUENCY λ

This is the reciprocal of the average spacing d_a between discontinuities measured along a borehole core or along a scanline set out across a rock exposure. On the assumption that the discontinuity spacing values d will follow a distribution of negative exponential form (a form of Poisson distribution), Priest and Hudson (1976) showed how fracture frequency could be related not only to the specific 100mm characteristic RQD spacing, expressed as $d* = 0.1$ in metres, but also to any target spacing $d*$ in metres. The general equation is

$$RQD_{d*} = 100[1 + (d*/d_a)]\exp[-(d*/d_a)]$$

and the specific equation is

$$RQD_{(d* = 0.1)} = 100[1 + 0.1/d_a]\exp[-0.1/d_a].$$

These two equations thus offer a means of estimating rock quality in respect of fracture density simply by counting the number of discontinuity intersections in a core or along a scanline. They are, however, dependent on discontinuity spacing values d following a distribution of negative exponential form, and the relations are also sensitive to the length over which the discontinuity presence is counted (the borehole core or scanline length).

MASS FACTOR *j*

This discontinuity index was developed by Hobbs (1975) for the assessment of the foundation support capacity of weak rocks, particularly chalk. It is the ratio of the field deformation modulus E_f to the laboratory deformation modulus E_{lab}. There is a propensity for the discontinuity spatial density to increase as the volume of rock increases. Thus, the *j*-factor will be low when this is the case. When the rock is sensibly intact there will be no discontinuities to promote deformation and the *j*-factor will tend to unity.

VELOCITY INDEX

This index concerns the ratio of the field P-wave velocity c_{Pf} to the P-wave velocity c_{Plab} derived in the laboratory. The assumption here is that a P-wave front in the rock mass has to weave its way around discontinuities and that the greater the number of such obstacles the greater is the path length (and hence the lower the apparent velocity). There will be more and longer discontinuities in the mass than in a laboratory specimen, and so the simple velocity ratio, field-to-laboratory, should be less than or equal to unity. The squared ratio should be used since the rock stiffness, expressed by the deformation modulus E and responsive to the number of discontinuities present, is proportional to the wave velocity squared:

$$c_{Pf}^2 = E_f(1\text{-}\nu)/\rho(1+\nu)(1\text{-}2\nu)$$

where c_P is the P-wave velocity (m/s),
        ν is Poisson's ratio (unit),
and ρ is the rock density (Mg/m^3).

The dominant far-field frequencies within a wave transmitted through a rock mass are likely to be several orders of magnitude less than the frequencies propagated through a test specimen in the laboratory. Wave arrivals in the field are usually detected by geophone (velocity transducer), although accelerometers may

sometimes be used, the latter being particularly appropriate for very low frequencies in the region of 1Hz. Knowing the time elapsed from wave input to wave arrival and the distance between source and measurement, the wave velocity c_{Pf} can be calculated. Laboratory waves, propagated through the same right-circular cylindrical specimens that would be used for unconfined compressive strength testing, are usually generated and received ultrasonically by piezoelectric crystals (barium titanate or titanate-zirconate). Typical transmission frequencies would be from 50kHz up to 2MHz. Since the respective wave velocities will be frequency-dependent to a degree, this fact must be borne in mind when calculating a velocity index.

Although P-waves tend to be used in the field, for example in engineering seismic refraction surveys, there are logical grounds also for the adoption of shear (S) waves which can easily be generated at source in horizontal mode. Since water does not support the transmission of shear waves, inferences concerning the degree of saturation of the ground can be drawn from a comparison of P-wave and lower-velocity S-wave amplitudes at a particular measurement point. S-waves are less commonly generated for laboratory ultrasonic tests, although shear transmitters and receivers are available and it is possible to generate S-waves from P-waves for transmission and transform from S to P for reception by means of critical angle wedges fixed to the two ends of the right-circular cylindrical test specimens.

SUGGESTED DESIGN VALUES OF ROCK STRENGTH PARAMETERS

Table SI 10.2 suggests how RQD can be related to rock mass strength parameters for the purposes of design in the rock. The unconfined compressive strength of the *macroscopically intact* rock is determined in the laboratory.

Table SI 10.2 RQD and rock mass strength properties (after Kulhawy and Goodman, 1987)

RQD%	Rock mass properties		
	Unconfined compressive strength (UCS)	Cohesion value	Friction angle
0-70	0.33 UCS	0.1 UCS	30°
70-100	0.33-0.8 UCS	0.1 UCS	0°-60°

REFERENCES

Barton, N. (1976) Recent experience with the Q-system for tunnel support. In: *Proc. Symp. Exploration for Rock Engineering*, Z.T. Bieniawski (ed), **1**, 107-114, A.A. Balkema, Rotterdam.

Barton, N., Lien, R. and Lunde, J. (1974) Engineering classification of rock masses for the design of tunnel support, *Rock Mech.*, **6**(4), 189-236.

Bieniawski, Z.T. (1974) Geomechanics classification of rock masses and its application in tunnelling, *Proc. 3rd Int. Congress on Rock Mechanics, Int. Soc. for Rock Mechanics*, Denver, Colorado, USA, **IIA**, 27-32.

Bieniawski, Z.T. (1979a) *Tunnel Design by Rock Mass Classifications*, Technical Report GL-79-19 for the Office, Chief of Engineers, US Army, Washington DC 20314, USA under Purchase Order No. DACW39-78-M-3114.

Bieniawski, Z.T. (1979b) The geomechanics classification in rock engineering applications, *Proc. 4th Int. Congress on Rock Mechanics*, Montreux, Switzerland, **1**, 41-48.

Bieniawski, Z.T. (1984) *Rock Mechanics Design in Mining and Tunneling*, A.A. Balkema, Rotterdam, 272pp.

Bieniawski, Z.T. 1989) *Engineering Rock Mass Classifications*, Wiley, New York, 251pp.

Deere, D.U. (1963) Technical description of rock cores for engineering purposes, *Felsmechanik und Ingenieurgeologie*, **1**(1), 16-22.

Deere, D.U. and Deere, D.W. (1989) *Rock Quality Designation (RQD) after Twenty Years*, US Army Engineer Waterways Experiment Station Report GL-89-1, Vicksburg, Miss., USA, 67pp + 25pp Appendix.

Hatheway, A.W. (1990) Perspectives No. 6. Rock Quality Designation (RQD: a wonderful shortcut), *AEG News*, **33**(4), 28-30.

Hobbs, N.B. (1975) Factors affecting the prediction of settlement of structures on rock: with particular reference to the Chalk and the Trias. In: *Settlement of Structures* (British Geotechnical Society), Pentech Press, London, 579-610.

Kulhawy, F.H. and Goodman, R.E. (1987) Foundations in rock. In: *Ground Engineer's Reference Book*, F.G. Bell (ed), Butterworths, London, 55/1-55/13.

Priest, S.D. and Hudson, J.A. (1976) Discontinuity spacings in rock, *Int. J. Rock Mech. Mining Sci. and Geomech. Abstr.*, **13**, 135-148.

Wickham, G.E., Tiedemann, H.R.T. and Skinner, E.H. (1972) Support determination based on geologic predictions. In: *Proc. 1st North American Rapid Excavation and Tunneling Conf.*, Chicago, Ill., USA, **1**, 43-64, AIME, New York.

Wickham, G.E., Tiedemann, H.R.T. and Skinner, E.H. (1974) Ground support prediction model - RSR concept. In: *Proc. Rapid Excavation and Tunneling Conf.*, 691-707, AIME, New York.

11

SYMBOLS USED ON BOREHOLE LOGS

SYMBOL	DEFINITION
U	U100 undisturbed soil sample (100mm diameter, 450mm long)
U^+ or $\mho$ or U^0	Undisturbed sample not recovered
U*	Full penetration of sampler not obtained
<u>U</u>	Open tube 40mm diameter, 300mm long
D	Disturbed bag or jar sample of soil
B or BD	Bulk disturbed sample of soil
P_a	Piston sample (100mm diameter, 600mm long)
P_b	Piston sample (250mm diameter, 300mm long)
Thick vertical line	Length over which sample taken
Narrow vertical open rectangle	Rotary core sample taken
w (or W)	Water sample
*	No recovery
_____	Borehole depth
- - -	Casing depth
S	Standard penetration test (SPT). In the borehole record the depth of the test is that at the start of the normal 450mm depth of penetration. The number of blows to achieve the standard penetration of 300mm (the *N* value) is shown after the test index letter, but the seating blows through the initial 150mm penetration depth are not reported unless the full penetration of 450mm is

not achieved. In the latter case, the symbols below are added to the test index letter:

S^+	Seating blows only
S^{++} or $S^\pm$	Blow count includes seating blows
S*	No penetration
S_+	Split spoon sampler sank under its own weight

[The Standard Penetration Test is usually completed when the number of blows reaches 50. For tests which achieve the full penetration of 450mm, the depth at which the test procedure is commenced is given in the depth column on the borehole record, whilst for those tests not achieving full penetration the actual penetration is noted after the blow count, ie. S93/125 = 125mm penetration. If a sample is not recovered in the split spoon sampler a disturbed sample is taken on completion of the test drive. Both are given the same depth as the top of the SPT test drive.]

C	Cone penetration test (with 60° cone fitted)
C	Dynamic cone penetration test (CPT)

[A test is usually conducted in coarse granular soils using the same procedure as for the SPT. The bulk disturbed sample taken is given the same depth as the top of the CPT drive.]

v	*In situ* vane shear test
V	Vane shear strength (MPa) Nat./Remoulded
J	Borehole jacking test
p	Standpipe or piezometer tip
K	Permeability test
Narrow vertical hatched rectangle	Different hatching to denote piezometer seal: upper seal - to response length to - lower seal
Cr	Core recovery (%) <u>or</u>
TCR	Total core recovery (% of core run) <u>and</u>
SCR	Solid core recovery (% of core run)
RQD	Rock Quality Designation
+	Seepage of groundwater, <u>or</u>
Inverted triangle (open)	Groundwater encountered
Inverted triangle (infilled, solid)	Standing level of the groundwater

An alternative symbol system as follows can be used.

Inverted triangle (infilled, solid)	Water level, am
Inverted triangle (open)	Water level, pm
Rhombohedron (open, point down)	Standing water level following initial strike

(Casing depths and the blows necessary to drive the casing will need also to be entered on the borehole log.)

The following symbols may be used to indicate that a laboratory test has been performed on the retrieved borehole material at the point shown on the log.

SYMBOL	DEFINITION
mc <u>or</u> w	Natural (*in situ*) moisture content
n	Porosity
ρ	Density
γ	Unit weight
c or c_u	Undrained shear strength
f	Angle of shearing resistance
lv	Laboratory shear vane value (soil)
LL	Liquid limit (soil)
PL	Plastic limit (soil)
UCS	Unconfined compressive strength of rock
I_s	Point load strength

When composite soils are described, the following terms are used to indicate the various proportions of the subsidiary constituents present:

'with a trace'	up to 10% (approximately)
'with a little'	up to 25% (approximately)
'with some'	25 to 40% (approximately)
'and'	approximately 50%

A statement, such as the following, would normally be included in the site investigation report:

'The consistencies of clays given in this report are based on both visual inspection of the samples and results of strength tests when carried out.'

12

DETERMINATION OF UNCONFINED COMPRESSIVE STRENGTH OF ROCK

INTRODUCTION

Unconfined (or 'uniaxial') compressive strength (UCS) is perhaps the parameter most used for engineering design in rock, and the specification for its determination is rigorous. Design may involve the support capacity of the rock as a material, or alternatively its excavation. The highest and the lowest possible material strengths are required for the former and the latter, respectively. Any failure to prepare the test specimens and to conduct the test properly can lead to low value results. Not only will the basis for design in the rock then be affected, but comparisons of strength between different rock types will be invalid. Test results on the same type of rock should be repeatable within narrow margins.

The specification often adopted is that suggested by the International Society for Rock Mechanics (Brown, 1981).

ISRM METHOD

Apparatus

(a) A suitable machine of sufficient capacity and complying with national standards.
(b) The two loading faces of the testing machine to be parallel with one another, and any spherical seat locked if not complying with (d) below.
(c) The steel platens bearing on the test specimen at both ends shall be of diameter $D + 2mm$, where D is the specimen diameter. Platen thicknesses shall be at least 15mm or $D/3$. Surface flatness should be better than 0.005mm.
(d) One of the platens shall incorporate a spherical seat placed on the upper end of the specimen. The seat shall be lightly lubricated so that it locks in position when loaded on the specimen.

Procedure

(a) Test specimens shall be right-circular cylinders, having a height-to-diameter ratio of 2.5-3.0, a diameter preferably of not less than NX core size, approximately 54mm. The specimen diameter shall be at least 10 times the size of the largest grain in the rock.
(b) The specimen ends shall be flat to 0.02mm and shall not depart from perpendicularity to the axis of the specimen by more than 0.001 radian (about 3.5 minutes) or 0.05mm in 50mm.
(c) The specimen sides shall be smooth and free from abrupt irregularities and straight within 0.3mm over the full length of the specimen.
(d) No capping or end surface treatments other than machining shall be applied to the specimen.
(e) The diameter of the specimen shall be measured to the nearest 0.1mm by averaging two diameters measured at right angles to each other at about the upper height, the mid height and the lower height of the specimen. The average diameter shall be used for calculating the cross-sectional area. The height of the specimen shall be determined to the nearest 1.0mm.
(f) Samples shall be stored for no longer than 30 days, in such a way as to preserve the natural water content, as far as possible, and be tested in that condition. This moisture content shall be reported in accordance with 'Suggested method for determination of the water content of a rock sample', Method 1, ISRM Committee on Laboratory Tests, Document No. 2, First Revision, December 1977.
(g) Load on the specimen shall be applied continuously at a constant stress rate such that failure will occur within 5-10 min. of loading; alternatively, the stress rate shall be within the limits of 0.5-1.0MPa/s recorded in

newtons (or kilonewtons and meganewtons where appropriate) to within 1%.

(h) The number of specimens tested should be determined from practical considerations but at least five are preferred.

Calculation

The unconfined compressive strength of the specimen is calculated by dividing the maximum load, carried by the specimen during the test, by the original cross-sectional area.

Reporting of results

(a) Lithologic description of the rock.
(b) Orientation of the axis of loading with respect to specimen anisotropy, bedding planes, foliation, etc.
(c) Source of sample, including: geographic location, depth and orientations, dates and method of sampling and storage history and environment.
(d) Number of specimens tested.
(e) Specimen diameter and height.
(f) Water content and degree of saturation at time of test.
(g) Test duration and stress rate.
(h) Date of testing and type of testing machine.
(i) Mode of failure, eg. shear, axial cleavage, etc.
(j) Any other observations or available physical data such as specific gravity, porosity and permeability, citing the method of determination for each.
(k) Uniaxial compressive strength for each specimen in the sample, expressed to three significant figures, together with the average result for the sample. The pascal (Pa) or its multiples shall be used as the unit of stress and strength.
(l) Should it be necessary in some instances to test specimens that do not comply with specifications as stated above, these facts shall be noted in the test report.

DISCUSSION

There is no British Standard for unconfined compressive strength testing. BS 5930 (British Standards Institution, 1981), which is under revision, states that tests on rock have not reached 'the same degree of standardization as those on soil', and recommends the procedures outlined by Hawkes and Mellor (1970) as a test specification. In the Remarks column of Table 5 (Schedule of laboratory tests on rock), BS 5930, page 74, an important statement is that 'Its (UCS) value is limited to giving an indication of the upper limit value'. This statement strongly infers the existence of so many factors in the test serving to produce lower values that such values should be discounted rather than averaged in determining the unconfined compressive strength. Thus, only the highest value derived from a test sequence should be used. The alternative would be to specify a limited scatter of the results from a specified number of tests. This could best be achieved by assuming the distribution of test strengths to be normal and defining a tolerable probability. But such a procedure is unlikely to be adopted under normal commercial test conditions. It should be noted that the ISRM specification requires the unconfined compressive strength for each specimen in the sample to be expressed to three significant figures, together with the average result for the sample.

The Hawkes and Mellor test specification is similar not only to the ISRM method given above but also to the ASTM specification. The specifications differ principally in their approach to moisture content, the influence of which is acknowledged by the Geological Society of London (Anon, 1977, p365): 'The strength of rock material determined in unconfined compression is dependent on moisture content of the specimen, anisotropy of the material, and the test procedure used.' Both the ISRM Suggested Method (Procedure (f)) and the ASTM specification recommend (the latter quite strongly) testing at the natural moisture content of the rock, although the ISRM specification allows rocks to be tested saturated or oven dry, with such conditions being noted in the report on the work. Hawkes and Mellor recommend testing in a saturated or 'air dry' state, the latter being defined as a moisture content represented by the surface adsorption and capillary condensation of moisture from the surrounding air, ideally of controlled relative humidity.

Hawkes and Mellor quote various reference sources to show that moisture in rocks reduces strength in two ways. The first is the effect of very small quantities of water which create a surface charge, or zeta potential, at capillary surfaces. This effect has been demonstrated subsequently by various workers. The second is a pore pressure effect in saturated or partially saturated rocks when the rate of loading in the UCS test exceeds the rate at which water can drain from the capillary pores. It is suggested that the former reduces the strength by about 10%. The latter can reduce the strength by about 50% or more. Unfortunately, and in total contrast to the considerations pertaining to the testing of soils, pore pressure effects are rarely taken into account explicitly during the testing of rock specimens.

It follows from the above discussion that it is important, when testing rocks at natural moisture content (Procedure (f) above) to control the rate of loading so that drainage can then occur conformably. It is also possible that for some saturated or partially saturated rocks the specified rates of loading in the ISRM Suggested Method (Procedure (g)) and in the ASTM Specification are too high to allow drainage from a rock replete with a tortuous pore structure.

There is, therefore, some merit in adopting, as recommended by BS 5930:1981, the Hawkes and Mellor approach whereby 'it is suggested that loading rates should be selected on the basis of a logical consideration of the test material and the application intended'.

Procedure (d) above permits no end treatment to the specimens other than machining and lapping. With certain types of rock, such as breccias which contain quite large particles of differing hardness, some 'plucking' of material during end preparation may occur. Load from the compression test machine is therefore taken by a reduced cross-sectional end area of specimen, which increases the local stresses and so promotes premature failure and a lower reported unconfined compressive strength than would otherwise be the case.

For a discussion on the variability of strength test results on rock, reference may also be made to the paper by Rohde and Feng (1990).

UNCONFINED COMPRESSIVE STRENGTH AND POINT LOAD STRENGTH

Point load tests, in which a rock specimen in the form of either a core, cut block or irregular lump (*see* Panek and Fannon, 1992) is compressed

between a pair of 'spherically truncated conical platens' (60° cone angle; 50mm tip radius) and induced to fracture in an essentially brittle tensile manner, do not require the rigorous specimen end preparation demanded for unconfined compression tests, and are thus relatively simple (and relatively inexpensive) to conduct. The basic aim for any suite of rocks must be to correlate the point load strength to uniaxial compressive strength so that the latter, which tends to be the fundamental parameter used in design both for support and excavation, can be inferred from the more extensive point load strength results. Reference may be made to the relations given in Table 4.13 of Section 4.4.2 in the main text (Part One).

Specific requirements of the test for different types of specimen are given in the ISRM Standard (Anon, 1985). The following points in particular should be noted:

(a) For core specimens the length/diameter ratio should be greater than 1.0 when testing across a diameter and should be between 0.3 and 1.0 when testing along an axis. Blocks or lumps should be of size 50 ± 35mm, with a depth/width ratio between 0.3 and 1.0 and preferably about 1.0. The other dimension should be at least 0.5 times the width.

(b) There should be at least 10 tests per sample.

(c) Point load strength $I_s = P/D_e^2$ where P is the applied platen load, $D_e^2 = D^2$ (diameter squared) for cross-diameter loading, and $D_e^2 = 4\pi A$ for axial, block and lump tests, where A is the minimum cross-sectional area of a plane through the platen contact points and is equal to the width/diameter ratio.

(d) The load P should be increased steadily such that failure occurs within 10-60 seconds.

(e) Point load strength varies as a function of D, so a *size correction* must be applied. The size correction to I_s is defined as the value of I_s that would have been measured by a diametral test with $D = 50$mm (typically an NX core of 54mm diameter). This I_s value is expressed as $I_{s(50)}$. A suite of tests should be conducted over a range of D or D_e values and then a graph of log P versus log D_e^2 prepared. A value of P_{50} corresponding to $D_e^2 = 2500$mm^2 can then be obtained by interpolation (and, if necessary, by extrapolation), and the size-corrected point load strength index calculated as $P_{50}/50^2$. When neither a 50mm diameter specimen nor a size correction procedure is practicable, then the Standard suggests applying the formula $I_{s(50)} = F \times I_s$, where $F = (D_e/50)^{0.45}$. If the specimen size does approximate to the 50mm standard then the approximate expression $F = (D_e/50)^{1/2}$ may be used.

(f) If the rock is anisotropic, and particularly if there is a plane of obvious weakness, tests must be performed in directions both parallel to and perpendicular to the planes of anisotropy, and the results reported accordingly.

(g) The ISRM Standard notes that, on average, UCS is 20 to 25 times I_s ($I_{s(50)}$), but can vary between 15 and 50, especially for anisotropic rocks. $I_{s(50)}$ is approximately 0.80 times the direct (uniaxial) tensile strength or Brazilian tensile strength.

According to ISRM recommendations (Anon, 1985), point load specimens should be tested and reported at a moisture content 'that is appropriate to the project for which the data are required'. However, because at degrees of saturation above 50% the rock strength is less affected by small changes in moisture content, tests in the range 50% to 100% saturation would be recommended unless tests on dry rock are specifically required. As was noted above, unconfined compression tests should ideally be conducted at natural moisture content but may be tested saturated or oven dry provided that such conditions are stated in the test report. Together these recommendations pose several problems:

- The latitude allowed for point load tests and UCS tests in respect of moisture content is not really satisfactory. It is not clear how any reported moisture content can be correlated with strength changes and used in design.
- Because of the need to correlate point load strength with UCS, both tests on the same rock should be performed at the same moisture content.
- Ideally, both point load and UCS tests should

be conducted at the natural moisture content of the rock. This requires:

(a) No water to be used in the hole drilling process.
(b) Rapid and effective core sealing, before placement in the core box, using cling film which will allow visual inspection of the core.
(c) Testing of the rock immediately after removal of the seal.

- If the rock is supplied in block form, then cores must be taken for UCS tests and, preferably at the same diameter, for point load tests. It may be possible to core some blocks dry, using compressed air, but in most cases water will be necessary for core barrel cooling and cutting removal. If dry coring is used, then the cores at the time of the test will not be at natural moisture content. If wet coring is used, the final moisture content will also be unknown.

The requirements would seem to be as follows:

- Both point load and UCS tests should be conducted at as near as possible to the same moisture content.
- Test specimens for both sets of test should be either completely dry or completely wet or at some easily achieved target moisture content, as recommended below.

As discussed above, strength is very dependent on moisture content, and if a wholly or partially saturated specimen is dried out in an oven its mechanical strength is then enormously increased. On the other hand, the strength of a specimen that has been saturated is severely depleted from that at natural moisture content. Many contractual claims for extra payment arise from the failure of plant satisfactorily to excavate a rock. The contractor will often attempt to show that the actual strength of the rock (adopting the unconfined compressive strength as an indicator and tending to ignore other factors such as discontinuity spatial density distribution) is greater than that noted in the site investigation report. For this reason it is probably preferable in a standard site investigation report that is related substantially to rock excavation to report the strengths 'on the high side'. This would imply that rock should not be tested wet, but should be dried out before testing. However, the strength may then be unrealistically high and perhaps encourage, at higher cost to the client, the use of heavier duty plant. There is a strong argument to be advanced for storing both point load and UCS specimens for 5 to 6 days in a humidity oven prior to testing. In this way test specimens that are wet from laboratory coring and/or end preparation will shed water and dry specimens will gain some moisture before testing takes place. There may be a problem that under commercial pressures the necessary 5 or 6 days

rest time in the humidity chamber may not be achievable.

It is recommended that all rock specimens be stored in a humidity chamber after preparation and prior to testing, the period in the chamber being 5 to 6 days or as near as possible to that time, with the actual time being reported. A separate small specimen of rock taken from the UCS specimen after preparation should be similarly stored in the humidity chamber for assessment of moisture content before testing takes place.

REFERENCES

Anon (1977) The description of rock masses for engineering purposes, *Quart. Jl. Engrg Geol.*, **10**, 355-388.

Anon (1985) International Society for Rock Mechanics Commission on Testing Methods, suggested method for determining point load strength, *Int. J. Rock Mech. Min. Sci. & Geomech. Abstr.*, **22**(2), 51-60.

British Standards Institution (1981) *Code of Practice for Site Investigations*, BS 5930:1981, London.

Brown, E.T. (ed) (1981) *Rock Characterization, Testing and Monitoring. ISRM Suggested Methods. Suggested Method for Determination of the Water Content of a Rock Sample*, p83.

Suggested Method for Determination of the Uniaxial Compressive Strength of Rock Materials, p113, 114. Pergamon Press, Oxford.

Hawkes, I. and Mellor, M. (1970) Uniaxial testing in rock mechanics laboratories, *Engrg Geol.*, **4**, 177-285.

Panek, L.A. and Fannon, T.A. (1992) Size and shape effects in point load tests of irregular rock fragments, *Rock Mech. and Rock Engrg*, **25**(2), 109-140.

Rohde, J. and Feng, H. (1990) Analysis of the variability of unconfined compression tests of rock, Technical Note, *Rock Mechanics and Rock Engineering*, **23**(3), 231-236.

13

RISING, FALLING AND CONSTANT HEAD PERMEABILITY TESTS

DETERMINATION OF THE COEFFICIENT OF PERMEABILITY

Soil permeability, or the ability of a soil to transmit a fluid, is not only a function of the size and pattern of the individual particles, but also, and more importantly, the actual stratification which can cause the flow properties to depend upon direction (permeability anisotropy). It is difficult to obtain suitable specimens from the field for the purposes of permeability assessment, and so measurements are usually determined *in situ*. This is done by applying a head difference between the water in a borehole or piezometer and that in the surrounding ground. The flow resulting from this difference in hydraulic pressure is measured and expressed as a permeability - or, more accurately, a hydraulic conductivity - value.

In the case of more permeable soils, such as sands and gravels, variable head tests may be appropriate for determining the permeability properties. These test procedures, together with their advantages and limitations, are provided in BS 5930 (British Standards Institution, 1981). Those tests which are carried out by increasing the pressure in an open borehole or piezometer (by introducing water into the hole casing or piezometer tube so that the water head exceeds the water level in the ground) are known as falling head or outflow tests. Those in which the water pressure is decreased by removing water so that the level in the borehole or piezometer tube is below that in the ground are known as rising head or inflow tests. In either of these cases during the tests, readings are taken of water head at frequent intervals of time.

The standard Darcy equation is normally used for the assessment of permeability:

$$Q = kAh/l \qquad (1)$$

where

Q is the volume flow rate through the soil (dimension $L^3 T^{-1}$),

A is the cross sectional area of flow; i.e. sample c/s/a in a laboratory test (L^2),

h is the head loss (L),

l is the flow length; ie. length of sample in a laboratory test (L),

and k is the coefficient of permeability (LT^{-1}).

This equation, however, is inadequate in that it takes no account of the conditions pertaining at the intake zone. Permeability k in units of metres per second may more satisfactorily be derived from the expression developed by Hvorslev (1951):

$$k = [A/6F(t_2 - t_1)][\ln(h_1/h_2)] \times 10^{-7} \text{m/s}.... \qquad (2)$$

where

A is the internal cross-sectional area of the borehole in units of mm^2,

F is the intake factor for the test zone in units of metres,

and h_1 and h_2 are the pressure heads (excess water head in borehole or piezometer over groundwater table) in units of metres, respectively at times t_1 and t_2.

Hvorslev (1951) provides a means of evaluating the intake factor F for different geometries of the test zone. A summary of the values for the more common-occurring geometries is given in BS 5930. For probably the simplest case of that of a borehole, diameter D ($=2r$, where r is the radius) open at its base and cased to full depth in uniform soil, shape factor

$$F = 2.75D. \qquad (3)$$

For the case of a borehole cased through impermeable soil with the base of the casing coincident with the top of an impermeable stratum,

$$F = 2D. \quad ....(4)$$

In cases where inflow/outflow tests are carried out through a filter of cylindrical form at the base of a borehole or piezometer, if the diameter of the filter is D (mm) and its length is L (mm), then if it can be assumed that the soil is hydraulically uniform it is proposed by Hvorslev that

$$F = 2\pi L/[\ln\{(L/D) + (1 + (L/D)^2)^{0.5}\}] \quad ....(5)$$

with the expression being approximately correct for ratios of L/D greater than about 4.

There is a simpler expression, again proposed by Hvorslev (1951), for determining the coefficient of permeability, k (m/s), using the concept of basic time lag of the installation, T

$$k = A/FT. \quad ....(6)$$

A curved plot of time t against head h may produce scatter of data points. One way of correcting for this is to plot the head ratio h/h_0 on a logarithmic scale against lapsed time t (minutes), where h_0 is the head at the beginning of the test. This should produce a straight line plot, and from it the basic time lag T corresponds to the time (minutes) at which the head ratio

$$h/h_0 = e^{-1} = 0.37. \quad ....(7)$$

This basic time lag, T, is thus the time that would be required for equilibrium to be achieved if the initial flow rate were maintained throughout the test. Interpretation of the test requires knowledge of the initial head h_0, but this may not always be available. In such cases, trial values of h_0 may be assumed and the semi-log plot, $\ln(h/h_0)$ versus time (linear scale), drawn. Any errors in the evaluation of h_0 will be reflected as a curved plot, in which case an iterative procedure may be used with other trial values of h_0 until such time as a straight line plot is obtained.

Table SI 13.1 represents an attempt to express in an integrated manner the shape factor effects and also to take account of possible permeability anisotropy of the soil.

ABSOLUTE COEFFICIENT OF PERMEABILITY

It should further be noted that k is not an absolute property of the soil, but rather a property of the soil/fluid system. Permeability k depends upon the fluid density and viscosity, and will vary with temperature. The Darcy coefficient of permeability, k, relates to the absolute coefficient of permeability, K (in dimensions of L^2), which is independent of the fluid characteristics, in the manner of Equation 8 below:

$$k = Kc_w/\mu \quad ....(8)$$

where

k is the permeability in, say, cm/s (dimension LT^{-1}),

K is the absolute coefficient of permeability in, say, cm^2 (L^2) (1 Darcy is equivalent to 9.88×10^{-9}cm^2),

$c_w = \rho g$ in, say, g/cm^2.s^2 ($ML^{-2}T^{-2}$),

ρ is the fluid density in, say, g/cm^3 (ML^{-3}),

g is gravitational acceleration in, say, cm/s^2 (ML^{-2}),

and μ is the dynamic viscosity in, say, g/cm.s ($ML^{-1}T^{-1}$).

[Note that absolute permeability is the permeability of a soil given a fluid with a viscosity of 1 centipoise having a flow of 1 cm^3/s through a sample 1cm^2 cross-sectional area with a pressure gradient of 1 atmosphere/cm.

Expressing dynamic viscosity (μ) in Poise (P),

1P = 0.1kg/m.s = 10^4g/cm.s.

Also, 1P = 0.1N.s/m^2 (newton seconds per metre squared).

So, 1g/cm.s = 0.1N.s/m^2.]

For water in soil at 20°C and at atmospheric pressure,

ρ = 1.0g/cm^3

g = 980.667cm/s^2

μ = 1.0019 centipoise = 10.019×10^{-3}kg/m.s

= 10.019g/m.s

= 0.10019g/cm.s.

Table SI 13.1 Coefficient of permeability calculation

| Horizontal permeability (k_h) equations | | Shape Factors (F) | |
Rising and falling head	Constant head	Open holes & piezometers	Tube wells
$k_h = \dfrac{\pi a^2}{F(t_2-t_1)}ln\dfrac{h_1}{h_2}$	$k_h = \dfrac{Q}{F(h_2-h_1)}$	**A** $\quad F = \dfrac{2\pi L}{ln[\alpha+(1+\alpha^2)^{0.5}]} = F'L$	**A** $\quad F = \dfrac{5.5a}{m} = F'a$

Symbols			
		B $F = \dfrac{2\pi L}{ln[2\alpha+(1+4\alpha^2)^{0.5}]} = F'L$	**B** $F = \dfrac{4.0a}{m} = F'a$
a	: radius of casing in which water level has been observed		
t_1,t_2	: two arbitrary observation times	A : assuming that the permeable ground extends infinitely in all directions	
h_1,h_2	: corresponding water levels	B : permeable material is limited in vertical direction by impermeable barrier immediately below well	
F	: shape factor, dependent on geometry of borehole and porous medium and has dimension of length	F' values for open holes and piezometer	
L	: length of piezometer head or length of section in borehole	k_h/k_v	
r	: radius of piezometer head or section of open hole		
m	: square root of ratio between horizontal and vertical permeabilities, ie $(k_h/k_v)^{0.5}$		
α	: quantity mL/2r		
h_2-h_1	: water level, measured from rest water level, maintained during constant head test.		

F' values for open holes and piezometer k_h/k_v

L/2r	1	2	5	10
01	7.13-4.35	5.48-3.56	4.07-2.85	3.36-2.47
02	4.35-3.00	3.56-2.58	2.85-2.18	2.47-1.94
05	2.72-2.10	2.37-1.88	2.02-1.65	1.82-1.52
10	2.10-1.70	1.88-1.56	1.65-1.40	1.52-1.30
20	1.70-1.43	1.56-1.33	1.40-1.21	1.30-1.14
50	1.36-1.19	1.27-1.11	1.16-1.03	1.09-0.97
100	1.19-1.05	1.11-0.99	1.03-0.92	0.97-0.88

F' values for tube wells k_h/k_v

L/2r	1	2	5	10
01	5.50-4.00	3.89-2.83	2.46-1.79	1.74-1.26

The same shape factors apply to rising, falling and constant head tests.

Now $K = $ (Darcy/0.101325) x 10^{-9}. Therefore,

$k = (10^{-9}$ x $980.665/0.101325$ x 0.10019 Darcy) cm/s
$= (9.66$ x 10^{-4} Darcy) cm/s,

or, expressing this approximately,

1 millidarcy = 10^{-8}m/s permeability.

Thus, at 20°C and atmospheric pressure, an absolute permeability of 1 Darcy would be an effective permeability of 9.66 x 10^{-4}cm/s with respect to water and would be 6.42cm/s with respect to air using Equation 8 above.

LUGEON TYPE PERMEABILITY TESTS

These tests are usually conducted in boreholes in sections 3m to 5m long as drilling proceeds. The test length depends upon the degree of rock fracturing. The bottom of the hole defines one end of the test section and the packer position the upper end. If the test is performed after completion of drilling, then a double packer system may be used.

The water (phreatic) level needs to be known before each test. This can be determined by the use of an electrical (audible bleeper) probe lowered from the top of the hole. The (single) packer can then be lowered to position and inflated in order to render the watertightness of the test section.

Generally, the tests are conducted at several pressure stages, say up to five. Water is pumped in and the pressure increased to p_1. Flow is maintained constant, and is registered over a specific period of time, say 10 minutes. This operation is repeated at a higher pressure, p_2, with monitoring over the same period, and again at the highest pressure, p_3 Thereafter, the pressure is decreased over two stages, with $p_4 = p_2$ and $p_5 = p_1$. A maximum pressure of $1MN/m^2$ or, in the case of a rock, a pressure that would not exceed half that to cause hydrofracture would be used.

One lugeon unit is equal to a flow of 1 litre of water per minute per metre of borehole length at an excess pressure of $1MN/m^2$. In the case of a regularly fissured rock, 1 lugeon unit is equivalent to a permeability of approximately $10^{-7}m/s$.

From Darcy's law, $v = ki$. Assuming steady state conditions, k can be calculated from

$$k = (Q/h)[\ln(r_e/r_o)/2pL] \quad \text{...................}(9)$$

where

Q = flow rate (m^3/s),

h = excess hydraulic head (m),

r_e = radius of influence of seepage (m),

r_o = borehole radius (m),

and L = length of the section (m).

For most rock masses, the ratio r_e/r_o lies

somewhere between 100 and 1000. The ratio $\ln(r_e/r_o)/2pL = L/C$, where C (metres) can be obtained from a nomograph, is a geometrical cut-off.

DEFINITIONS

1.) **Piezometer**: A filter is sealed into a particular stratum by means of seals placed above and below the filter section. For a piezometer at the base of a borehole, the lower seal is normally omitted. For bentonite seals, the balls of bentonite should be dropped down the hole wrapped in (old!) nylon stockings so as to prevent the particles from breaking up in the water. Alternatively, very thin polythene bags can be used for containment and be punctured in the ground by means of a tamper. If bentonite/cement type grouts are used, the bentonite must first be mixed with water before cement is added to the mixture, otherwise the grout will segregate or remain in suspension. Mixing is best done by means of a mechanical mixer.

The single-entry ('Casagrande') types of piezometer simply indicate the water level. More complicated double-entry types indicate the pressure on a gauge or manometer. Casagrande 'drive-in' type piezometers can be used in soft-to-firm soils without the need for a borehole.

(a) *Hydraulic*: This type is suitable for saturated soils, but a conical fine-grained ceramic element which restricts air entry from the soil may be substituted when dealing with partially saturated material. The lines themselves require careful de-airing. There are twin tubes for flushing the piezometer, which is of small diameter in order to minimise the response time. Measurements are by manometer, Bourdon gauge or electrical transducer. It is only possible to measure positive pressures, and so careful positioning is important. (A -5m head is measurable on very carefully de-aired lines.)

(b) *Pneumatic*: This type can be useful for reading pore pressures without the need for extensive de-airing (in fact, this type cannot be de-aired). The pore pressure on one side of a diaphragm is balanced against a pneumatic pressure established on the other side using a portable unit, but diaphragm flexibility can affect valve operation. There are high and low air entry ceramic filters (as with a hydraulic piezometer). If long-term negative pore pressures in partially saturated soils are to be measured, such a diaphragm type of piezometer may be unsuitable. These piezometers cannot be used for *in situ* permeability tests.

(c) *Electrical*: These comprise vibrating wire types and hydraulic/pneumatic types having an electrical output. They permit remote output, but are expensive and cannot be used for *in situ* permeability tests. It is not possible to de-air them, and there can be changes in transducer calibration over long periods.

(d) *Installation*: All details of installations should be recorded. No dirt should be allowed to enter the pipe or piezometer tip. The bottom end of the standpipe should be plugged (a plastic plug, for example) to keep soil out if the pipe scrapes the side of the borehole as it is lowered in. Filter material should be suitably graded, but kept free from clay and silt. For preliminary guidance, a filter of clean, well-graded sand and gravel with only a small proportion of fine-to-medium sand is suitable for soils with some clay or silt content. For a fine sand soil, the filter should comprise coarse sand or coarse sand with gravel, with not more than a few per cent medium sand.

When backfilling a lined borehole, the casing should be withdrawn as the filling is being placed so that the casing shoe is just below the top of the backfill. This is especially important when placing clay seals, and in this case the lining tubes should never be pulled above the top of the clay backfill.

Whenever an instrument is installed in a clayey or silty soil, the supervising geotechnical engineer or engineering geologist may require some water to be poured into the top of the pipe in order to reduce the time needed for the water level to reach an equilibrium level and for the

purpose of making a permeability test. Clean water must always be used.

2.) **Standpipe**: This is an installation where the upper seal is absent or where the filter section covers more than one aquifer. The standpipe comprises UPVC or HDPE tubing not less than 19mm internal diameter. There is a plug at the bottom of the tubing and the lower 1m or so is perforated by holes less than 5mm in diameter at intervals of about 75mm, or by an equivalent area of slots of equivalent cross-sectional area but not greater than 5mm wide. There will normally be a pea gravel (6-10mm diameter) filter placed at the bottom of the hole, or a similar material can be used with the approval of the Engineer. The reduced level to the top of the filter must be recorded. On occasions the depth of the hole may exceed the depth to which the standpipe tubing and filter are to be placed. In such a case the hole must be backfilled with impermeable material up to the base of the filter. The tubing is then lowered down the hole to the top of the filter and the hole then filled around the tube with gravel or similar material to within 1m of the top of the hole. There is normally a steel cover for the tube at the top of the hole, the cover being set in concrete and containing an air vent.

3.) **Casagrande-type filter tips for a standpipe piezometer**: These filter tips are made from ceramic material having a permeability $k = 3 \times 10^{-4}$m/s and a pore size of 60μm. Tip sizes are, typically, 50mm diameter x 127mm long and 50mm diameter x 254mm long. These piezometers also de-air themselves (diameter >12mm). Response times are in minutes ($k = 10^{-6}$m/s for silts, fine sands) or months ($k = 10^{-10}$m/s for clays). They are generally too slow to measure construction pore pressures.

4.) **Well-point-type filter tips**: These are an alternative to the Casagrande-type tips and are normally constructed from perforated rigid PVC tubing with a porous plastic insert, 19mm in diameter and length 0.3m or 1m. This type can be used for both piezometer and standpipe installations.

GROUNDWATER OBSERVATIONS

Observations should be made during the boring process and in the longer term after piezometer installation. It is useful to note the routine observations that should be made as a borehole is sunk:

(a) Depth at which groundwater is first encountered.
(b) An approximate indication of the rate of water inflow to the borehole, using descriptive terms such as 'rapid rise of *n* metres in *t* minutes', or 'slow seepage'.
(c) Depth at which the groundwater is sealed off by the borehole lining tubes.
(d) Depth or depths at which groundwater is encountered again.
(e) Depth to the water level in the borehole at the start of a new morning shift before boring continues.
(f) Depth to the water level in the borehole on completion of drilling.

It is good practice, when the opportunity presents itself, to leave the borehole open for several days after completion of the drilling and for the water level to be recorded in the morning and in the evening each day. The depth to the bottom of the borehole should be checked at the time that these measurements are made because the water level could be affected by some collapse at the sides of the borehole after withdrawal of the casing. All groundwater observations must be submitted with the daily borehole record. There may be circumstances when a decision has to be taken whether to install standpipes or piezometers for longer-term recording in the borehole. A decision on this will depend upon the type of construction on the site, and so it is usually possible to make a specification to cover the point before the ground investigation begins. In the case of a tunnel site investigation it should generally be policy to install standpipes, and sometimes piezometers, as a minimum requirement in all or most of the boreholes, in which case standpipes and/or piezometers should be purchased well in advance so that they are ready on site for installation by experienced personnel at the right time.

REFERENCES AND BIBLIOGRAPHY

British Standards Institution (1965) *Specification for Water Well Casing*, BS 879:1965, London.
British Standards Institution (1981) *Code of Practice for Site Investigations*, BS 5930:1981, London.
British Standards Institution (1983) *Code of Practice for Test Pumping Water Wells*, BS 6316:1983, London.
Clark, L. (1977) The analysis and planning of step drawdown tests, *Quart. J. Eng. Geol.*, **10**, 125-144.
Clark, L. (1979) The analysis and planning of step drawdown tests: a clarification, *Quart. J. Eng. Geol.*, **12**, 124.
Houlsby, A.C. (1976) Routine interpretation of the Lugeon water-test, *Quart. J. Eng. Geol.*, **9**, 303-314.
Hvorslev, M.J. (1951) *Time Lag and Soil Permeability in Ground-water Observations*, Bulletin No. 36, Waterways Experimental Station, Vicksburg, Mississippi, USA.
Klinkenberg, L.J. (1941) The permeability of porous media to liquids and gases, *Drilling and Production Practice*, 200-213.
Lugeon, M. (1933) *Barrages et Géologie*, Dunod, Paris; Librairie de l'Université, Lausanne.
Oliveira, R. and Graça, J.C. (1987) *In situ* testing of rocks. In: *Ground Engineer's Reference Book*, F.G. Bell (ed), Butterworths, London, 26/1-26/28.

14

ROCK EXCAVATABILITY VIA 'Q'

Kirsten (1982) has applied the Q-system concept (Barton *et al.*, 1974; Barton, 1976) to the estimation of rock excavatability (Table SI 14.1):

Excavatability Index, $N = M_s(RQD/J_n)J_s(J_r/J_a)$

where

M_s = Mass Strength Number - the effort needed to excavate the material as if it were homogeneous, unjointed and dry. M_s approximates to the unconfined compressive strength of rock in MPa.

RQD = Rock Quality Designation.

J_n, J_r and J_a are Q-system parameters.

J_s = relative ground structure number. Its value is a function both of the gradients of two intersecting joint sets, which are assumed to be orthogonal, relative to the direction of cutting and of the ratio of the two joint set spacings. Values range from 0.37 to 1.5, and for intact material, $J_s = 1$.

Abdullatif and Cruden (1983) produced a graph of Q against RMR for digging, ripping and blasting, based on investigations at 23 sites. The graph showed that the rock mass can be dug up to an RMR value (*see* Bieniawski, 1989) of 30 and ripped up to an RMR value of 60. Rock masses rated as 'good' or better by RMR must be blasted.

Many of the publications relating to the general field of rock excavatability are concerned with 'rippability', often in an open excavation setting and Coal Measures rocks. Some of these publications, together with the thesis by Glossop on tunnelling, are listed below in the Bibliography.

REFERENCES (*see also References to Barton and Bieniawski in the main text*)

Abdullatif, O.M. and Cruden, D.M. (1983) The relationship between rock mass quality and ease of excavation, *Bull. Int. Assoc. Engrg Geol.*, no. 28, 183-187.

Barton, N. (1976) Recent experiences with the Q-system of tunnel support design, *Symp. on Exploration for Rock Engrg.*, Johannesburg, **1**, 107-117.

Barton, N. Lien, R. and Lunde, J. (1974) Engineering classification of rock masses for the design of tunnel support, *Rock Mech.*, **6**(4), 189-236.

Bieniawski, Z.T. (1989) *Engineering Rock Mass Classifications*, Wiley Interscience, New York, 251pp.

Kirsten, H.A.D. (1982) A classification system for excavation in natural materials, *The Civil Engineer in South Africa*, **24**, 293-308.

Table SI 14.1 Excavatability Index

Range of N	Rippability
1 < N < 10	Easy ripping
10 < N < 100	Hard ripping
100 < N < 1000	Very hard ripping
1000 < N < 10 000	Extremely hard ripping/blasting
N > 10 000	Blasting

BIBLIOGRAPHY

Braybrooke, J.C. (1988) The state-of-the-art of rock cuttability and rippability prediction, *Fifth Australian New Zealand Conf. on Geomechanics*, 13-42.

Glossop, N.H. (1982) The influence of geotechnical factors on tunnelling systems, PhD thesis, University of Newcastle upon Tyne.

Hadjigeorgiou, J. and Scoble, M.J. (1990) Ground characterization for assessment of ease of excavation, *Proc. 4th Int. Symp. on Mine Planning and Equipment Selection*, Calgary, R.K. Singhal and M. Vavra (eds), A.A. Balkema, Rotterdam, 323-331.

MacGregor, F., Fell, R., Mostyn, G.R., Hocking, G. and NcNally, G. (1994) The estimation of rock rippability, *Quart. J. Engrg Geol.*, **27**, 123-144.

Singh, R.N., Denby, B. and Egretli, I. (1987) Development of a new rippability index for coal measures excavations, *Proc. 28th U.S. Symp. on Rock Mech.*, Tucson, Arizona, A.A. Balkema, Boston, 935-943.

Smith, H.J. (1986) Estimating rippability by rock mass classification, *Proc. 27th U.S. Symp. on Rock Mech.*, Tuscaloosa, Alabama, AIME, New York, 443-448.

Weaver, J.M. (1975) Geological factors significant in the assessment of rippability, *The Civil Engineer in South Africa*, **17**, 313-316.

15

WELL-POINT DEWATERING

There are three vacuum well systems:
- vacuum well-points
- deep wells with vacuum
- ejector (eductor) wells

Vacuum well-points comprising slotted PVC plastic pipes, 38mm up to 152mm diameter and about 200mm long (or sometimes longer), are often gravel packed for greater efficiency. The screened or slotted sections of a well-point should have holes or slots fine enough to keep soil or backfill out but sufficiently open to let the water enter freely. Such systems offer a flexible dewatering facility for fine sands and silts where the quantities of water to be pumped out are not large. A typical pumping capacity of a 38mm diameter riser pipe is approximately 1 litre/s.

The well-point, with a metal riser pipe, is hammered or jetted into the ground 1 or 2m below tunnel invert for a maximum lift depth of about 4 to 6m below header level. This tends to limit its application to small diameter shallow tunnels. Well-points, at spacings from 0.8m up to 3m or more and set to one side of or on both sides of the proposed tunnel, are connected through swing joints to header pipes connected with an air separation chamber that removes air from the system so that the suction pumps (usually combined centrifugal and vacuum) do not lose their 'prime'. (The purpose of the vacuum pump is to increase the vacuum developed by the centrifugal pump, and so ensure that it is self-priming.) Pumping depresses the groundwater table via a series of overlapping cones of depression and allows excavation to take place in sensibly dry working conditions. Water discharge is at a sufficient distance from the working site to prevent back-flows to the dewatered ground.

If a depth greater than 4 to 6m needs to be dewatered, then several stages of groundwater lowering up to 18-24m or more may be needed. Alternatively, deep, gravel-packed pumped wells may be used. These deep wells comprise a perforated well liner surrounded by a suitable gravel/sand filter, with an electro-submersible pump at the bottom of the hole. They tend to be relatively widely spaced, say 10 to 30m with depths around twice tunnel depth, and are capable of developing vacuums up to about 95kPa to pump relatively large flow rates. The pumping capacity, in theory, is limited only by the power and size of the submersible pump, but in the case of fine soils the ground becomes a significant determinant.

Suitable filter design to reduce loss of fines from the surrounding soil is important, otherwise attributable settlement might be induced by dewatering. A general formula that can be adopted for the design of filters may be ascribed to Bertram (1940) and is written as

$$\frac{D_{15(filter)}}{D_{85(soil)}} < 4 \text{ to } 5 < \frac{D_{15(filter)}}{D_{15(soil)}}$$

<------------------------->
Criterion for prevention
of piping

<------------------------->
Ensures that filters will
be more permeable than
protected soils

In this formula, D_{15} means the size at which 15% of the particles are finer and D_{85} means the size at which 85% of the particles are finer. This information is obtained from grading curves which are drawn up from the results of sieving tests.

There are other filter design criteria, such as the specification that the 50% size of a filter should be less than 25 times the 50% size of a soil in which it is placed. According to the US Army Corps of Engineers (Anon, 1955) the piping ratio: [D_{15}(of filter)/D_{85}(of soil)] criterion for protecting medium- to highly plastic clay soils permits the D_{15} size of the filter to be as high as 0.4mm. When the protected soil contains a high gravel fraction then according to Sherard *et al.* (1963) the filter should be designed on the basis of the grading curve of the portion of the material which passes a 25mm sieve. So, the permeability of the filter is approximately proportional to $(D_{15})^2$. Reference may also be made to Cedergren (1987).

Should the tunnel be too deep for complete dewatering to be achieved economically, one option is to lower the water table partially in order to tunnel under the reduced head with a tolerable internal compressed air pressure or with a slurry shield or earth pressure balance shield.

Well-point jetting involves forcing water under pressure down the inside of the riser pipe and out through orifices or jetting tips. Special filters may be needed to prevent fine sands and silts from clogging the tips, and may also be required for clay soils which are prone to erosion as a result of dispersion and de-flocculation mechanisms and for which standard filters are unprotective (Sherard *et al.*, (1972).

Synthetic filter fabric membranes may sometimes be used instead of a graded granular soil. These filters may be formed of materials such as PVC, polypropylene, polyester, nylon, and so on, but some can be affected by alkalis in the soil, others by acids or fuel oils. Most of them deteriorate in the (long-term) presence of ultra violet light, so they need to be stored away carefully before use.

The question of geotextile filter criteria has been addressed by Schober and Teindl (1979), McGown *et al.* (1982), Giroud (1982) and Hoare (1987).

If the flow of water from each well-point is small (say, less than 1 litre/s), jet ejector pumps can be used. In this case the well is pumped by a nozzle and venturi device positioned at depth within the well and driven by water from high pressure supply pumps at ground level. Water at high pressure (greater than 750kPa) is pumped through the supply pipe to a nozzle within the ejector casing. The supply water emerges from the nozzle at high (around 40m/s) velocity, entraining the surrounding groundwater out of the well through the return riser. An ejector has the capability of pumping both water and air, and provided that the capacity of the ejector exceeds that of the well the ejector will automatically develop a vacuum in the well if the well casing and the surrounding (sand) filter are sealed. There are also twin-pipe ejectors capable of

giving higher induced flow rates from a larger diameter hole.

Although these ejector dewatering systems are relatively inefficient, being prone to clogging and requiring cleaning and maintenance because of nozzle wear, they can produce vacuums up to 95kPa in the well and a lift of 30m. They can also operate at very low flow rates but cannot handle large flow rates. Ejectors could be particularly appropriate for tunnelling works under high water table/silty ground conditions where the other logical options would be compressed air, full-face machine tunnelling, or even ground freezing. Each jet ejector can be powered by a separate high-pressure centrifugal pump or by a high-pressure ring main serving several pumps. Further information on vacuum dewatering systems can be obtained from the papers by Preene and Powrie (1994) and Powrie and Preene (1994a), and on time-drawdown behaviour from Powrie and Preene (1994b).

Drawdown itself could lead to differential settlements at the foundations of adjacent property (*see* Hsi *et al.*, 1994). In the ancient centres of some cities (Amsterdam, for example, where the groundwater table may be only one or two metres below ground surface), the possibility of negative skin friction effects and pile drawdown, together with the rotting of timber piles when exposed to the atmosphere, must be taken fully into account. One method suggested sometimes in the literature for partially overcoming the problem is to drill recharge holes between the tunnel and the property, and to circulate the abstracted water through them. The foundation support capacity of the water to the sides of the tunnel is maintained in this way by means of a steepened drawdown surface.

Neglecting the effects of well overlaps, assuming soil homogeneity and permeability isotropy, the standard Theim equation can be used to establish water quantities requiring to be pumped in order to achieve the requisite drawdown. Let H be the height of the water table above the base of the soil stratum to be dewatered, h the height from the base of the stratum to the point of maximum drawdown at the well, b the thickness of the aquifer, r_0 the radius of the drawdown depression, r_w the radius of the well, k the ground permeability (isotropy), and Q the flow rate in units of volume per unit time from the well, then the radial water flow in a confined aquifer is

$$Q_c = 2\pi k b (H - h)/2.30 \log(r_0/r_w),$$

and, in the case of an unconfined aquifer,

$$Q_u = \pi k (H^2 - h^2)/2.30 \log(r_0/r_w)$$

in order to achieve the required drawdown (which is $H - h$) at the well. Reference would naturally be made to standard textbooks for further reading on the theory of this subject and the in-house guidance manuals of reputable site investigation firms will tend to contain much more extensive coverage on the subject than will be found in many textbooks. It is always preferable, of course, to seek advice from specialist groundwater lowering contractors, since the ideal ground conditions required to satisfy the theory will rarely pertain in practice.

A dominant factor controlling the choice of well-pointing rather than ground freezing, grouting, or even the use of compressed air, is the soil particle size. The applicability of the different ground improvement options, based on readily available charts, is summarised broadly below in terms of approximate particle size.

Groundwater lowering:
 Well-points and pumped wells: 210μm - 2mm
  (medium and coarse sand)

Sump pumping: 600μm - 2mm (coarse sand)
Eductors for vacuum enhanced pumping:
 63μm - 425μm
 (fine to medium sand)

Freezing: $<63\mu$m - 2mm (silt to coarse sand)

Grouting:
 Cement-based grouts: 1mm - 10mm
  (coarse sand to medium gravel)
 bentonite grouts: 420μm - 6.3mm
  (coarse sand to fine gravel)
 silicate grouts and resin grouts: 100μm - 2mm
  (fine to coarse sand)
 chemical grouts: 20μm - 600μm
  (coarse silt to medium sand)

Compressed air: $<63\mu$m - 2mm
 (silt to coarse sand)
 (above 2mm size there are heavy air losses)

These figures, however, tend to be somewhat conservative; for example, a cement-based grout would have a D_{90} of about 30μm and could inject a D_{10} soil of 1mm particle size. However, there are some cements with a D_{90} of 10μm which can inject a D_{10} soil of 0.33mm particle size (midway in the sand band). In all instances when a vacuum dewatering system is contemplated for ground engineering works, advice needs to be sought from a specialist contractor.

REFERENCES

Anon (1955) *Drainage and Erosion Control - Sub-surface Drainage Facilities for Airfields*, Part XIII, Chapter 2, Engineering Manual, Military Construction, US Army corps of Engineers, Washington, D.C.

Bertram, G.E. (1940) *An Experimental Investigation of Protective Filters*, Grad. School of Engrg Publ., Harvard Univ., 267.

Cedergren, H.R. (1987) Drainage and dewatering techniques. In: *Ground Engineer's Reference Book*, F.G. Bell (ed), Butterworths, London, 29/1 - 29/21.

Giroud, J.P. (1982) Filter criteria for geotextiles, *Proc. 2nd Int. Conf. on Geotextiles*, Las Vegas, 1, 103-108.

Hoare, D.J. (1987) Geotextiles. In: *Ground Engineer's Reference Book*, F.G. Bell (ed), Butterworths, London, 34/1 - 34/18.

Hsi, J.P., Carter, J.P. and Small, J.C. (1994) Surface subsidence and drawdown of the water table due to pumping, *Géotechnique*, 44(3), 381-396.

McGown, A., Kabir, M.H. and Murray, R.T. (1982) Compressibility and hydraulic conductivity of geotextiles, *Proc. 2nd Int. Conf. on Geotextiles*, Las Vegas, 2, 167-172.

Powrie, W. and Preene, M. (1994a) Performance of ejectors in construction dewatering systems, *Proc. Inst. Civ. Engrs, Geotech. Engrg*, 107(3), 143-154.

Powrie, W and Preene, M. (1994b) Time-drawdown behaviour of construction dewatering systems in fine soils, *Géotechnique*, **44**(1), 83-100.

Preene, M. and Powrie, W. (1994) Construction dewatering in low-permeability soils: some problems and solutions, *Proc. Inst. Civ. Engrs, Geotech. Engrg*, **107**(1), 17-26.

Schober, W. and Teindl, H. (1979) Filter criteria for geotextiles, *Proc. 7th Europ. Conf. Soil Mech. Found. Engrg*, Brighton, **2**, 121-129.

Sherard, J.L., Decker, R.S. and Ryker, N.L. (1972) Piping in earth dams of dispersive clays, *Proc. ASCE Speciality Conf., Performance of Earth and Earth-supported Structures*, Purdue University, **1**(1).

Sherard, J.L., Woodward, R.J., Gizienski, S.F. and Clevenger, W.A. (1963) *Earth and Earth-Rock Dams*, Wiley, New York.

16

ESTIMATION OF THE FORM OF A LONG-TERM (GROUND LOSS PLUS CONSOLIDATION) FREE-GROUND TRANSVERSE (NORMAL TO THE TUNNEL CENTRE LINE) SURFACE SETTLEMENT TROUGH

Definition of the form of the surface settlement trough is necessary if the criticality of superadjacent buildings and buried services is to be estimated. The truncated form of this supplementary information is taken from the paper by Attewell (1988) which, together with the book

Figure SI 16.1 indicates in diagrammatic form how a surface settlement trough develops above, to the sides of and in front of an advancing tunnel face. Assume that both the short-term and long-term transverse surface settlement troughs are of inverse bell-shape configuration and can

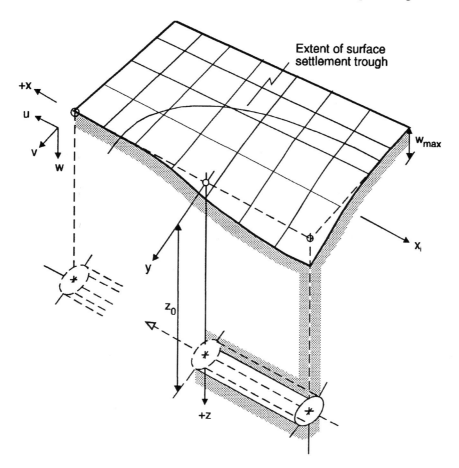

Figure SI 16.1 Development of settlement trough caused by tunnelling in soil

by Attewell *et al.* (1986), should be read for further information. The field measurements from which the following empirical equations are derived come, in the main, from open, hand-driven shield tunnels.

reasonably be approximated by normal probability curves characterised by a mean (maximum settlement w_{max}) and a standard deviation (i_y). The characteristics of a transverse normal probability settlement curve are shown in Figure

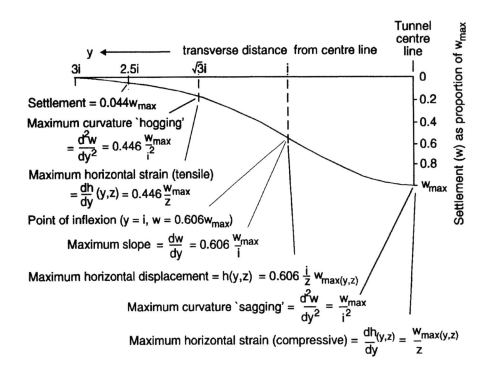

Figure SI 16.2 Characteristics of a single tunnel transverse
settlement semi-profile as defined by a normal probability curve

(Note that ground without a sufficient degree of cohesion will not support tensile strain)

SI 16.2 and the stages in the definition of the settlement are as given below.

STEP 1

Estimate the *short-term* (ground-loss) i_{ys} as indicated by Attewell *et al.* (1986, pp64, 65). The following empirical relations between i_{ys} and the tunnel depth have been suggested by O'Reilly and New (1982) from case history reviews of British tunnels for which adequate ground settlement records are available:

$$i_y = 0.43(z_0 - z) + 1.1 \text{ metres}$$
for cohesive soils $(3\text{m} < z_0 < 34\text{m})$ .... (1)

$$i_y = 0.28(z_0 - z) - 0.1 \text{ metres}$$
for granular soils $(6\text{m} < z_0 < 10\text{m})$ .... (2)

where z_0 is the depth from ground surface to the tunnel axis (centre) and z is the depth from ground surface to the point of settlement interest, often a pipeline. If it is ground surface settlement that is of concern, then z is equal to zero and i_y can be written as i_{ys}. On the other hand if it is a buried pipeline that is of concern, then z_0 is the depth from ground surface to the pipeline. (It should be

noted that the original O'Reilly and New equations did not incorporate the z term which stems from the analyses of Attewell and Woodman, 1982.) The trough width seems to be independent of tunnel diameter. In the case of a two-layer medium:

$$i_y = 0.43(z_A - z) + 0.28(z_B - z) + 1.1 \text{ metres}$$
for tunnels in clay overlain by sands
.........................(1a)

$$i_y = 0.28(z_A - z) + 0.43(z_B - z) - 0.1 \text{ metres}$$
for tunnels in sand overlain by clay
.........................(1b)

where z_A is the depth of the tunnel axis (spring-line) beneath the interface between the two lithological bands and z_B is the thickness of the surface layer of soil.

As an alternative expression to those in equations (1) and (2), Leach (1985) analysed data from 23 tunnels constructed by different methods (no-shield, shield, shield in free air, mini-tunnel, shield in compressed air) and suggested the following relations:

$$i_y = 0.57 + 0.45(z_0 - z) \pm 1.01 \text{ metres} \quad ....(3)$$

for those sites where consolidation effects are generally considered to have been insignificant, and

$$i_y = 0.64 + 0.48(z_0 - z) \pm 0.91 \text{ metres} \quad \text{.........} \quad (4)$$

for those sites where consolidation effects are deemed to have been significant. The same comments as above apply to the z term.

Because equation (4) is less comprehensive than the equations suggested below, it is recommended that the later equations be used.

Note that rather than accepting reported values of i_y (i_{ys}) for the above equations, if surface settlement values w as a function of transverse distance y from the tunnel centre line are known, then in order to obtain a value for i_y the following procedure can be applied. Plot the logarithm of the recorded settlements (log w) against the square (y^2) of the transverse distance and draw a regression line through the points. Maximum settlement (w_{max} at y = 0) can then be defined by the intercept of the regression line with the axis y^2 = 0, and the value of i_y is specified by the fact that i^2 is the value of y^2 where $w/w_{max} = 0.606$.

STEP 2

Estimate the *short-term* (ground-loss) surface settlement volume V_{ss} as given by Attewell *et al.* (1986, pp60-64). One method of estimation for tunnels in *cohesive soils* is by means of the equation:

$$V_{ss}\% = 1.33 \times (OFS) - 1.4 \quad \text{..........................} \quad (5)$$

for 1.5 < OFS < 4. OFS, the Simple Overload Factor, or Stability Ratio, is defined as

$$OFS = \left(\frac{\gamma z_0 + q - \sigma_i}{c_u} \right) \quad \text{..............................} \quad (6)$$

where γ is the soil unit weight, q is the estimated magnitude of any surface surcharge loading, σ_i is any tunnel internal support pressure (usually applied by compressed air as a temporary support measure, and c_u is the soil undrained shear strength. Reference should be made to Figures 2.11(a) and 2.11(b) in Attewell *et al.* (1986) for further guidance.

Mitchell (1983, p99) has also given a volume loss equation:

$$V_t = \left(\frac{c_u}{E_u} \right) \exp \left[\frac{(\gamma z_0 - \sigma_i)}{2c_u} \right] \quad \text{.....................} \quad (7)$$

In this equation V_t is equivalent to V_{ss} and E_u, the undrained soil deformation modulus, is generally about 200 to 700 times the value of c_u for soft soils. Mitchell has also suggested that for very sensitive soils or for poor construction control (for example, excavating ahead of the tunnelling shield, and poor jacking techniques at the shield) the ground loss estimate should be increased by a factor of about 3.

In the case of *cohesive soils* it is assumed that the ground losses at the tunnel (say V_t) are transferred entirely to the ground surface to form the settlement depression volume V_{ss}. For these soils, typical values of V_{ss} will range from about 0.5% to 2.5% of the tunnel face area, depending on the stiffness of the soil and the speed at which the initial support is erected. A typical, and not over-conservative, value of 1.5% could reasonably be taken.

For *stiff fissured clays*, tunnelled with and without a shield, V_t (V_{ss}) would normally be between 1% and 2%. *Glacial tills* often contain silt/sand/gravel lenses containing water at artesian or sub-artesian pressure. Although the clay soil itself may be stiff, the presence of the lenses may dictate the adoption of compressed air temporary support measures. During tunnelling of such deposits in free air, V_t (V_{ss}) would probably be between 2% and 2.5%, but with compressed air support in the tunnel they could be reduced to 1.5% or 1%. *Silty clays*, having undrained shear strengths probably in the range 10 to $40kN/m^2$ and being shield-driven under compressed air, would tend to incur ground losses (V_t V_{ss}) in the range 2% to 10%. When tunnelling in *granular soils* above a water table a V_t (V_{ss}) range of 2% to 5% is appropriate, but ground losses in these soils are dependent upon *in situ* density and, in contrast to those in cohesive soils, are much more dependent upon operator experience and skill. If compressed air is used to control the stability of a granular soil tunnelled below the water table, then a V_{ss} range of between 2% and 10% will apply. In both cases a trial value of 5% should be adopted for preliminary calculations, but trial values should be selected on the basis of site investigation SPT (standard penetration test) evidence. In the case of *man-filled ground*, estimation of V_{ss} is rendered uncertain because of the usually quite variable composition and compaction of the fill. Ground losses of about 17% were estimated for recent household/industrial waste fill in the north of England (Dobson *et al.*, 1979). For trial purposes, a lower value of about 8% should be adopted for an old (and better compacted) fill comprising natural ground material, a value of 10% to 12% for an old established industrial fill, and a figure of 15% for a recent loose industrial or household waste fill.

STEP 3

From i_{ys} and V_{ss} evaluate the *short-term* (ground-loss) maximum (tunnel) centre-line surface settlement, $w_{\max s}$.

$$w_{\max s} = \frac{V_{ss}}{\sqrt{2\pi}\,i_u} \qquad\qquad (8)$$

STEP 4

After Hurrell (1985) and Attewell *et al.* (1986), estimate *long-term* maximum centre-line settlement, $w_{\max t}$, from

$$w_{\max t} = 2w_{\max s} \cdot A \cdot OFS \qquad (9)$$

where

$$A = 0.39\left[1 - \left(\frac{w_{\max s}}{100}\right)\right] \qquad (10)$$

STEP 5

If the ground permeability is considered to be *isotropic*, then the long-term i_{yt} is expressed as

$$i_{ut} = \frac{i_{us}}{\exp\left[\dfrac{-D^2}{2i_{us}^{\,2}}\right]} \qquad (11)$$

where D is the tunnel excavated diameter.

STEP 6

From $w_{\max t}$ and i_{yt} the estimated *form* of the long-term surface settlement trough may be defined.

STEP 7

The *volume* of the *long-term* surface settlement trough is then

$$V_{st} = \sqrt{2\pi}\,.i_{ut}\,.w_{\max t} \qquad (12)$$

STEP 8

If the ground permeability is *anisotropic*, say $k_h > k_v$, then using a flow net procedure, let the following relation apply:

$$K = \sqrt{\frac{k_h}{k_v}} \qquad\qquad (13)$$

STEP 9

After Hurrell (1987), let the potential ultimate spread of the *long-term* settlement trough, y_t, be

$$y_t = 2Kz_0 \qquad (14)$$

STEP 10

Also after Hurrell (1987), the *anisotropic permeability* case standard deviation $i_{yt(a)}$ for the *long-term* surface settlement trough is estimated as

$$i_{ut(a)} = \frac{y_t}{3} \qquad (15)$$

STEP 11

As in equation (7) above, the volume of the *long-term* settlement trough is

$$V_{st} = \sqrt{2\pi}\,.i_{ut(a)}\,.w_{\max t} \qquad (16)$$

STEP 12

Definition of the complete form of the *long-term* transverse settlement trough is then

$$w(y)_{t(i)} = w_{\max t}\exp\left[\frac{-y^2}{2i_{ut}^{\,2}}\right] \qquad (17)$$

for *permeability isotropy*.

$$w(y)_{t(a)} = w_{\max t}\exp\left[\frac{-y^2}{2i_{ut(a)}^{\,2}}\right] \qquad (18)$$

for *permeability anisotropy*.

COMMENT

It is stressed that the above equations only produce *estimates* of the possible ground settlements for hand-shield tunnels. Settlements in particular cases may vary significantly from the estimates. Equation 10 is based on only four measurements of $w_{\max s}$ and $w_{\max t}$, together with the associated OFS values, and even with the caution expressed above its use should be restricted to ground loss settlement values between about 6mm and 63mm, a range which encompasses the observed values which generate the equation.

An alternative method of estimating settlement-time relations using a hyperbolic model has been adopted by Fang *et al.* (1993) for the Taipei Mass Rapid Transit. It would be expected that slurry-shield tunnels would reduce substantially the ground settlement, but this is not always the case. Indeed, the pressurised slurry may actually cause ground heave ahead of the face. Construction techniques such as the New Austrian Tunnelling Method (*see* the main text, Part One of the book and Supplementary Information 17) might be expected to allow rather more settlement to take place than with a shield tunnel, but this is by no means confirmed under circumstances of single tunnel construction. Measurements on three monitoring sections of the Heathrow Airport NATM trial tunnel showed that the ground losses contributing to settlement were only 1.1% to 1.5% of the tunnel excavated volume, which is much the same in London Clay as for shield tunnelling in that material. In the light of the collapses at Munich and at the actual tunnel construction at Heathrow Airport, judgement on the *general* applicability of the NATM in soil must be suspended at the time of writing. Also, there may have been special circumstances such as the pre-disturbance of the London Clay towards its boundaries at Heathrow Airport, the proximity of other works such as the large access shaft and junctions which create stress concentrations in the ground, the local presence of old piled foundations, fuel oil contamination of the clay from an old oil depot, and remedial work on the concrete lining at invert which may, singly or jointly, have promoted the collapse. Before this civil engineering set-back it did seem that there was sufficient confidence in the method for it to be used carefully in sensitive locations.

Induced settlement by shield tunnelling can be offset by the use of a continuous grouting system through the tail of the shield in order to fill automatically the annulus between the lining ring and the surrounding soil as soon as the shield advances. It can also be reduced by adopting the technique of compensation grouting (*see* also Supplementary Information 8 in Part Two of the book).

Settlement is the vertical component of ground movement and the one most referred to when assessments of tunnelling effects on structures are being made. However, as noted in Section 6.3.1 of the main text, Part One of the book, it is important to understand that the induced horizontal displacements, the vertical and horizontal strains, and the induced curvatures of the ground, all as a function of depth, are also important parameters affecting above-ground and in-ground structures. Reference needs to be made to the publications cited above and to the more recent paper by New

and O'Reilly (1991). The question of ground movements associated with adjacent and super-adjacent tunnels is considered in Attewell (1978) and the equations that can be used for estimating ground movements caused by the excavation of twin tunnels in close proximity can be found in New and O'Reilly (1991).

Estimation of ground movements is really a means of inferring structural movements on the assumption that there is full transfer of displacement from soil to structure. Patently, this will not usually be the case, and so the estimates will generally be on the pessimistic side. For analyses of above-ground structural, and below-ground pipeline response, reference may be made to the book by Attewell *et al.* (1986).

It has been usual to resort to the damage criteria published by Skempton and McDonald (1956), Polshin and Tokar (1957), Bjerrum (1963), Grant *et al.* (1974) and Burland and Wroth (1975). There are three main criteria, as indicated in Figure SI 16.3. *Angular distortion*, in simple terms, is the amount that a structure or a structural component deflects as a ratio of the length over which the deflection takes place. Whole (rigid) body rotation (α) of the structure is removed from any such assessment. Angular distortion is thus related to differential settlement along the base of the structure, or the curvature of the ground if it can be assumed that there is close contact between the structure and the ground. The magnitude of horizontal distortion varies as a function of location within a settlement trough. *Deflection ratio* can be regarded as the ratio of the maximum normal separation between two structural components originally in contact and the length over which the separation takes place. It is also a function of a differential settlement. *Horizontal distortion* is really horizontal strain within a structure, being the ratio of the amount of induced relative horizontal movement and the length over which that movement has taken place. Horizontal distortion can be both compressive and tensile: compressive towards the centre of a settlement trough where the tendency is towards the sagging mode of deformation and tensile towards the outer limbs of a settlement trough where the deformation mode is one of hogging. Structures and structural components towards the limbs of a settlement trough should be regarded as being most prone to damage since they are there subjected to both angular distortions of magnitudes which depend on the actual location and the steepness of the settlement trough (itself a function of the size of the tunnel, its depth, and the nature of the ground and groundwater), and tensile horizontal distortions.

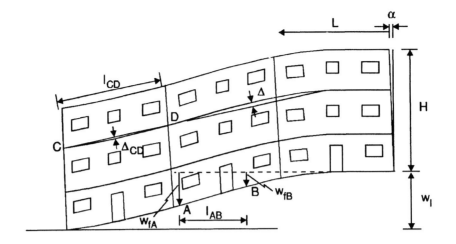

Definition of angular distortion and deflection ratio

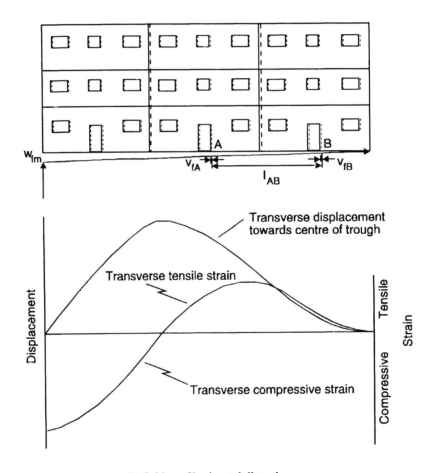

Definition of horizontal distortion

Figure SI 16.3 Styles of building deformation

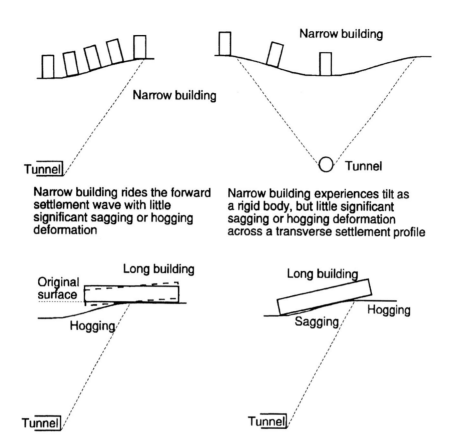

Narrow building

Narrow building

Narrow building rides the forward settlement wave with little significant sagging or hogging deformation

Narrow building experiences tilt as a rigid body, but little significant sagging or hogging deformation across a transverse settlement profile

Long building

Original surface

Hogging

Tunnel

Long building

Sagging   Hogging

Tunnel

Progressive deformation of a long building in a forward settlement trough

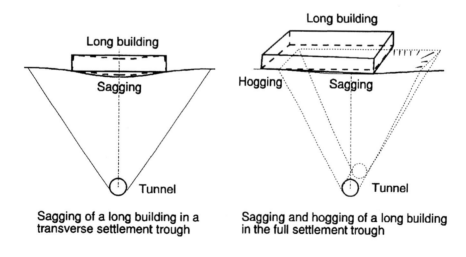

Long building

Sagging

Tunnel

Long building

Hogging   Sagging

Tunnel

Sagging of a long building in a transverse settlement trough

Sagging and hogging of a long building in the full settlement trough

Figure SI 16.4 Idealised settlement behaviour of narrow and long buildings

The proneness of a building to damage is also dependent on its length relative to the settlement trough size parameters. As indicated in Figure SI 16.4, narrow buildings are better able to 'ride' a settlement trough, mainly experiencing rigid body tilt, than are longer buildings.

Table SI 16.1 Criteria for building damage under conditions of self-weight settlement

(a) Angular distortion				
Structure	**Limit**		**Notes**	**Reference**
Load-bearing walls or panel walls in frame structures	1/300	(0.003)	Cracking likely	Skempton and McDonald (1956)
Load-bearing walls or panel walls in frame structures	1/150	(0.007)	Structural damage probable	Skempton and McDonald (1956)
Load-bearing walls or panel walls in frame structures	1/500	(0.002)	Design criterion against cracking	Skempton and McDonald (1956)
Frames with diagonals	1/600	(0.002)	Danger	Bjerrum (1963)
Buildings generally	1/500	(0.002)	Safe limit for no cracking	Bjerrum (1963)
Panel walls	1/300	(0.003)	First cracking	Bjerrum (1963)
Panel walls	1/150	(0.007)	Considerable cracking of panel and brick walls	Bjerrum (1963)
Buildings generally	1/150	(0.007)	Danger of structural damage	Bjerrum (1963)
Flexible brick walls, L/H > 4	1/150	(0.007)	Safe limit	Bjerrum (1963)
	Sand and hard clay	**Plastic clay**		
Column foundations, steel and reinforced concrete structures	1/500 (0.002)	1/500 (0.002)		Polshin and Tokar (1957)
Column foundations for end rows of columns with brick cladding	1/150 (0.007)	1/1000 (0.001)		Polshin and Tokar (1957)
Column foundations for structures where auxiliary strain does not arise during non-uniform settlement of foundations	1/200 (0.005)	1/200 (0.005)		Polshin and Tokar (1957)
(b) Deflection ratio				
Plane load-bearing brick walls	1/3333 (0.003)	1/250 (0.004)	For multi-storey buildings at L/H < 3	Polshin and Tokar (1957)
Plane load-bearing brick walls	1/2000 (0.005)	1/1500 (0.0007)	For multi-storey buildings at L/H > 5	Polshin and Tokar (1957)
Plane load-bearing brick walls	1/1000 (0.001)	1/1000 (0.001)		Grant *et al.* (1974)
(c) Horizontal distortion				
Load-bearing walls/continuous brick cladding	1/2000 (0.0005)	1/2000 (0.0005)	On structural damage criteria set of cracking	Burland and Wroth (1975)

These criteria and also their empirical quantification are based on slowly developing movements, which is in contrast to the (ground and structural) dynamic movements generated by tunnelling. Assessments, as in Table SI 16.1, based on the former type of movements may thus be somewhat pessimistic when applied to the latter type.

REFERENCES

Attewell, P.B. (1978) Ground movements caused by tunnelling in soil, *Proc. Conf. Large Ground Movements and Structures*, Cardiff, Pentech Press, 812-948.

Attewell, P.B. (1988) An overview of site investigation and long-term tunnelling induced settlement in soil. In: Engineering Geology of Underground Movements (*Proc. 23rd Annual Conf. of the Engineering Group of the Geological Society, Nottingham, Sept. 1987, Lead Lecture Session 2: Ground Movements due to Tunnelling), Geol. Soc. Eng. Group Special Publication No. 5*, F.G. Bell, M.G. Culshaw, J.C. Cripps and M.A. Lovell (eds), 55-61.

Attewell, P.B. and Woodman, J.P. (1982) Predicting the dynamics of ground settlement and its derivatives caused by tunnelling in soil, *Ground Engrg*, **15**(8), 13-22, 36.

Attewell, P.B., Yeates, J. and Selby, A.R. (1986) *Soil Movements Induced by Tunnelling and their Effects on Pipelines and Structures*, Blackie, Glasgow, 325pp.

Bjerrum, L. (1963) Discussion Session, *Proc. European Conf. on Soil Mech. and Found. Engrg*, Wiesbaden, Germany, **III**, 135-137.

Burland, J.B. and Wroth, C.P. (1975) Settlement of buildings and associated damage, *Proc. Conf. on Settlement of Structures*, British Geotechnical Soc., Pentech Press, London.

Dobson, C., Cooper, I., Attewell, P.B. and Spencer, I.M. (1979) Settlement caused by driving a tunnel through fill, *Proc. Midland Geotech. Soc. Symp. on the Engineering Behaviour of Industrial and Urban Fill*, Birmingham, E41-E50.

Fang, Y.-S., Lin, S.-J. and Lin, J.-S. (1993) Time and settlement in EPB shield tunnelling, *Tunnels and Tunnelling*, **25**(11), 27-28.

Grant, R., Christian, J.T. and Vanmarke, E.H. (1974) Differential settlement of buildings, *J. Geotech. Eng. Div., ASCE*, **100**(GT9), 973-991.

Hurrell, M.R. (1985) The empirical prediction of long-term surface settlements above shield-driven tunnels in soil, *Proc. 3rd Conf. on Ground Movements and Structures*, Cardiff, J.D. Geddes (ed), Pentech Press, London, 161-170.

Hurrell, M.R. (1987) Personal communication.

Leach, G. (1985) Pipeline response to tunnelling. Unpublished paper presented to the North of England Gas Association, January 1985.

Mitchell, R.J. (1983) *Earth Structures Engineering*, Allen and Unwin, Boston.

New, B.M. and O'Reilly, M.P. (1991) Tunnelling induced ground movements, predicting their magnitude. *Proc. 4th Int. Conf. Ground Movements and Structures*, Cardiff, Pentech Press, 671-697.

O'Reilly, M.P. and New B.M. (1982) Settlements above tunnels in the United Kingdom - their magnitude and prediction, *Proc. Tunnelling '82*, M.J. Jones (ed), IMM, London, 137-181.

Polshin, D.E. and Tokar, R.A. (1957) Maximum allowable non-uniform settlement of structures, *Proc. 4th Int. Conf. on Soil Mech. and Found. Engrg,*, London, **1**, 402-405.

Skempton, A.W. and McDonald, D.H. (1956) The allowable settlement of buildings, *Proc. Inst. Civ. Engrs*, Part III, **5**, 727-768.

17

NEW AUSTRIAN TUNNELLING METHOD

INTRODUCTION

The NATM system of tunnelling was defined in 1980 by the Austrian National Committee on 'Underground Construction' of the International Tunnelling Association in the following terms:

The New Austrian Tunnelling Method (NATM) is based on a concept whereby the ground (soil or rock) surrounding an underground opening becomes a load bearing structural component through activation of a ring-like body of supporting ground.

No doubt triggered by the NATM tunnel failures at London's Heathrow Airport, Kovári (1994) has produced a succinct critique of the NATM system, pointing out that the concept of the ground becoming part of the support system is by no means unique to the NATM. The 'ground ring', the concept of which for different tunnel shapes and for twin tunnels is considered by Kovári in his paper, is said to be ambiguous in meaning. It is believed that the NATM is not merely a method but rather a universal collection of knowledge and skill. Kovári demonstrates that the concepts of ground self-support were established and published decades before Rabcewicz, Müller and Pacher were publishing the system concepts and definitions. He particularly questions the validity of published ground response curves purporting to describe the interaction of the ground and the primary support that is applied. Reference may be made to, for example, Brady and Brown (1993, p297, Fig. 11.3) for an indication of the style of, and explanation of, ground support curves which are plotted in support pressure (ordinate axis) p and radial displacement (abscissa) δ space. In Pacher's NATM ground response curve the radial pressure decreases as inward relaxation movement takes place but then increases as the 'supporting ring' of ground develops strength. The lining characteristic line intersects the ground response curve at the point where the radial pressure is at its lowest, thus implying the

need for *minimum* primary lining support. However, although for the assumed *ground* characteristic curve this does say something about the required strength of the *lining* it says nothing about the *stiffness* and the *time* of placement of the *lining* support which are dependent variables in the determination of the ground-lining interaction.

Notwithstanding the above discussion, Kovári states that there is no evidence, theoretically or empirically, for the existence of a ground response curve of a form in which the radial ground stress (support pressure) around a tunnel opening begins to increase with increasing radial movement towards the tunnel opening once that pressure has minimised. Rather, as he notes, plasticity theory suggests that the ground characteristic (support) curve (radial ground pressure) will tend to decline continuously with increasing ground movement inwards. Only if a body of rock in a state of 'detachment failure' in the roof of a tunnel falls from the parent rock will the radial pressure in the ground around the tunnel increase, so creating an upward reversion in the ground characteristic curve of the form indicated by Pacher for the NATM.

OUTLINE OF ONE POSSIBLE SUPPORT DESIGN IN ROCK

There are obviously many variations on NATM-style support systems, these depending on the *in situ* nature of the tunnelled rock (its intrinsic strength and its discontinuous character), the tunnel depth and the prevailing earth pressures, and the tunnel size. Here an attempt is made to suggest one possible sequence of measures that would be modified according to the circumstances that apply to a particular case.

Primary support

Primary shotcrete: 25-50mm thick
Welded wire mesh (WWM): This acts as reinforcement for the later shotcrete shell. The

mesh may be all steel, plastic-covered steel, or some approved equivalent. There may be one or two layers, separated by thin layers of shotcrete, but the first layer must be located as close to the rock as possible, subject to protection by shotcrete from the rock. Individual sheets should be overlapped by at least 200mm.

Shotcrete: This must be of 100mm minimum thickness, uniform over the whole tunnel cross-section, and fully enclose the WWM and lattice girders if used. Exposed ends of dowel rods on the inside face of the tunnel should be covered by a minimum shotcrete thickness of 30mm. If any part of the shotcrete shell becomes deformed or cracked it needs to be replaced to its original thickness.

Lattice girders: These act to provide immediate support to the ground, control the tunnel shape, support the WWM while it is installed, reinforce the concrete lining shell, and offer support to any forepoling that might be required. The girders need to be set vertically and perpendicular to the centre line of the tunnel, and the invert design must take account of their placement.

Fibre or steel wire reinforced concrete: This may be used in place of WWM. Plain shotcrete is particularly necessary at the rock contact if a protective waterproof membrane is used. The thickness of the concrete will depend on the nature of the fibres and their length. Provided that a good quality concrete is used, fibre reinforced shotcrete is satisfactory for initial support even when it is exposed to saline groundwater.

Dowel rods: These would be typically 25mm diameter, 2.5-4m long steel bars with a screw thread, patch plate, washer and nut for tensioning purposes inserted into boreholes and fully grouted up using a cement grout or epoxy resin cartridges (in the latter case the separate resin and hardener cartridges being placed first in the hole and then mixed through the action of insertion of the dowel rod). Alternatives include bolts in the form of expansion shells (such as Atlas Copco Swellex bolts), split rods and wedges (for example, Ingersoll-Rand split sets), and perfo-anchors.

Additional support

Inserted ahead of the tunnel face, at an upward and outward angle of about 15°, could be the following means of applying further support to the tunnel face:

Rebar spiling: 25mm diameter steel bars spaced at 100-250mm could be used when there is a tendency to overbreak but when the individual blocks of rock are sufficiently wide not to dilate between adjacent spiles.

Sheets (metal): Interlocking sheets can be driven outside the excavation line in weaker, ravelling ground of a frictional nature.

Pipes: 25-40mm diameter perforated steel pipes can be driven into frictional ground ahead of and outside the excavation cross-section. Then, after insertion, they are shotcreted up under pressure to (ideally) form a cylindrical mass of stabilised ground around each pipe.

Monitoring

The measurement of support performance and rapid change of design and design support implementation are essential features of the NATM. There must be provision for monitoring the deformation of the primary lining and of the surrounding rock mass, of the radial and circumferential stresses within the primary lining, and of the stresses acting on the bolting system.

Deformation: Convergence can be measured by installing Demec-type points on the intrados of the tunnel lining and bolts driven into the rock. Triple points in a triangular configuration would be used for the lining strain measurement. Extensometers would be used to monitor strain in the surrounding rock mass. Each monitoring cross-section would contain at least three sets of three-point extensometers fixed radially in boreholes.

Stress: A necessary requirement with stress measurement is that the presence of the device should not affect the *in situ* stress being measured. Stress in the concrete primary lining can be monitored by means of stiff Glötzl-type hydraulic pressure cells filled with mercury and equipped with a bourdon pressure gauge. Pressure between the lining and the rock can be monitored with the same type of cells but filled with a suitable oil. One possible arrangement at each monitoring cross-section would comprise, say, seven pairs of pressure cells installed either side of the tunnel centre line (with two pairs on the centre line at crown and invert) at angles of 45°, 90°, 135°, 225°, 270° and 315°. Bolt tensions would be monitored via strain gauges (and appropriate circuitry) cemented at several points along the rod.

Groundwater pressure: Electrical piezometers, with suitable protection for the connections, would be most appropriate for this pressure behind the lining.

Installation of instruments

Installations have to be by experienced personnel. There needs to a full description of the ground geology (petrology and structure) at each monitoring cross-section, as derived from observation prior to placement of the primary shotcrete shell and, where appropriate, from the forward probing observations (*see* below). A record needs to kept of the instruments at each monitoring cross-section together with the thickness of the lining. Convergence bolts and points can usually be installed about 5m behind the face. Before stress cells are mortared carefully into place against the surface of the rock the surrounding area must be drained. A zero reading is then taken, to be followed by a second reading after the shotcrete is placed. The cells then need to be re-pressurised after the shotcrete has hardened. Plots of the re-pressurising curves are needed to confirm that the cells are functioning correctly. Instruments need to be read continuously for about the first 5 days after placement.

Readings

If convergence measurements indicate more than a target figure, say 5mm over each 24 hours, by the third day following installation of the monitors, the support system is then in need of strengthening by means of additional rock bolts and a thickening of the shotcrete lining. Daily readings of the deformation measurements should continue until the movement stabilises. If the deformation does not exceed another target figure, say 2mm over 24 hours, reading intervals can then be increased to, say, 2 days. As the deformation decreases to below the resolution of the measurement system, there can be a further increase in the period between the readings.

Readings of the stress gauges should follow the same procedure.

Forward probing

With suitable protection at the drill rods against high water pressures, which is especially important in karstic rock, forward probing could be used to acquire advance information on the ground and groundwater from about 10-20m ahead of the tunnel face, depending on the rate of advance. The geological and geotechnical information would be back-correlated to the measurements acquired at the monitoring cross-sections. Open hole drilling would most likely be used, and this would require interpretation of the cuttings, flushing water volume and colour returns, flushing water losses, drilling rates (for checking against the petrology of the cuttings), and the actual distance probed.

ADDITIONAL COMMENT

As noted in the main text, when fully implemented, the NATM system of primary (temporary) support depends fundamentally upon the use of Peck's observational method (Peck, 1969). This means that the method of support should be capable of responding immediately to the results of ground deformation monitoring, and this in turn means that the style of works contract must readily accommodate such on-the-spot changes. The New Engineering Contract, discussed in Section 2.8 of the main text, is particularly suited to this flexibility requirement.

Many NATM-type constructions are really of hybrid form since there is no built-in provision for the design to be changed as the work proceeds. The project tends to be standardised in respect of a few designs of reinforced concrete shells determined at the outset of the work for each section of the tunnel on the basis of the ground investigation information.

A fundamental requirement is for rapid placement of the primary support system. With some TBM systems, bolting and shotcreting may only take place some distance behind the face - 20m, for example, in the case of the Evinos-Mornos water transfer tunnel in Greece - and the bolting pattern may then be constrained by the actual space available at the TBM. The type of bolt needs to be chosen with this in mind. Another drawback is the time taken for the installation of the mesh reinforcement. A further requirement will be for a rapid hardening (strengthening) shotcrete. There are two basic systems of shotcrete application. In the currently more common dry system the cement and accompanying aggregates (sand) are mixed and projected to the delivery nozzle at which point the water in the form of a fine spray and any accelerating additives are added just before the mix is projected onto the newly exposed ground surface. The operator adjusts the water and the additive to produce the necessary degree of cohesion in the mix so that it sticks to the surface with minimum rebound and will be self-supporting on a vertical surface without sagging. In some cases there can be a rebound of up to 50% by weight of material, particularly of the coarser aggregate fraction. In the wet system, all the shotcrete constituents are pre-mixed and pumped to the delivery nozzle together, with only air being added separately.

Ingredients are added to the concrete to enhance its cohesive properties, reduce rebound, and accelerate the setting and hardening of the shell layer. Some have health and environmental implications. With the dry system, liquid accelerator can be added at the nozzle with the water mix. Sodium aluminate accelerators, which are very caustic, are tending to be supplanted by calcium aluminate accelerators. Microsilica (fine pozzolanic material) when used in shotcrete can reduce the amount of rebound and increase the thickness of the layer that can be applied in a single pass. A layer more than 200mm thick can be sprayed to even an overhead surface with little or no accelerator being used.

Compared with the wet process the dry process generates more dust and rebound, and a somewhat more variable final product, but it is more suitable for delivering over a longer distance. On the other hand, the use of the wet process seems to be becoming more widespread due to the consistency of the product. A stabiliser can be added to the wet mix to inhibit hydration, and a combined activator and accelerator added at the delivery nozzle to neutralise the stabiliser and promote rapid setting of the shotcrete.

Shotcrete setting times of less than two minutes and unconfined compressive strengths of 20MPa can be achieved with suitably designed materials. One problem that applies to all shotcretes, both wet and dry mix, concerns the exothermic nature of the reaction. Because considerable heat is generated as the high cement content hydrates, individual layer thicknesses can usually be no more than about 150mm.

Shotcrete thickness can be controlled by using distance pins drilled into the rock. For more accurate control high breaking strain fish-line can be used to interconnect the pins.

When a TBM is tunnelling through squeezing ground it may be necessary to provide early shotcrete support at the actual face rather than delaying the operation for some 10m or 20m of advance. The TBM motors and cylinders may then become covered with the shotcrete, removal of which will delay advance. Squeezing ground will also create problems for the installation of any pre-sized steel rings specified as part of the primary support system. Such problems would normally be anticipated by the provision of cutterhead gauge cutters which can be set to give a substantial overcut beyond the extrados diameter of the shield.

In the example given above of the application of the NATM to rock, three methods of 'additional support' were indicated. When the NATM is applied to a tunnel in a clay soil potentially substantial ground movements may take place at the exposed face proper unless such additional support is inserted to stiffen the soil. (The stiffness of the overconsolidated London Clay, for example, decreases as the shear strain induced by the excavation increases.) The most usual methods of applying this additional support may be summarised as:

- Overlapping spiles, which can be of breakable material such as fibre-reinforced plastic, so that no tunnelling time is lost having to retrieve them.
- Longitudinal ground anchors stressed and grouted up before tunnelling takes place.
- Dewatering ahead of the tunnel face in order to reduce the porewater pressure in the soil by means of well-points, vacuum well-points or electro-osmosis.
- Compressed air in the tunnel to decrease the simple overload factor (stability ratio).

Another possibility is to use jet grouted columns to reduce face movements. Other methods such as freezing and chemical grouting are unlikely to be applicable on technical and/or economic grounds.

The possibility of incorporating a protective waterproof membrane is mentioned in the example above. Difficulties with the use of high density polyethylene (HDPE) sheeting arise in its mounting (clamps or toroidal 'doughnut'-shaped fixtures) and the testing of circumferential welds. These operations slow down progress at a time when the shotcrete ring needs to be closed quickly. An alternative is the use of a hot spray solvent-free epoxy polyamid two-component coating applied in layers of 500 - 700μm per layer to form a continuous lining without joints or welds and one that is both amenable to testing with a spark gun and relatively easily repairable. A possible addition to either of these water-proofing systems is a ('polyfelt') geosynthetic membrane (a needle punched endless fibre structure) for protecting the waterproofing from puncture by sharp-edge stones or of the shotcrete. The membrane needs to have a high chemical resistance against alkaline substances and a high transmissibility to allow free drainage of seepage water.

A waterproof welded membrane may sometimes be placed between the primary shotcrete lining and the final concrete lining in order to intercept any water passing from the ground through shrinkage and stress-induced cracks in the primary lining. There would be provision for the intercepted water to be collected by a suitable arrangement of pipework located at the sides and base of the tunnel. In all cases there would be provision, through the incorporation of suitable joints and sealing, for the final concrete lining to respond to shrinkage and creep without unacceptable cracking.

For the purposes of contract pricing, the NATM support systems may be classified within the context of reference ground conditions, particularly when those conditions are likely to include collapsing ground, swelling ground, and anticipated substantial water inflows. The simplest category might include bolting alone; the next would require arches, mesh and shotcrete; the next would include bolts, mesh and an increased thickness of shotcrete (perhaps an increase to 75mm from 50mm in the previous category); and a final category would require the installation of steel rings, mesh and thicker (say 100mm) shotcrete. The contractor would price for each of these categories of primary support over the lengths of drive for which the site investigation deemed that they would be appropriate, and there would be provision for the contractor to be paid on a remeasurement or target cost basis for the level of primary support that needed to be used.

In a comment on the use of the NATM in London Clay (as part of the continuing commentary on the tunnel failure at London Heathrow Airport), Barton (1994) has suggested Q-values of between 0.01 and 0.1 which would imply rib (reinforcing bar) reinforced (or lattice girder reinforced) steel fibre reinforced shotcrete of about 150mm to 250mm thickness as a final primary support for the best quality end of his Q-value range. He has suggested robot-applied steel fibre-reinforced wet process shotcrete at production rates of 5 to 25m^3 per hour instead of the mesh reinforced shotcrete used at Heathrow, stressing the point that primary rib-reinforced support is preferable to mesh reinforcement which takes longer to install. A Swedish robotic shotcreting unit lined out a 2.5km long, 3.4m diameter gas pipeline TBM tunnel in granite on Hong Kong Island over a period of 5 weeks.

As an alternative to the use of the NATM method, reference may be made to the paper by Barton and Grimstad (1994) which describes the Norwegian Method of Tunnelling (NMT). It is said that this method is most appropriate for tunnels driven by drill and blast in jointed rock which tends to overbreak and is often characterised geotechnically by a rock mass classification system such as the Q-system.

Finally, it is now recognised in the industry that as a result of the failures at London Heathrow Airport the NATM has an image problem. Because of this, coupled with the fact that there are so many variants on the general theme, there is a growing feeling that the name should be dropped in the UK and another name substituted.

REFERENCES

Barton, N. (1994) Updating the NATM, *Tunnels and Tunnelling*, Dec. 1994, 40.

Barton, N. and Grimstad, E. (1994) Rock mass conditions dictate choice between NMT and NATM, *Tunnels and Tunnelling*, Oct. 1994, 39-42.

Brady B.H.G. and Brown, E.T. (1993) *Rock Mechanics for Underground Mining*, 2nd Edition Chapman and Hall, London, 571pp.

(ISBN 0 412 47550 2.)

Kovári, K. (1994) Erroneous concepts behind the New Austrian Tunnelling Method, *Tunnels and Tunnelling*, Nov. 1994, 38-42.

Peck, R.B. (1969) Advantages and limitations of the observational method in applied soil mechanics, (Ninth Rankine Lecture), *Géotechnique*, 19(2), 171-187.

18

WATER POLLUTION LEGISLATION

INTRODUCTION

With increasing industrialisation more and more toxic effluents have added to the urban sewage that has often been allowed to enter the river systems in a virtually untreated state. Before the mid-1970s, for example, there were almost 200 outfalls discharging chemical and other waste into the River Tyne. Such discharges created 'dead' rivers by affecting the biochemical oxygen demand (BOD) as determined from a standard water treatment test for the presence of organic pollutants. (BOD is a measure of the oxygen required by microbes which reduce waste materials to simple compounds.) In common with many other major conurbations located on major rivers, Newcastle upon Tyne now has a new sewerage system, major features of which are the interceptor sewers along both banks and a treatment works.

Modern pollution prevention legislation began in Britain with the River Boards Act 1948, which recognised 34 River Boards throughout England and Wales and conferred upon them responsibility for pollution control. The Rivers (Prevention of Pollution) Act of 1951 introduced the prohibition of the use of a natural river course for the disposal of polluting matter. It did not help the cause of pollution a great deal since it dealt only with the control of new discharges. The Rivers (Prevention of Pollution) Act 1961 extended control to the discharge of existing effluents. However, only those causing gross pollution and generating sufficient public complaints were dealt with.

River Boards were superseded by 27 River Authorities in 1965 with the introduction of the Water Resources Act of 1963. This Act was intended to promote the conservation and proper use of water resources in England and Wales. A major reorganisation took place following the Water Act of 1973, with 10 regional Water Authorities being formed in 1974. The principal legislation remained the Rivers (Prevention of Pollution) Acts of 1951 and 1961. The Water Authorities were responsible for the water supply, sewage disposal and river management.

Upon privatisation of the water industry in Britain under the Water Act 1989 the National Rivers Authority (NRA) was established on 1 September 1989 in England and Wales. As a statutory, regulatory body, independent of Government and industry, the NRA is currently responsible for overseeing water resources and the environment, more specifically, for pollution prevention, land drainage and fisheries.

The main responsibilities of the NRA are:

- monitoring of water quality and its pollution control, with regulation of discharges to water

- management of public water supply

- protection against flooding

- maintenance and improvement of fishery resources

- promotion of recreational activities.

The NRA, soon to be part of the new Environment Agency, is answerable to the Secretary of State for the Environment. It is also overseen by the Secretary of State for Wales and the Minister for Agriculture, Fisheries and Food (MAFF). The NRA is self-financing and exercises its authority throughout ten regions based on the river catchments of England and Wales. As one example, the Northumbrian region of the NRA covers 3600 square miles and is responsible for regulating the water supply to 2.6 million people in and around the catchment areas of the Rivers Tyne and Wear, Coquet and Tees. Water supply by statutory undertakings in 1984-85 for domestic and industrial requirements in the Northumbrian Region was 1.02 million m³/day, or just over 220 million gallons per day.

The Control of Pollution Act of 1974 paved the way for public involvement in pollution issues, although the main provisions of Part II of the Act were not introduced until 1984/85. COPA provisions relating to pollution issues have been re-enacted by the Environmental Protection Act 1990. The Government intends that the duties of the NRA, Her Majesty's Inspectorate of Pollution (HMIP), and the local authorities in respect of waste disposal regulation, will be incorporated into a new UK Environmental Agency, but with ratification of the legislation probably not before 1996.

EUROPEAN COMMUNITY LAW

Official European Community Directives set out rules and standards and are used by all Member States to ensure that water quality standards and effective control against pollution is strictly adhered to. However, EU Directives are not enforceable through UK law. Member States are required to take action to comply with and ensure proper implementation of the Directives.

The European Community has agreed more than 200 measures on environmental protection, ensuring that there are common environmental standards throughout western Europe. Several major Directives having direct implications on pollution are considered below.

76/464/EEC, 1976, Control of Dangerous Substances Directive

This Directive concerns pollution caused by certain dangerous substances discharged into the aquatic environment. It aims to eliminate or reduce pollution of territorial waters by identifying 129 dangerous compounds which are likely to be of risk to the environment. These compounds are compiled in two lists: thus the dangerous substances Directive distinguishes between particularly dangerous compounds (List I, or Black List) and less dangerous compounds (List II, or Red List) on

the basis of their toxicity, bioaccumulation and persistence (Agg and Zabel, 1989).

Annex (Council Directive 76/776/EEC)

List I of Families and Groups of Substances

List I contains the individual substances which belong to the families and groups of substances enumerated below, with the exception of those which are considered inappropriate to List I on the basis of low risk of toxicity, persistence and bioaccumulation. Such substances which with regard to toxicity, persistence and bioaccumulation are appropriate to List II are to be classed in List II.

1. Organohalogen compounds and substances which may form such compounds in the aquatic environment.
2. Organophosphorous compounds.
3. Organotin compounds.
4. Substances which possess carcinogenic, mutagenic or teratogenic properties in or via the aquatic environment.
5. Mercury and its compounds.
6. Cadmium and its compounds.
7. Mineral oils and hydrocarbons.
8. Cyanides.

(A carcinogen is a substance that causes cancer in animal tissue; a mutagen is a substance that causes genes in an organism to mutate or change; a teratogen is a substance that causes physical birth defects in the offspring following exposure of the pregnant female to that substance.)

The Department of the Environment and the Welsh Office (Circular 20/90 and Circular 34/90, respectively) have specified that, on the basis of their intrinsic properties, the three pesticides brooxynil, bromoxynil octanoate and chlorpyrifos should be regarded as List I substances for the purposes of EC Directive 80/68/EEC.

List II of Families and Groups of Substances

List II contains the individual substances and the categories of substances belonging to the families and groups of substances that are given below and which could have a harmful effect on groundwater.

1. Metalloids and metals and their compounds which are listed below:

1. Zinc	11. Tin
2. Copper	12. Barium
3. Nickel	13. Beryllium
4. Chrome	14. Boron
5. Lead	15. Uranium
6. Selenium	16. Vanadium
7. Arsenic	17. Cobalt
8. Antimony	18. Thallium
9. Molybdenum	19. Tellurium
10. Titanium	20. Silver

2. Biocides and their derivatives not appearing in List I.
3. Substances which have a deleterious effect on the taste and/or odour of groundwater, and compounds liable to cause the formation of such substances in such water and to render it unfit for human consumption.
4. Toxic or persistent organic compounds of silicon, and substances which may cause the formation of such compounds in water, excluding those which are biologically harmless or are rapidly converted in water into harmless substances.
5. Inorganic compounds of phosphorus and elemental phosphorus.
6. Fluorides.
7. Ammonia and nitrites.

In the Directive, List I substances are controlled by limit values (or Uniform Emission Standards, UES) and Environmental Quality Standards (EQS) published in the 'daughter' Directives. The limit values are determined by (a) the maximum concentration of a substance which is permissible in a discharge, and (b) where appropriate the maximum quantity of such a substance expressed as a unit weight of the pollutant per unit of the characteristic element of the polluting activity, for example, the unit of weight per unit of raw material or per product unit. An EQS is that concentration which must

not be exceeded if a specified use of the aquatic environment is to be maintained. Only four Directives have been published concerning cadmium, HCH (Lindane) and mercury. List II substances are controlled by the EQS of individual states (Agg and Zabel, 1989). The UK is the only state to have set EQS for all List II metals, but the government has been accused of being lax on the issue of limit values.

The Department of the Environment in the UK has published a document (DoE, 1988a) to provide a unified system for the control of dangerous substances, and has issued a 'red list' containing 23 compounds. The provisions of the Control of Pollution Act 1974 meet the requirements of this Directive in the UK but, as noted above, this has been re-enacted by the Environmental Protection Act 1990.

80/68/EEC, December 1979, Protection of Groundwater against Pollution Caused by Certain Dangerous Substances

This Directive sets a framework for the reduction or elimination of pollution of groundwater by controlling the discharge of certain specified (listed) substances. Member states were required to comply with the Directive within two years of its notification, that is, by 19 December 1981. Under Article 3, member states are required

'...to take the necessary steps to:
(a) prevent the introduction into groundwater of substances in List I, and
(b) limit the introduction into groundwater of substances in List II so as to avoid pollution of this water by these substances'.

Discharge of List I substances needs to be controlled only if there is evidence of toxicity, persistence or bioaccumulation, and List II substances only if they could have a harmful effect on the groundwater (DoE, 1982a). Exemptions from the Directive include, under Section 8(b) Circular 4/82...

'...discharges found to contain substances in List I or List II in a quantity and concentration so small as to obviate any future danger or deterioration in the quality of the receiving water'.

Until re-enactment by the Environmental Protection Act 1990, implementation of the groundwater Directive has been brought about in the UK by the exercise of powers provided under the Control of Pollution Act 1974. Further provision for the specification of all underground water was brought about by the Control of Pollution (Underground Water) (Specification) Regulations 1984. Interestingly, there is a lot of debate in the European Community concerning the definition of groundwater - the EU Directives define 'groundwater' to mean 'all water which is below the surface of the ground in the saturation zone and in direct contact with the ground or subsoil'. The United Kingdom legislation, however, defines groundwater as 'any water which is found in the saturated zone', but does not include water in the unsaturated zone. Thus, the legislation within the UK clearly does not encompass the requirements set out in the EU Directive concerning groundwater contamination.

80/778/EEC, July 1980, Quality of Water Intended for Human Consumption

This Directive was adopted and implemented in July 1985 by administrative means. Member States were allowed a period of two years to bring into force any necessary legislation and a further three years to ensure that waters comply with the provisions of the Directive. This Directive ensures the achievement and maintenance of high standards for drinking water by setting out maximum admissible concentrations (MACs) for various parameters (including pesticides). MAC is defined as the concentration in water below which a substance is not expected to cause or indirectly result in an identifiable effect harm-ful to the health of a statistically representative sample of the population concerned (DoE, 1982b). The MAC values for the various parameters are listed in Table SI 18.1.

Table SI 18.1 List of parameters with maximum admissible concentrations in drinking water (DoE, 1989a);Annex 1 (Council Directive 80/778/EEC)

Parameter	Expression of results	Guide level	MAC	Comments
Odour	Dilution no.	0	2 at 12°C 3 at 25°C	
Taste	Dilution no.	0	2 at 12°C 3 at 25°C	
Chlorides	Cl mg/litre	25		Approximate concentration above which effects might occur: 200mg/litre
Dissolved	%O_2 saturation			>75% except for groundwater
Phenols	C_6H_5OH		0.5	Excluding natural phenols which do not react to Cl
Surfactants	.µg/litre		200	
Pesticides	µg/litre		0.1	Substances considered separately
HCH	µg/litre µg/litre µg/litre		0.5 0.1 0.02	Substances in total Inland surface waters Estuary and marine waters

Under Article 7 of the Directive, steps must be taken to ensure that standards for water intended for human consumption must meet the MAC. The Directive is not, however, based on any toxicological evidence and the British Government has been pressing the EU for a review of the pesticide parameter. The Government would like to see it replaced with limits for individual pesticides or groups of compounds which are more closely related to the toxicity of the individual pesticides and so more clearly define the health risk. Current advisory levels in the UK are calculated from published toxicological data with a wide margin of safety. It is proving expensive to monitor water supplies for complex substances the MACs of which are less than one part per billion (DoE, 1988b).

Legislation

According to Circular 85/18, by definition of Parliament, pollution is defined as...

'...the discharge by man, directly or indirectly, of substances or energy into the aquatic environment the results of which are such as to cause hazards to human health, harm to living resources and to aquatic ecosystems, damage to amenities or interference with other legitimate uses of water'.

The wording of this definition is changed slightly in Circular 20/90, Appendix 1:

'...the discharge by man, directly or indirectly, of substances or energy into groundwater, the results of which are such as to endanger human health or water supplies, harm living resources and the aquatic ecosystem or interfere with other legitimate uses of water.'

'Pollution of the environment' is defined more generally in the Environmental Protection Act 1990 as:

'pollution of the environment due to the release (into any environmental medium) from any process of substances which are capable of causing harm to man or any other living organisms supported by the environment.'

In the Act there are general definitions of 'environment', 'harm', and 'process' (including 'prescribed process' and 'prescribed substance').

The Control of Pollution Act 1974 required that applications for consent to make a discharge of effluent to waters must undergo Public Inquiry. Registers must be kept by the Water Authorities (Companies), listing all applications and consents for discharge under Clause 9 of the Pollution Act 1974 (DoE, 1985)...

'...discharges of effluent..., are permitted provided that consent is obtained and that any conditions attached to that consent are observed.'

These registers should be made available to the general public. Clause 50 of the same circular lists the following instructions:

'...Under Section 41, Water Authorities have a duty to maintain registers recording particulars of:
(i)   applications for consent,
(ii) consents given,
(iii) samples of effluent and receiving water,
(iv) exemption certificates,
(v)   notices to abstain from certain agricultural practices'.

The Water Act 1989 superseded the Water Act 1973 in Great Britain and in turn has been superseded by the consolidation of water law in 1991. When introduced in September 1989 the Water Act facilitated the transfer of water and sewage services to the private sector and set up regulatory controls for price and the levels of service that would have to be provided for the public. The Act also placed water quality and environmental standards on a statutory basis through statutory water quality objectives (SWQOs). Various bodies were established under the Act in addition to the National Rivers Authority: the Office of Water Services (Ofwat), the Water Services Company (WSC) and the Consumer Services Committees (CSCs). Under the terms of 1989 Act it is now an offence to cause...

'...any poisonous, noxious or polluting matter, or any solid waste matter to enter any controlled waters.'   (Controlled waters are classed as rivers, lakes, estuaries, coastal waters and ground-water.)

The consolidation of water legislation in 1991 brought together the Water Resources Act, the Water Industry Act, the Statutory Water Companies Act and the Land Drainage Act 1991. Amendments to the Water Act 1989 are set out in the Water Consolidation (Consequential Provisions) Act 1991. The Water Resources Act 1991 ratified the position of the NRA under the 1989 Act in respect of management of rivers, basins and the water environment in England and Wales, and responsibility for seeing that the SWQOs are maintained for all controlled waters. It also confirmed the responsibility of Her Majesty's Inspectorate of Pollution (HMIP) for the operation of integrated pollution control (IPC) as given in the Environmental Protection Act of 1990. The Water Industry Act of 1991 re-enacted the part of the 1989 Water Act concerned with the privatisation of water and sewage services and the establishment of Ofwat, with the Secretary of State for the Department of the Environment exercising control and supervision of water and sewage services. The Secretary of State is also ultimately responsible for checking the environmental impact of decisions that he makes when exercising his management of the water environment.

Additional to the above legislation are the Control of Substances Hazardous to Health Regulations (1988). Reference may be made to the booklet *COSHH Assessments* published by the Health and Safety Executive (1989).

The UK Government subscribes to the 'polluter pays' principle, whereby polluters are not subscribed to enable them to meet pollution control standards, nor are the costs of dealing with polluting emissions from unknown sources borne by the public. The discharger has to meet the monitoring cost of the National Rivers Authority and to pay an administrative charge for the consent (DoE, 1989b).

The Water Supply (Water Quality) Regulations 1989 sets out the standards for UK drinking waters. Part IV of the Regulations prescribes that water shall be sampled each year in order to determine the microbiological parameters. There is no requirement under British legislation to monitor the quality of groundwater at source, nor is there any legislation specifying the quality standards of raw groundwater.

91/271/EEC, Urban Waste Water Directive

As a result of this Directive, the main requirements of which are given in Table SI 18.2, the water industry in Britain could be faced with £1bn expenses for the improvement of environmental standards. Out of a sum of £24bn which is expected to be spent by the water

Table SI 18.2 Main requirements of the EU Urban Waste Water Directive, 91/271/EEC

Date for compliance	Size of community (number of inhabitants)	Receiving waters	Requirement
1988	>10 000	Sensitive[1]	Collection
2000	>15 000	All	Collection and secondary treatment
2005	2000 - 15 000 10 000 - 15 000 2000 - 10 000	All All Freshwater and estuaries	Collection Secondary treatment Secondary treatment
Less sensitive areas[2]	10 000 - 150 000	Coastal waters	Primary treatment

[1] Mainly freshwater lakes, estuaries and coastal waters which are eutrophic (a lake-type habitat with gently sloping shores and a wide belt of littoral vegetation) or surface sources of drinking water containing higher than permitted levels of nitrate.

[2] Relaxation of the requirement is on the understanding that comprehensive studies, submitted to the EC Commission, show that the environment is not adversely affected.

companies (England and Wales) on water and sewerage up to the end of the century, up to £7bn would be needed to meet the waste water Directive. Spending on water-industry capital works in Scotland was planned to rise from £138 million for the year ending 31 March 1991 to £165 million in 1992 and £206 million in 1993. This, however, only represents an annual increase of about 10%.

The Directive lays down the minimum requirements for the treatment and collection of urban waste water and sewage sludge, and recommends that, wherever appropriate, sludge should be processed and re-used. Incorporation of the objectives of this Directive into UK legislation will be part of a progressive move away from the UK's traditional 'assimilative capacity' approach to pollution control and towards the more rigorous method of standard emission limits. The basic approach of the Directive is to set standard emission limits on effluent discharge from sewage treatment works, but in more sensitive areas more treatment will be needed to satisfy specific environmental requirements, to be assessed using criteria set down in the Directive. Thus, according to the Directive, the concentration of faecal coliforms in bathing water should not exceed 2000/100ml. In raw sewage it averages 10^7/100ml and primary treatment only halves this concentration. In order to meet the requisite standards for bathing in sea water long sea outfalls for sewage are no longer sufficient for beach protection, and sewage will have to be at least partially treated. There is an agreement that dumping in the North Sea will end by the year 1998, and this means that sewage that would have been disposed of by sea outfall will need to be incinerated or disposed of in landfill sites, although some of it could be treated for use as fertiliser. Incineration seems to be the most popular option in the proposals of regional water companies, but there have been problems based on environmental grounds (ever-increasing constraints of air pollution). Thus Northumbrian Water's proposal for two co-incineration plants - sludge with domestic and perhaps toxic waste - went to public enquiry after opposition and was eventually rejected by the Secretary of State. (Somewhat controversially it is claimed that toxic waste has a higher calorific value than most other wastes so its inclusion would reduce the overall cost of burning high

water content sludge.) North West Water has been dumping 2 million tonnes of sludge out at sea (51% of its total quantity) and plans to build a £100 million incineration plant, while Yorkshire Water is gradually reducing its annual 5400 tonnes dump of sludge in favour of incineration as the best practical environmental option (BPEO). It has built a successful incinerator at Bradford and one is planned for the Hull area. Planning permission was received in April 1992 for a new sludge incinerator at its Knostrop Sewage Treatment Works, the generated electricity being used to help power the works. Severn Trent Water also has incinerators under construction or constructed at Coleshill and Roundhill sewage treatment works, and Thames Water is planning to use incineration as the BPEO for disposing of sewage sludge from Greater London. In contrast, Southern Water's Horsham works was the first Simon-N-Viro plant to be operational in the UK, the plant pasteurising the sludge to remove the heavy metal problem and producing usable soil. Somewhat similarly, a South West Water pilot plant near Plymouth composts organic municipal waste with sewage sludge in a windrow composter to produce substitute soil for land reclamation. Wessex Water also expected a £10 million Swiss-Combi bio-drying plant at Avonmouth, near Bristol to be operational by 1994. It is designed to convert 300 000 tonnes of wet sludge annually from the Bristol area into an agricultural peat substitute, the energy for the plant coming from methane produced by an existing plant at Avonmouth and with oil as a back-up fuel. The sludge enters the plant as 96% water, is centrifuged to reduce the volume by 85%, then enters a drier to reach a temperature of 450°C, to be followed either by bagging or recycling to change the size of granular product (similar to freeze-dried coffee, having a final volume only 7% of that of the original). Heat from steam emitted during the drying process is recovered in a condenser and used to warm the sludge digesters. Operation at slightly below atmospheric pressure allows air to be drawn into the plant from outside to prevent smells escaping to the atmosphere. Thames Water is also manufacturing topsoil substitute from sludge and subsoil.

In contrast to the established reed bed

technology, used for example by Southern Water at Gravesend sewage treatment works in Kent and which is particularly relevant not only to suitable areas of the western world but also to emergent Third World countries, the several newer methods of treating sewage are particularly appropriate. For example, there is the bioprecipitation process for removing heavy metals, the Simon-N-Viro process mentioned above which produces a contamination-free soil and any heavy metals rendered insoluble, the VarTech treatment (aqueous phase oxidation) process, and the Zerofuel sludge incineration process. More recently, at a demonstration plant at Histon in Cambridgeshire, Anglian Water has been investigating membrane technology as a means of upgrading the quality of effluent discharged from treatment works. Here, the membranes are capable of removing particles as small as 0.1 micron (μm), which means that some bacteria can be caught in the filter. Preventing the filters from clogging up can pose a problem, and this adds to the expense of the system. In another system, ultra-violet light is being used as a disinfectant at a treatment works at Bellozanne in Jersey. Space is at a premium on the island, and treatment must be at a high level because marine disposal is into an almost closed bay which permits minimal dispersal. UV light tubes lie in the effluent, the light attacking coliform bacteria and pathogens that were largely unaffected by earlier stages of treatment. The plant is built under an IChemE form of lump sum contract for £4 million, and for information on this form of contract reference should be made to Section 2.7 in the main text.

A further added benefit of the 1991 Directive is that its tighter regulations on effluent will stimulate moves towards more 'at source' controls on effluent production.

REFERENCES

Agg, A.R. and Zabel, T.F. (1989) EC Directive on the Control of Dangerous Substances (76/464/EEC): Its impact on the UK water industry, *J. IWEM*, 3, 436-442.

Commission of the European Communities (1976) *Council Directive on Pollution Caused by Certain Dangerous Substances Discharged into the Aquatic Environment of the Community* (76/464/EEC), OJ L129, 4 May 1976.

Commission of the European Communities (1979) *Council Directive on the Protection of Groundwater against Pollution Caused by Certain Dangerous Substances* (80/68/EEC).

Commission of the European Communities (1980) *Council Directive on the Quality of Water Intended for Human Consumption* (80/778/EEC).

Commission of the European Communities (1991) *Council Directive on Urban Waste Water Treatment* (91/271/EEC), OJ L135, 21 May 1991.

Department of the Environment (1982a) *EC Directive on the Protection of Groundwater against Pollution Caused by Certain Dangerous Substances* (80/68/EEC), Joint Circular DoE 4/82 and the Welsh Office 7/82.

Department of the Environment (1982b) *EC Directive Relating to Quality of Water Intended for Human Consumption* (80/778/EEC), Joint Circular DoE 20/82 and the Welsh Office 33/82.

Department of the Environment (1985) *River Quality in England and Wales*. A Report of the 1985 Survey. HMSO, London.

Department of the Environment (1988a) *Inputs of Dangerous Substances to Water*. Proposal for a Unified System of Control. HMSO, London.

Department of the Environment (1988b) *Assessment of Groundwater Quality in England and Wales*, HMSO, London.

Department of the Environment (1989a) *White Paper on Pesticides in Water Supplies*, WP 18/1989, HMSO, London.

Department of the Environment (1989b) *Water and the Environment: The Implementation of European Community Directives on Pollution Caused by Certain Dangerous Substances Discharged into the Aquatic Environment*. Joint Circular DoE 7/89 and the Welsh Office 16/89.

Department of the Environment (1990) *EC Directive on Protection of Groundwater against Pollution Caused by Certain Dangerous Substances* (80/68/EEC): *Classification of Listed Substances*. Joint Circular DoE 20/90 and the Welsh Office 34/90.

Health and Safety Executive (1989) *COSHH Assessments: A Step-by-step Guide to Assessment and the Skills Needed for it*, First

BIBLIOGRAPHY

Bascombe, A., Crathorne, B., Frost, R.C., Hall, T., Hedgecott, S., Hunt, D.T.E., Johnston, D., Miller, D.G., Pike, E.B. and Zabel, T. (1990) *Water Quality*, Water Research Centre Report No. CO 2515-M.

Control of Substances Hazardous to Health Regulations 1988.

Department of the Environment (1985) *Digest of Environmental Protection and Water Statistics*, Government Statistical Service No. 8, HMSO, London.

Department of the Environment (1985) *Water and the Environment. The Implementation of Directive 76/464/EEC on Pollution Caused by Certain Dangerous Substances Discharged into the Aquatic Environment of the Community.* Joint Circular DoE 18/85 and the Welsh Office 37/85.

Department of the Environment (1989) *Drinking Water Quality in Public Supplies: an Explanation of the Water Act 1989 and the Water Supply (Water Quality) Regulations 1989*, HMSO, London.

Department of the Environment (1989) *Environment on Trust*, HMSO, London.

Department of the Environment (1990) *Guidance on Safeguarding the Quality of Public Water Supplies*, HMSO, London.

Environmental Data Services Ltd (1992) *Dangerous Substances in Water: a Practical Guide*, ENDS Report, 1-30.

Health and Safety Executive (1990) *Occupational Exposure Limits 1990, Guidance Note Environmental Hygiene (EH) 40/90*, 32pp. (Also includes lists of COSHH documents and Guidance Notes in the Environmental Hygiene and Medical Series.)

House of Commons (1990) *Contaminated Land, Environment Committee First Report*, Volumes 1, 2 and 3, HMSO, London.

Howarth, W. (1988) *Water Pollution Law*, Shaw and Sons Ltd., London, 608pp.

Control of Pollution Act 1974.

Environmental Protection Act 1990, Chapter 43, HMSO, London, 235p.

River Boards Act 1948.

Rivers (Prevention of Pollution) Act 1951

Rivers (Prevention of Pollution) Act 1961.

Water Act 1973.

Water Act 1983.

Water Act 1989.

Water Resources Act 1963.

Water Supply (Water Quality) (Amendment) Regulations 1989, Statutory Instrument 1989, No. 1384.

Water Supply (Water Quality) Regulations 1989, Statutory Instrument 1989, No. 1147.

Published 1988, Second Impression 1989, HMSO, London (ISBN 0 11 885470 4), 43pp.

19

ANAYSIS OF PRESSURE TUNNEL IN ROCK

Figure SI 19.1 shows in simple cross-section form the conditions that apply to the analysis of the problem of a pressure tunnel in rock subjected to a uniform state of *in situ* stress.

The following primary variables are used in the analysis, first of a thick-walled concrete cylinder that is subjected to unrestrained dilation when internal hydrostatic pressure is applied, and, second, when any dilation at the extrados of the concrete cylinder is resisted by contiguous, relaxed and broken (perhaps grouted) rock:

r is the radius from centre of circular, concrete-lined pressure tunnel to any point of interest (in the lining or the rock mass)
r_i is the radius to the inside concrete surface
r_a is the radius to the outside concrete surface
$(r_a - r_i)$ is the concrete thickness
$a = r_a / r_i$
σ_i is the water pressure in the tunnel
σ is the external (to the tunnel) pressure (due to stress field in the rock)
δ_{bi} is the displacement of a point on the inner (water) side of the concrete lining
δ_{ba} is the displacement of a point on the outer (rock contact) side of the concrete lining
v_b is Poisson's ratio of concrete
v_g is Poisson's ratio of the rock mass adjacent to the lining extrados
m_b is v_b^{-1} (Poisson's number) for the concrete
m_g is v_r^{-1} (Poisson's number for the rock)
E_b is Young's modulus for the concrete

E_g is Young's modulus for the rock mass adjacent to the concrete lining
σ_r is the radial stress
σ_θ is the circumferential (hoop) stress
ε_θ is the hoop strain

First, considering the problem as plane strain axisymmetric about the centre of the tunnel, development from the biharmonic equation leads to the standard equations for the radial and hoop stresses in an elastic annulus:

$$\sigma_r = \frac{r_i^2 r_a^2 (\sigma_i - \sigma)}{r_a^2 - r_i^2} \cdot \frac{1}{r^2} + \frac{\sigma r_a^2 - \sigma_i r_i^2}{r_a^2 - r_i^2} \quad ......(1)$$

$$\sigma_\theta = \frac{-r_i^2 r_a^2 (\sigma_i - \sigma)}{r_a^2 - r_i^2} \cdot \frac{1}{r^2} + \frac{\sigma r_a^2 - \sigma_i r_i^2}{r_a^2 - r_i^2} \quad ..(2)$$

Equations 1 and 2 show that for σ_r positive, σ_θ is negative. Thus for a compressive positive water pressure inside the tunnel, a concrete lining goes into hoop tension. If there is no external rock pressure to resist lining dilation associated with the hoop tension then the concrete must support this tension on its own. An initial conservative approach is to design the lining on the basis of this condition, that is, no inward compressive pressure external to the tunnel lining.

So, re-writing equations 1 and 2 for $\sigma = 0$,

$$\sigma_r = \frac{r_i^2 r_a^2 \sigma_i}{(r_a - r_i^2)} \cdot \frac{1}{r^2} - \frac{\sigma_i r_i^2}{(r_a^2 - r_i^2)} \quad ....(3)$$

$$\sigma_\theta = \frac{-r_i^2 r_a^2 \sigma_i}{(r_a^2 - r_i^2)} \cdot \frac{1}{r^2} - \frac{\sigma_i r_i^2}{(r_a^2 - r_i^2)} \quad ....(4)$$

Suppose that there is a 200m head of water creating pressure in the tunnel. The value of σ_i is then 200m x 9.81kN/m³ which is approximately 2MN/m².

Let $r_i = 5$m and $r_a = 5.25$m. This means that the lining thickness is 250mm. The hoop stress in the concrete is then 39.02MN/m² tensile on the

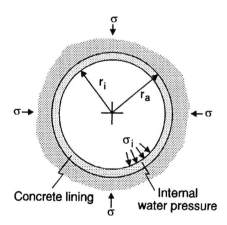

Figure SI 19.1 Assumed pressure conditions, internal and external, in a circular concrete-lined tunnel in a hydrostatic stress field

extrados (outer fibre) and 30.37MN/m^2 tensile on the intrados (inner fibre). Clearly, for this size of tunnel, lining thickness, internal pressure and unreinforced concrete the stress conditions in the concrete are unacceptable. Reinforcement would be included in the concrete under normal circumstances and there would also be some supporting pressure at the extrados from the grouted annulus and surrounding rock.

The above elastic analysis takes no account of the constraint to dilation offered by the surrounding rock. Charles Jaeger (1979) has shown that expressions can be derived for the hoop stresses at the inner face ($\sigma_{r(\theta)i}$) and the extrados ($\sigma_{r(\theta)a}$) of the lining:

at the inner face, $\sigma_{ri} = \sigma_i$

$$\sigma_{r(\theta)i} = -\left[\frac{r_a^2 + r_i^2 - 2kr_a^2}{r_a^2 - r_i^2}\right]\sigma_i \quad \text{.............(5)}$$

and

at the extrados, $\sigma_{ra} = k\sigma_i$

$$\sigma_{r(\theta)a} = -\left[\frac{2r_i^2 - k(r_a^2 + r_i^2)}{r_a^2 - r_i^2}\right]\sigma_i \quad \text{.............(6)}$$

In these equations, k is the proportion of the internal pressure σ_i that is transferred to the lining extrados:

$$k = \frac{\sigma_a}{\sigma_i} = \frac{\dfrac{2r_i^2}{E_b(r_a^2 - r_i^2)}}{\dfrac{m_g + 1}{m_g E_g} + \dfrac{(m_b - 1)r_a^2 + (m_b + 1)r_i^2}{m_b E_b(r_a^2 - r_i^2)}}$$
$$\text{.........................(7)}$$

Say, for example, that the elastic modulus of the concrete E_b is 20GN/m^2 and that of the fractured rock E_g surrounding the lining is 2GN/m^2. Poisson's ratio v_b for the concrete can be taken as 0.25, and although, as is the case with the modulus, a Poisson's ratio value for fractured rock v_g is not strictly applicable, it is assumed to be 0.35. Using the same geometrical (r_i = 5m; r_a = 5.25m) and pressure (σ_i = approximately 2MN/m^2) values as before, the magnitudes of k, $\sigma_{r(\theta)i}$ and $\sigma_{r(\theta)a}$ are 0.52, -18.65MN/m^2 and -17.69 MN/m^2, respectively, the negative sign denoting tension.

As would be expected, the effect of outer restraint offered by the rock has been to reduce the overall level of hoop tension in the lining. It has also reduced the difference in the level of hoop tension between the outer and inner faces of the lining and actually placed the inner fibre at a slightly higher tension than the outer fibre, reversing the significant difference that existed earlier.

Concrete lining cracks

The concrete lining may contain cracks, generally in a radial direction, resulting perhaps from thermal gradients during curing and also perhaps from squeezing compressive pressures from the rock after construction and before the internal water pressure is applied in the tunnel. Assuming that the cracking is uniform, then a pressure

$$\sigma_r = \frac{r_i}{r_a}\sigma_i$$

is applied to the rock, and the stress on the rock surface is of the same magnitude in tension ($-\sigma_t = \sigma_r$). If the water pressure is sufficiently high and the concrete cracks sufficiently wide, then water will penetrate from the inside to the outside of the lining, so causing $\sigma_r = -\sigma_t = \sigma_i$.

Rock mass cracks

Discontinuities in a rock mass, generally tight and relatively impermeable as a result of geostatic stress, will be loosened in the vicinity of the tunnel. Let $r = r_d$ be the distance into the rock at which the construction-imposed disturbance ceases. For $r < r_d$ within the more open rock mass structure,

$$\sigma_r = \frac{r_i}{r_d}\sigma_i \quad \text{.................................(8)}$$

and because the cracks are open and are assumed not close under the stress imposed by the tunnel internal water pressure σ_i,

$$\sigma_t = 0 \quad \text{.....................................(9)}$$

assuming also that the cracks are not full of groundwater nor have they been penetrated by water from the tunnel through a cracked concrete lining. If water is present,

$$-\sigma_r \leq \sigma_t \leq 0.$$

Beyond the cracked rock annulus and into the (relatively) intact rock,

$$\sigma_r = -\sigma_t = \frac{r_i}{r_d} \frac{r_d^2}{r^2} \sigma_i = \frac{r_i r_d}{r^2} \sigma_i \quad \ldots\ldots\ldots(10)$$

There are two major unknowns in the foregoing analysis: the 'elastic' modulus for the rock E_g and the distance into the rock r_d from the centre of the tunnel. The variable r_d is considered later, but at present equations 1 and 2 can be re-written as

$$\sigma_r = \frac{(\sigma_i - \sigma)}{a^2 - 1} \cdot \frac{r_a^2}{r^2} + \frac{(a^2\sigma - \sigma_i)}{a^2 - 1} \quad \ldots\ldots\ldots(11)$$

$$\sigma_\theta = \frac{-(\sigma_i - \sigma)}{a^2 - 1} \cdot \frac{r_a^2}{r^2} + \frac{(a^2\sigma - \sigma_i)}{a^2 - 1} \quad \ldots\ldots(12)$$

Equation 12 reduces to

$$\sigma_\theta = \sigma_r - 2\left\{ \frac{(\sigma_i - \sigma)}{(a^2 - 1)} \cdot \frac{r_a^2}{r^2} \right\} \quad \ldots\ldots\ldots(13)$$

Take δ_b as the associated radial displacement. Then r becomes $(r + \delta_b)$ after straining, where δ_b is small compared with r.

So, circumferential (hoop) strain ε_θ

$$= \frac{\text{Increase in circumference}}{\text{Original circumference}}$$

$$= \frac{2\pi(r + \delta_b) - 2\pi r}{2\pi r} = \frac{\delta_b}{r} \quad \ldots\ldots\ldots\ldots(14)$$

Thus,

$$\varepsilon_\theta = \frac{\delta_b}{r} \quad \ldots\ldots\ldots\ldots\ldots\ldots\ldots\ldots\ldots\ldots\ldots(15)$$

We are taking compressions as positive.

So,

$$-\varepsilon_\theta = \frac{\delta_b}{r}$$

Resolving ε_θ in terms of the stresses in the concrete annulus,

$$\varepsilon_\theta = \frac{\sigma_\theta}{E_b} - \frac{v_b \sigma_r}{E_b} = -\frac{v_b \sigma_r}{E_b} + \frac{\sigma_r}{E_b} - \frac{2}{E_b}\left\{ \frac{(\sigma - \sigma_i)}{a^2 - 1} \cdot \frac{r_a^2}{r^2} \right\}$$

$$\ldots\ldots\ldots\ldots\ldots\ldots\ldots(16)$$

or

$$-\delta_b = \frac{v_b \sigma_r r}{E_b} + \frac{\sigma_r r}{E_b} - \frac{2}{E_b} r \left\{ \frac{(\sigma_i - \sigma)}{a^2 - 1} \cdot \frac{r_a^2}{r^2} \right\}$$

$$= \frac{1 - v_b}{E_b}\left\{ \frac{\sigma_i - \sigma}{a^2 - 1} \cdot \frac{r_a^2}{r} \right\} + \frac{1 - v_b}{E_b}\left\{ \frac{a^2\sigma - \sigma_i}{a^2 - 1} \cdot r \right\}$$
$$- \frac{2}{E_b}\left\{ \frac{\sigma_i - \sigma}{a^2 - 1} \cdot \frac{r_a^2}{r} \right\}$$

$$= \frac{-1 - v_b}{E_b}\left\{ \frac{\sigma_i - \sigma}{a^2 - 1} \cdot \frac{r_a^2}{r} \right\} + \frac{1 - v_b}{E_b}\left\{ \frac{a^2\sigma - \sigma_i}{a^2 - 1} \cdot r \right\}$$

$$= -\left(\frac{1 - v_b}{E_b} \right)\left\{ \frac{\sigma_i - a^2\sigma}{a^2 - 1} \right\} r - \left(\frac{v_b + 1}{E_b} \right)\left\{ \frac{\sigma_i - \sigma}{a^2 - 1} \cdot \frac{r_a^2}{r} \right\}$$

or

$$\delta_b = \frac{(1 - v_b)}{E_b}\left\{ \frac{\sigma_i - a^2\sigma}{a^2 - 1} \right\} r + \frac{(1 + v_b)}{E_b}\left\{ \frac{\sigma_i - \sigma}{a^2 - 1} \right\}\frac{r_a^2}{r}$$

$$\ldots\ldots\ldots\ldots\ldots\ldots\ldots(17)$$

This analysis is for the lining. Now consider the displacement of the inner surface of the rock mass surrounding the concrete lining (i.e. at the point $r = r_a$). For this purpose, consider the rock mass to be a cylinder of infinite radius.

Let E_g be the elastic modulus of the rock. For this purpose take equations 1 and 2 and set r_a equal to infinity.

Then

$$\sigma_r = \frac{(\sigma_i - \sigma)r_i^2}{r^2} + 0$$

$$\sigma_\theta = -\frac{-(\sigma_i - \sigma)r_i^2}{r^2} + 0$$

$$\sigma_r - \sigma_\theta = \frac{2(\sigma_i - \sigma)r_i^2}{r^2}$$

$$\delta_{b(r=\infty)} = \frac{r\sigma_\theta}{E_g} - \frac{r v_g \sigma_r}{E_g}$$

where v_g is Poisson's ratio of the rock.

So,

$$\delta_{b(r=\infty)} = \frac{-(\sigma_i - \sigma)r_i^2}{rE_g} + \frac{v_g(\sigma_i - \sigma)r_i^2}{rE_g}$$

$$= \frac{-(1-v_g)}{E_g} \cdot \frac{(\sigma_i - \sigma)r_i^2}{r} \quad \ldots \ldots \ldots (18)$$

In the case of this elastic rock cylinder, suppose that $\sigma_i = 0$. This means that the effect of the concrete is ignored. Radius $r = r_i$ at this inner surface of the rock cylinder. Also, r_i for the rock becomes r_a for the concrete when the equations for rock and concrete are balanced for displacement.

Thus, $\delta_b = \delta_{gi} = -\dfrac{(1-v_g)}{E_g} r_a \sigma \quad \ldots \ldots \ldots (19)$

where δ_{gi} is the rock-concrete surface contact, and v_g is Poisson's ratio for the rock. (Note that under firm operational contact pressures between concrete lining and the rock any discontinuities in the rock are likely to be tightened up, so a monolithic, quasi-elastic rock Poisson's ratio (and elastic modulus) is likely to be reasonably valid.)

Now turn to equation 17 to obtain displacement δ_{ba} of the extrados of the concrete lining in contact with the rock mass. It follows that δ_b can be written δ_{ba} and r becomes r_a.

So, from equation 17, setting $r = r_a$,

$$\delta_{ba} = \frac{(1-v_b)}{E_b} \left\{ \frac{\sigma_i r_a}{a^2 - 1} \right\} - \frac{(1-v_b)}{E_b} \left\{ \frac{a^2 \sigma r_a}{a^2 - 1} \right\}$$
$$+ \frac{(1+v_b)}{E_b} \left\{ \frac{\sigma_i r_a}{a^2 - 1} \right\} - \frac{(1+v_b)}{E_b} \left\{ \frac{\sigma r_a}{a^2 - 1} \right\}$$

$$= \frac{r_a \sigma_i}{E_b(a^2 - 1)} \underbrace{\{(1-v_b) + (1+v_b)\}}_{=2}$$
$$- \frac{r_a \sigma}{E_b(a^2 - 1)} \underbrace{\{a^2(1-v_b) + (1+v_b)\}}_{set = K_b}$$

So,

$$\delta_{ba} = \frac{r_a}{E_b(a^2 - 1)} \{2\sigma_i - K_b \sigma\} \quad \ldots \ldots \ldots (20)$$

The next step is to obtain the *radial displacement of the inner surface* of the concrete lining (i.e. the surface in contact with the water under pressure). This radial displacement is δ_{bi}, and r becomes r_i. So, from equation 17 again,

$$\delta_{bi} = \frac{(1-v_b)}{E_b} \left\{ \frac{\sigma_i - a^2 \sigma}{a^2 - 1} \right\} r_i + \frac{(1+v_b)}{E_b} \left\{ \frac{a^2(\sigma_i - \sigma)}{a^2 - 1} \right\} r_i$$

$$= \frac{r_i}{E_b(a^2 - 1)} \{ \sigma_i \underbrace{[(1-v_b) + a^2(1+v_b)]}_{set = C_b} \}$$
$$- a^2 \sigma \underbrace{[(1-v_b) + (1+v_b)]}_{=2}$$

Then,

$$\delta_{bi} = \frac{r_i}{E_b(a^2 - 1)} (C_b \sigma_i - 2a^2 \sigma) \quad \ldots \ldots \ldots (21)$$

The composite term factors K_b and C_b are a function simply of known terms: Poisson's ratio for the concrete lining and the ratio of the inner and outer lining radius.

The contact pressure σ_a between the rock and the concrete lining can now be obtained by equating the outer concrete displacement to the rock displacement, vis, $\delta_{ba} = \delta_{gi}$ (equating equations 19 and 20), and setting $\sigma = \sigma_a$ as follows:

$$\delta_{ba} = -\frac{(1-v_g)}{E_g} r_a \sigma_a = \frac{r_a}{E_b(a^2 - 1)} \{2\sigma_i - K_b \sigma_a\}$$

$$\sigma_a \left\{ \frac{K_b}{E_b(a^2-1)} - \frac{(1-v_g)}{E_g} \right\} = \frac{2\sigma_i}{E_b(a^2-1)}$$

$$\sigma_a \left\{ \frac{E_g K_b - E_b(a^2-1)(1-v_g)}{E_b E_g(a^2-1)} \right\} = \frac{2\sigma_i}{E_b(a^2-1)}$$

$$\sigma_a \left\{ \frac{E_g K_b - E_b(a^2-1)(1-v_g)}{E_g} \right\} = 2\sigma_i$$

$$\sigma_a = \frac{2\sigma_i}{K_b - \frac{E_b}{E_g}(1-v_g)(a^2-1)} \quad \text{...............(22)}$$

Now assume that there is good contact and no looseness at the lining extrados/rock interface (as achieved by grouting and checks on the efficiency of the grouting process). If this is indeed so, then the ratio of any displacement (radial), measured at the concrete lining inner surface caused by trial jacking operation, u_{bi} say, to the rock displacement δ_{gi} (equation 19 above) will be equal to the ratio δ_{bi} (equation 21) to δ_{ba} (equation 20).

So we can write down

$$\frac{u_{bi}}{\delta_{gi}} = \frac{\delta_{bi}}{\delta_{ba}} \quad \text{..............................(23)}$$

noting that the value of u_{bi} will be derived from an in-tunnel jacking test across the tunnel diameter with the concrete lining in place. From this test it should be possible to determine the elastic modulus E_g of the surrounding rock.

$$-\frac{u_{bi}}{\frac{(1-v_g)}{E_g}r_a\sigma_a} = \frac{\frac{r_i}{E_b(a^2-1)}(C_b\sigma_i - 2a^2\sigma_a)}{\frac{r_a}{E_b(a^2-1)}(2\sigma_i - K_b\sigma_a)}$$

$$= \frac{C_b\sigma_i - 2a^2\sigma_a}{a(2\sigma_i - K_b\sigma_a)}$$

So,

$$E_g = \frac{(1-v_g)r_a\sigma_a(C_b\sigma_i - 2a^2\sigma_a)}{u_{bi}a(2\sigma_i - K_b\sigma_a)} \quad \text{...........(24)}$$

During a jacking test there will be some 'slackness' to be taken up initially, so the load will be cycled. It will then be possible to separate out the elastic deformation from the plastic component.

Elastic modulus of concrete

As per BS 8110 Part 2: 1985 (British Standards Institution, 1985), and noting Table SI 19.1, E_c in terms of the cube strength is

$$E_{c,28} = K_0 + 0.2 f_{cu,28,}$$

Table SI 19.1 Elastic modulus of concrete related to cube strength (British Standards Institution, 1985)

$f_{cu,28}$	$E_{c,28}$	
	Mean value	Typical range
N/mm^2	kN/mm^2	kN/mm^2
20	24	18 to 30
25	25	19 to 31
30	26	20 to 32
40	28	22 to 34
50	30	24 to 36
60	32	26 to 38

where $E_{c,28}$ is the static modulus of elasticity at 28 days,

$f_{cu,28}$ is the characteristic cube strength at 28 days (in N/mm²),

and

K_0 is a constant closely related to the modulus of elasticity of the aggregate (which is taken as 20 kN/mm²) for normal-weight concrete).

Also, $E_c = 1.25\, E_{cq} - 19$,

where E_c is the dynamic modulus of elasticity .

It is said that such an estimate will generally be correct within ±4 kN/mm². Thus, it is possible to determine the value of E_b from a cube crushing test.

In order to test this value for E_g, the same values as in the earlier examples should be used, i.e. $r_a = 5.25$m.

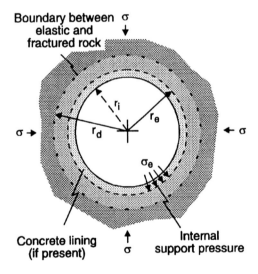

Figure SI 19.2 Idealised stress distribution in and around a circular tunnel in a hydrostatic stress field and surrounded by a zone of fractured rock

Extent of the fractured rock annulus

There has been earlier consideration of the extent of the zone of imposed rock fracturing around a tunnel, and because of the international use of the NATM method of tunnelling it is useful to extend the analysis. A starting point is an assumption that the strength of the rock mass is described by a Coulomb-Mohr criterion in the same manner as for intact rock. It is also assumed that there is a

well-defined interface between rock fractured as a result of inward relaxation on excavation of the tunnel and the rock that has not been fractured by construction. This situation is shown in Figure SI 19.2. Here, as in the earlier text, a tunnel lining is shown in place offering internal support. Alternatively, the support can be provided by anchorages. In the figure, r_d is the radial distance from the centre of the tunnel to the boundary between the 'intact' rock and the rock fractured by the construction operation. The distance r_e is from the centre of the tunnel to the extrados of the lining (the inner surface of the fractured rock), or if there was no lining present it would be the distance from the centre of the tunnel to the rock surface (equivalent to r_i in Figure SI 19.1).

As per Coulomb-Mohr, and analysing on a

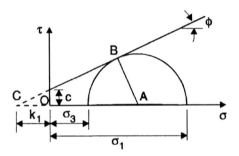

Figure SI 19.3 Mohr envelope for rock

total stress basis, reference to Figure SI 19.3 gives the expression,

$$\sin\phi = \frac{AB}{AC} = \frac{\frac{1}{2}(\sigma_1 - \sigma_3)}{k_1 + \frac{1}{2}(\sigma_1 + \sigma_3)} = \frac{\sigma_1 - \sigma_3}{2k_1 + \sigma_1 + \sigma_3}$$

Thus,

$$\sigma_1 - \sigma_3 = 2k_1 \sin\phi + (\sigma_1 + \sigma_3)\sin\phi.$$

Since

$$k_1 = c \cot\phi$$

then

$$(\sigma_1 - \sigma_3) = 2c \cos\phi + (\sigma_1 + \sigma_3)\sin\phi.$$

Now

$$\sigma_1 (1 - \sin\phi) = 2c \cos\phi + \sigma_3 (1 + \sin\phi)$$

and so

$$\sigma_1 = \frac{2c\cos\phi}{(1-\sin\phi)} + \frac{\sigma_3(1+\sin\phi)}{(1-\sin\phi)}$$

$$\sigma_1 = C_o + b\sigma_3 \quad\text{...(25)}$$

In equation 1, σ_1 and σ_3 are the major and minor principal stresses, respectively, and c and ϕ are the effective cohesion and effective friction, respectively, for the fractured rock annulus. If the strength of the fractured rock is taken to be frictional only, then the limiting state of stress is defined from equation 25 by

$$\sigma_1 = \sigma_3\frac{(1+\sin\phi_f)}{(1-\sin\phi_f)} = d\sigma_3 \quad\text{.......................(26)}$$

where ϕ_f denotes the friction angle of the fractured rock annulus.

As Brady and Brown (1993) point out, expressing the condition of static equilibrium in polar coordinates, for the axisymmetric condition there is only one differential equation of equilibrium which is satisfied throughout the problem domain:

$$\frac{d\sigma_r}{dr} = \frac{\sigma_\theta - \sigma_r}{r} \quad\text{.......................................(27)}$$

where σ_r and σ_θ are respectively the radial stress and circumferential stress.

From equations 26 and 27,

$$\frac{d\sigma_r}{dr} = (d-1)\frac{\sigma_r}{r} \quad\text{..................................(28)}$$

Integrating equation 28 and using the boundary condition $\sigma_r = \sigma_i$ when $r = r_e$,

$$\sigma_r = \sigma_e\left[\frac{r}{r_e}\right]^{d-1} \quad\text{...................................(29)}$$

$$\sigma_\theta = d\sigma_e\left[\frac{r}{r_e}\right]^{d-1} \quad\text{................................(30)}$$

These latter two expressions apply throughout the fractured zone and on its boundaries. Using the

fact that when $r = r_d$, the fractured rock is in equilibrium with intact elastic rock, the equilibrium radial stress σ_{rd} at the outer boundary r_d of the fractured annulus is

$$\sigma_{rd} = \sigma_e\left[\frac{r_d}{r_e}\right]^{d-1} \quad\text{.................................(31)}$$

or

$$r_d = r_e\left[\frac{\sigma_{rd}}{\sigma_e}\right]^{(d-1)^{-1}} \quad\text{.................................(32)}$$

Superimposing stresses gives the distribution of stress in the zone of intact elastic rock $(r > r_d)$:

$$\sigma_\theta = \sigma\left[1+\frac{r_d^2}{r^2}\right] - \sigma_{rd}\frac{r_d^2}{r^2} \quad\text{.....................(33)}$$

$$\sigma_r = \sigma\left[1-\frac{r_d^2}{r^2}\right] + \sigma_{rd}\frac{r_d^2}{r^2} \quad\text{.....................(34)}$$

At the boundary of the elastic zone with the crushed rock zone $(r = r_d)$,

$$\sigma_\theta = 2\sigma - \sigma_{rd} \text{ and } \sigma_r = \sigma_{rd}. \quad\text{.....................(35)}$$

This represents the limiting state of stress for uncrushed rock.

Substituting for $\sigma_\theta(\sigma_1)$ and $\sigma_r(\sigma_3)$ in equation 26 gives

$$2\sigma - \sigma_{rd} = b\sigma_{rd} + C_o$$

or

$$\sigma_{rd} = \frac{2\sigma - C_o}{1+b}. \quad\text{.......................................(36)}$$

Substituting equation 36 into equation 32,

$$r_d = r_e\left[\frac{2\sigma - C_o}{(1+b)\sigma_e}\right]^{(d-1)^{-1}} \quad\text{.........................(37)}$$

Equations 29, 30, 33, 34 and 37, with internal pressure σ_e, field stress σ and rock properties c and ϕ (σ_f) define the stress distribution and the crushed rock zone.

REFERENCES

Brady, B.H.G. and Brown, E.T. (1993) *Rock Mechanics for Underground Mining*, 2nd Edition, Chapman and Hall, London, 571pp.

British Standards Institution (1985). *Structural Use of Concrete*, Part 2. Code of Practice for Special Circumstances, BS 8110 Part 2: 1985, HMSO, London.

Jaeger, C. (1979) *Rock Mechanics and Engineering*, 2nd Edition, CUP, Cambridge, 523pp. (ISBN 0521 21898 5.)

20

GROUND VIBRATION AND NOISE FROM CIVIL ENGINEERING OPERATIONS

INTRODUCTION

Construction vibrations are generally of three different types:

(a) Impact vibration
(b) Continuous vibration
(c) Pseudo-continuous vibration.

British Standard 6472 (British Standards Institution, 1984) uses slightly different terminology without relating it to construction equipment (*impulsive*: rapid build-up to peak followed by damped decay, perhaps over several cycles, but can also comprise several short-duration cycles of less than 2 seconds total; *intermittent*: a string of vibration incidents, each of short duration, separated by intervals of much lower vibration magnitudes, but may also consist of bursts of vibration flanked by a more or less gentle rise and decay; *continuous*: which continues uninterrupted for a period of 16 hours).

Examples of the sources of vibration are:

(a_1) Impact (Instantaneous):
   Blasting (single shot and delayed action firing)

(a_2) Impact (Repetitive):
   Impact pile driving, pile extraction
   Dynamic compaction
   Demolition by drop ball, etc

(b) Continuous:
   Vibratory pile driver
   Large pumps
   Heavy static compressors
   Vibroflotation and vibrocompaction

(c) Pseudo-continuous:
   Heavy pneumatic breaker
   Heavy plant

There are two main sources of vibration from tunnelling works: blasting in rock and driven sheet piling (and extraction) for ancillary works such as manholes and storm sewage overflow chamber construction.

Measurements of ground vibration at different distances from the source are usually recorded by velocity transducers of the coil/permanent magnet type (geophones) which produce a voltage signal proportional to the velocity of the surface upon which they are mounted, carefully calibrated (usually by dynamic back-to-back testing on a vibrator using an accurate accelerometer), and recording through suitable instrumentation onto paper, magnetic tape, or floppy disk. Accelerometers may be used instead of velocity transducers to record vibration, and they are particularly appropriate for monitoring signals having very low frequencies of the order of 1 or 2 hertz. The signal conditioning requirements may then, however, be more demanding.

Although relatively simple equipment, which provides basic information on vibration magnitude (three orthogonal components and a time-varying resultant), may be hired on a temporary basis by site staff, relatively expensive specialist equipment, which further processes the signal in order to derive more information from it (typically, the frequency spectrum of the signal), is available.

The data recording and processing system used by the writer and colleagues (*see* Selby and Swift, 1989) comprises up to 64 input channels (including 4 high speed), 4 A/D convertors, each with a multiplexor, and 2Mbytes of RAM, a keyboard, two 5.25inch floppy disk drives, and a back-lit screen, all housed in a substantial box suitable for demanding site conditions. Power is provided either from site mains or from a portable generator. Post-processing facilities include time-based vector resolution of the radial (R), transverse (T) and vertical (V) components of the vibration signal at any one monitoring station, fast Fourier transform for frequency spectrum presentation, integration and also differentiation of the signals, printing of the peak particle velocities or accelerations, and colour plotting of one or more of the time-based vibration records in either single or stacked mode. In operation, a specially written data capture program is loaded into the recorder and the correct function of the transducers verified. Data capture parameters, for example trigger channels, trigger levels, rate of sampling, and record duration (including pre-trigger capture), are then entered. The RAM is then cleared for data capture and files of vibration data recorded on demand using an auto-triggering facility. Files are then transferred to floppy disk for storage, a careful manual record being made of file codes, hammer type and operation, pile type, toe depth and the several geophone recording distances from the feature of concern (in the case of piling), or the type of explosive, charge configuration (number of holes, explosive weight per hole, delay settings, spacings, burdens), and standoff distance in the case of blasting.

For civil engineering assessment purposes it is usually the vibration *velocity* of the ground that is measured and then checked against national codes and/or published criteria. The velocity at which the ground moves should not be confused with the speed at which the wave or waves which create the disturbance travel from the disturbance source. Ground velocity is expressed in units of millimetres or micrometres ($1\mu m = 1mm \times 10^{-3}$) per second whereas wave velocity (celerity) is measured in units of metres per second. The ratio between the two is a measure of dynamic strain in the ground. Ground *acceleration* is measured by accelerometers in units of 'g', or in centimetres (or millimetres) per second squared, and is usually processed by integration into units of millimetres (or μm) per second. If it is assumed that the motion is sinusoidal and that the ground velocities and accelerations are carried at the same frequencies, then the maximum velocity is simply the maximum acceleration divided by $2\pi f$, where f is the frequency in hertz (Hz). The onset of vibration-generated damage in buildings, and also human susceptibility to vibration, becomes independent of frequency when expressed in terms of velocity.

During the period of a transient vibration, the sense of the ground movement tends to be a mixture of up and down, forwards and backwards, and sideways. This movement can be resolved by a combination of three geophones or accelerometers mounted at right angles, for example, one vertical (v), one horizontal and aligned to the source of the disturbance (radial, r) and one horizontal and at right angles to r (transverse, t). The resultant ground vibration velocity, v_{res}, at any instant of time, $v_{res}(t)$ is then the square root of the sum of the squares of the three orthogonal vibration velocities $[v_v(t) + v_r(t) + v_t(t)]^{\frac{1}{2}}$. It is this maximum v_{res} value that should be reported and checked against limit criteria for vibration, although Vuolio (1990), while concluding that particle velocity is the best description for limiting potential damage to structures, also concludes that the most practical description is the *vertical component* of the peak particle velocity. If the resultant *is* used, it is sometimes computed from the individual maxima along the three axes of vibration independent of time, but this method is not recommended since it produces an over-conservative result.

Vibration traces, magnitude-ordinate against time-abscissa, are of sinusoidal character. It is the distance from the baseline (zero to peak) that should be calibrated to and expressed as the vibration velocity. Alternatively, a root mean square (peak-to-peak/root two) value may be reported, but it must be clearly stated which has been used. The relation between zero-to-peak and root mean square vibration velocity is $v_{0\,to\,peak} = v_{rms}/0.707$. As noted below, root mean square values may be used when expressing attenuation levels.

Measurements made on the ground are usually assumed to be free from interference due to the presence of a building. Sometimes measurements are made inside a building. In either case an assessment of a building's vulnerability, or a person's sensitivity, is what is required.

SCALED DISTANCE LAW

There are many factors that control the amplitude of vibration experienced by a building or person. Of these factors the most important are the amount of energy (W) that is put into the ground at source, the distance (r) from the source to the receiver (person or building), and the nature of the ground.

In the case of excavation (quarrying, trenching, tunnelling, and so on), the *energy* is often introduced by means of explosives. Although the explosives have different densities (strengths) it is usual in any assessment to ignore these differences and to enter a value of weight (force: newtons or kilograms force) *per delay* if delayed action blasting is used. Most of the explosive force should be absorbed by the process of fragmentation (provision of new free surface energy), and so only a small fraction of the total charge weight contributes to transmitted vibration. Similarly, in the case of driven piling most of the nominal input energy goes into advancing the pile, some is lost at the head of the pile and some is absorbed in other parts of the driver-pile system. Energy distribution in driven piling is discussed by Attewell (1986a). Notwithstanding these losses, and as with explosives, energy is usually referred to the input source (the driver) as quoted by the manufacturers, in this case in units of joules (newton metres) or alternatively kilojoules (kilonewton metres).

There are analogies between the explosive and driven pile vibration sources. In both cases dynamic energy is transmitted to the ground over a finite period of time, usually along a vertical column (sticks of explosive on the one hand and the length of the pile on the other). Strictly, therefore, a point source of disturbance cannot be assumed. This does not create a problem in, say, quarry blasting, when measurements and assessments of vibration are usually taken several hundred metres from the blast. Pile vibration measurements, on the other hand, are taken only a few, or at the most a few tens of metres from the pile, and so the source characteristics should loom large in the assessment.

Vibration in the form of ground waves undergoes geometrical divergence as the transmission distance increases. This means that the dynamic energy density in a wavefront decreases with distance travelled. If the wave expands uniformly in three dimensions in the ground from an assumed point source (such as from the toe of a driven pile), then its *energy* amplitude at any point of measurement on the wavefront decreases as the square of the distance (r) travelled according to the formula for the surface area of a sphere ($4\pi r^2$), and its vibration *velocity* amplitude decreases directly in proportion to the distance travelled. If a wavefront expands outwards into the ground as a result of dynamic friction along the shaft of a driven pile, then the wavefront is of quasi-cylindrical configuration and the vibration *energy* amplitude decreases linearly as the distance, as per the formula $2\pi rh$ where h is the depth of the pile. The vibration *velocity* amplitude decreases in proportion to the square root of the distance from source. A better understanding of the wave transmission mechanics can be gleaned from the simple explanation in Attewell and Farmer (1973). Of special interest are the styles of reflection and refraction of the wave trains as they interact with layers of ground having different acoustic impedances (ρc), and particularly when they interact with the ground surface.

It will now be apparent that the depth of a pile toe can enter into the transient vibration assessment and estimation. However, the influence of h is not confined to the style of the transmitted wave. It also affects the distance parameter. For example, on entry to the ground the pile toe is sensibly coincident with the ground surface, and so the reference distance r is unambiguous. After penetration, the depth h may be similar to, or could be even greater than r, and this might be significant when assessing vibration levels close to the pile. If it is assumed that most of the vibration energy stems from the pile toe, then the distance parameter to be used for assessment purposes should really be the direct distance (r) from the toe to the measurement point, that is, $R = (r^2 + h^2)^{1/2}$.

In addition to the purely geometrical energy reduction with distance the vibration level undergoes a progressive attenuation with distance (r and R) as a result of solid friction and other mechanisms in the ground. Clay soils tend to attenuate ground vibrations more rapidly than do granular soils, but soils and rocks do not have natural frequencies as such. These are the frequencies at which attenuation effects are less dominant. However, soils and rocks do seem to transmit certain frequencies more intensely than others. The approximate frequencies related to different types of ground are:

Silts and peats	5 - 10Hz
Clays	15 - 25Hz
Sands	30 - 35Hz
Sandstones and limestones	30 - 40Hz
Igneous rock	around 240Hz

These frequencies may be compared with typical natural frequencies for buildings, which range between about 0.1Hz for high rise buildings to about 10Hz for low rise buildings. Within buildings, floors and slabs will tend to resonate at natural frequencies of about 10Hz to 30Hz, windows at about 10Hz to 100Hz, and plaster ceilings at about 10Hz to 20Hz.

FREQUENCY

A signal recorded conventionally in the time domain can be expressed in the frequency domain by a fast Fourier transform. The distribution of the signal amplitude among the frequencies that compose it can then be examined. A frequency graphical (x) axis is often scaled in octaves or one-third octaves, with the vibration amplitude on the y-axis. An octave is a frequency band where the highest frequency is twice the lowest frequency. (The name 'octave' stems from the fact that an octave covers eight notes of the diatomic scale.) The one-third octave covers a range where the highest frequency is $2^{1/3}$ (or 1.26 times) the lowest frequency. Octave and third-octave passband frequencies, with nominal centre frequencies, within the general range of civil engineering interest are given in Table SI 20.1.

ATTENUATION

Attenuation of sound, or vibratory disturbance in a solid, is expressed as units (nepers) per unit length:

$$a = (1/2r) \log_e(I_0/I_r) \quad ....(1)$$

where the initial *intensity* I_0 has decreased to I_r after distance r as a result of frictional mechanisms. In terms of decibels (dB) per unit length, this value is $8.686a$.

The decibel expresses amplitude of vibration (displacement, velocity, or acceleration) relative to some particular reference value. Under the decibel unit, therefore, the recorded vibration is expressed as a *vibration level*, since it gives the *level* of the vibration amplitude relative to the reference.

Table SI 20.1 Octave and third-octave passband frequencies

Nominal centre frequency (Hz)	Third-octave passband (Hz)	Octave passband (Hz)	Nominal frequency (Hz)	Third-octave passband (Hz)	Octave passband (Hz)
1.25	1.12-1.41		40	35.5-44.7	
1.60	1.41-1.78		50	44.7-56.2	
2.00	1.78-2.24	1.41-2.82	63	56.2-70.8	44.7-89.1
2.50	2.24-2.82		80	70.8-89.1	
3.15	2.82-3.55		100	89.1- 112	
4.00	3.55-4.47	2.82-5.62	125	112 - 141	89.1- 178
5.00	4.47-5.62		160	141 - 178	
6.30	5.62-7.08		200	178 - 224	
8.00	7.08-8.91	5.62-11.2	250	224 - 282	178 - 355
10.00	8.91-11.2		315	282 - 355	
12.50	11.2-14.1		400	355 - 447	
16.00	14.1-17.8	11.2-22.4	500	447 - 562	355 - 447
20.00	17.8-22.4		630	562 - 708	
25.00	22.4-28.2		800		708 - 891
31.50	28.2-35.5	22.4-44.7	1000	891 -1120	708 -1410

Two sounds or, more generally, vibrations having intensities I_1 and I_2 differ in intensity level by m decibels, where

$$m = 10 \log_{10}(I_1/I_2). \quad \dots\dots\dots\dots\dots\dots (2)$$

Intensity is difficult to measure, so it is usual to express amplitude differences in terms of sound (vibration) *pressure*, in which case

$$m = 20 \log_{10}(p_1/p_2), \quad \dots\dots\dots\dots\dots (3)$$

p_1 and p_2 being the root mean square pressures of the two sounds (vibrations).

In terms of vibration velocity, equation 2 can be written

$$\text{velocity level} = 10 \log_{10}(v_{rms}^2/v_{ref}^2) \text{ dB}$$
$$\text{relative to } v_{ref},$$

which, in terms of equation 3, becomes

$$\text{velocity level} = 20 \log_{10}(v_{rms}/v_{ref}^2) \text{ dB re } v_{ref}$$

For convenience, v_{ref} would usually be taken as unity. Then, for example, if the measured root mean square velocity is 1.5mm/s, the vibration *level* is described as

$$\text{velocity level} = 20 \log_{10}(1.5/1)$$
$$= 3.52 \text{dB re 1mm/s}.$$

Should the r.m.s. velocity double, the velocity level then increases by $20 \log_{10}2 = 6$dB, and if there is a 10-fold increase in the r.m.s. velocity the velocity level then increases by $20\log_{10}10 = 20$dB.

Ground quality factor Q can be expressed as

$$Q = \pi f/\alpha c \quad \dots\dots\dots\dots\dots\dots\dots (4)$$

where f is the wave frequency and c is the phase velocity. Equation 4 shows that for constant frictional ground conditions (Q^{-1}), the spatial attenuation coefficient α (in units of metres^{-1}) should be proportional to frequency (Attewell and Ramana, 1966) except where scatter and relaxation phenomena may exist. Richart and Woods (1987) give the values for α shown in Table SI 20.2.

However, for civil engineering works, where r and R are relatively small compared with geophysical seismic distances, the influence of solid friction on the amplitude of a vibration transient is secondary to that of geometrical dispersion. A number of values for the spatial attenuation coefficient α are also given in Attewell (1986a).

Both the geometrical divergence and the solid friction attenuation mechanisms operate coincidently. However, an understanding of the transmission mechanics is not really necessary for practical vibration assessment, since if tolerable vibration levels can be pre-specified then estimates of actual vibration can be made using the above-noted parameters in a so-called scaled distance law.

The most widely used scaled distance law is

$$v = K(W^{1/2})/r)^n . \quad \dots\dots\dots\dots\dots (5)$$

Table SI 20.2 Values of attenuation coefficient α (after Richart and Woods, 1987)

α (metres^{-1})		Ground material type
5Hz	50Hz	
0.01 - 0.03	0.1 - 0.3	Weak or soft soils - lossy soils, dry or partially saturated peat, loose beach sand, organic soils, topsoil
0.003 - 0.01	0.03 - 0.1	Competent soils - sands, sandy clays, gravels, silts, weathered rock
0.0003 - 0.003	0.003 - 0.03	Hard soils - dense compacted sand, dry consolidated clay, consolidated glacial till, some exposed rock (pick needed to break)
<0.0003	<0.003	Hard competent rock - bedded rock, freshly exposed hard rock (difficult to break with hammer)

In equation 5, v is the vibration velocity, usually zero to peak resultant, recorded at ground surface, and K and n are so-called 'site constants', being dependent upon the ground conditions and the form of the imposed disturbance. In the case, for example, of blasting, the site constants can often be established by single hole trial blasts and then be applied to commercial (production) blasts.

This law is sometimes presented with an inversion of the parameters inside the brackets:

$$v = K(r/W^{1/2})^{-n}. \qquad (6)$$

GROUND VIBRATION FROM DRIVEN PILING

The draft of the British Standard BS 5228: Part 4 on pile vibrations (British Standards Institution, 1992a) in its recommendations for ground vibration assessment drew strongly on the original equations proposed by Attewell and Farmer (1973):

$$v = 1.5(W^{1/2}/r) \text{ mm/s,} \qquad (7)$$

where

v, in mm/s, is read as zero to peak resultant,
W, the source energy per blow (or per cycle), is in joules (N.m), and
r, the horizontal distance from source to receiver, is in metres.

Expressing W in units of *kilojoules* the above equation 7 is then

$$v = 47.43(W^{1/2}/r) \text{ mm/s.} \qquad (7a)$$

Thus, knowing the notional energy at source from a piling hammer, and being able to measure the distance r, an estimate can then be made of the vibration velocity for comparison with nationally recommended upper limits. Attewell and Farmer have stated, however, that equation 6 encompasses all the measured data points that it describes, and so for predictive purposes it is likely to be generally conservative. A more realistic equation from the Attewell and Farmer

paper for driven piling is

$$v = 0.75(W^{1/2}/r) \text{ mm/s zero-to-peak resultant} \qquad (8)$$

Expressing W in the now more usual units of *kilojoules*, the above equation 5 is then

$$v = 23.72(W^{1/2}/r) \text{ mm/s} \qquad (8a)$$

with the other units the same as in the above equation 7.

These equations encompass different types of ground, and therefore serve only as 'rough and ready' guides to possible pile vibration amplitudes. However, this is often all that is required on a busy working site. No account is taken of the depth of pile penetration h in order to derive a full toe-to-measurement point distance R, but again there is usually little time on site for this added detail.

More recent work performed by Attewell *et al.* (1989) has shown how the complex three dimensional vibratory motions can be presented two dimensionally, and has related the motion to structural effects. Further, Uromeihy (1990) has compiled the results of numerous site measurements made throughout Britain and has proposed more detailed equations which accommodate different piling and ground conditions. These are given in Table SI 20.3. The standoff distance measurement in the equations is R rather than r, and this does pose more observational requirements on the measurement staff. The control equation is

$$v = K(R/W^{1/2})^{-n} \qquad (9)$$

As an essential element in a knowledge-based (expert) system designed specifically for industry, Oliver (1990, unpublished work, personal communication) has derived site parameters for pile vibration estimation using the direct horizontal radial distance r. These particular parameters, given in Table SI 20.4 and used in conjunction with equation 5, may be used for estimating the vibration that could occur on a construction site.

Table SI 20.3 Attenuation coefficients for different piling conditions (after Uromeihy, 1990)

Hammer	Soil	Pile	Toe Depth	K	n
Diesel	Loose to medium dense	H	6 - 10m	39	-1.80
Diesel	Medium dense to dense	H	17 - 27m	13	-0.59
Drop		End bearing	7 - 12m	38	-1.97
Hydraulic	Medium dense	H	10 - 17m	51	-1.55
Hydraulic	Medium dense	H	19 - 24m	249	-2.27
Hydraulic	Very dense	H	20 - 30m	15	-0.41
Impact		Sheet	6 - 14m	33	-1.47
Vibrodriver		Sheet	6 - 10m	79	-1.95
Vibrodriver		End bearing	10 - 16m	87	-1.63

<u>Note</u>: From equation 6, v is in units of mm/s, W is in units of kilojoules, and R is in units of metres.

Table SI 20.4 Attenuation conditions for different piling conditions (after Oliver, 1990)

Hammer, Soil & Pile Type	K	n	Sample Size
Diesel, granular, sheet	0.54	0.92	15
Diesel, granular, bearing	0.61	0.93	30
Diesel, cohesive, sheet	0.12	1.33	30
Diesel, cohesive, bearing	1.25	0.79	20
Air, granular, sheet	0.66	0.92	15
Air, granular, bearing	2.90	0.39	10
Air, cohesive, sheet	1.26	0.60	30
Air, cohesive, bearing*	-	-	-
Drop, granular, sheet	1.28	0.75	20
Drop, granular, bearing	0.36	1.00	55
Drop, cohesive, sheet	0.24	1.16	35
Drop, cohesive, bearing	0.47	0.91	100
Vibrodriver, granular, sheet	0.82	1.00	30
Vibrodriver, granular, bearing	1.25	0.88	10
Vibrodriver, cohesive, sheet	0.79	1.07	80
Vibrodriver, cohesive, bearing	1.37	0.85	50

<u>Notes</u>:

* Small sample size renders these parameters meaningless.

From equation 5, v is in units of mm/s, W is in units of joules (Nm), and r is in units of metres.

For formulation of the above parameters, each digital record of vibration (radial, transverse and vertical components), taken at distances of 2, 4, 8, 12 and 20m from the driven pile at numerous piling sites throughout the UK, was characterised by cubic equations. Altogether, 120 sets of data were available for scanning in the system which employed Smartware II on an RM Nimbus microcomputer with enhanced memory and linked to a laserjet printer. The enquirer could use a 'loose' definition only on one of the system's formative parameters (for example, the term 'sheet piles' being input without any further qualification), in which case there would then be a large number (around 50) of selected records of vibration available for recall and display. On the other hand, a 'tight' definition on several parameters (for example, 'vibrodriver, steel H-piles, dense sandy gravel, toe depth 5-10m) would then recall only a small number of comparable records, say 5, or even result in zero recall, signifying a mismatch between input requirement and record availability.

Use of any or all of the above equations needed to be reserved for first estimate purposes and be supplemented when necessary by direct measurements. Where the soil was deemed to be a mixture of granular and cohesive, estimates of the relative volumes of the two fractions could be made and then a vibration velocity between the two calculated according to the composition.

If the requirement is for *a high level of confidence that a particular vibration level will not be exceeded*, then for *impact hammers* the original equation

$$v = 1.5(W^{1/2}/r) \quad \text{..............................(7)}$$

may be used. For *more probable* vibration velocity values, and again for *impact hammers*, the following equation from Attewell and Farmer (1973) can be used:

$$v = 0.76(W^{1/2}/r)^{0.87} \quad \text{....................(7b)}$$

The equivalent equations for *vibrodrivers* are:

$$v = 1.00(W^{1/2}/r)^{0.95} \quad \text{(upper bound) ....(10)}$$

$$v = 1.8(W^{1/2}/r) \quad \text{(more probable values)}$$
$$\text{...........................(10a)}$$

In equations 7, 7b, 10, 10a, vibration velocity v is expressed in units of millimetres per second, nominal input energy W (see manufacturer's literature) is in units of newton metres, and standoff distance r is in units of metres.

In equations 10 and 10a, nominal input energy W is the *energy per vibrodriver cycle*.

Further, more detailed, statistical analyses have been performed by Attewell *et al.* (1992a) on different sets of data, singly and combined, for impact hammers and vibrodrivers. In particular, when the original Attewell and Farmer (1973) data are combined both with the later Durham University data and with data provided by consulting engineers W.S. Atkins and Partners there is discernible curvature in the data point trends when the plots are in vibration velocity (v) versus scaled distance ($W^{1/2}/r$) space. Best fit quadratic equations were applied to the data sets, and then further assessments were directed towards the one-half and one standard deviation curves. It was concluded that it would be unreasonable to apply best-fit equations, since these would leave a 50% probability of the empirically recommended vibration velocity values being exceeded. It was concluded also that it would be equally unreasonable to adopt quadratic curve parameters that encompassed all the data points of each data set. Thus the recommendation was that data sets should be combined and that *one-half standard deviation quadratic curve* parameters should be adopted. This means that there is then a 32% probability of the predicted vibration velocity being exceeded in any particular case. The recommended equations are:

Impact Hammers

$$\log v = -0.296 + 1.38 \log (W^{1/2}/r)$$
$$- 0.234 \log^2(W^{1/2}/r) \quad \text{...............(11)}$$
(one-half standard deviation)

Vibrodrivers

$$\log v = -0.213 + 1.64 \log (W^{1/2}/r)$$
$$- 0.334 \log^2(W^{1/2}/r) \quad \text{...............(12)}$$
(one-half standard deviation)

As before, in these equations vibration velocity (v) is in units of mm/s, nominal energy (W) is in units of N.m (energy per cycle in the case of a vibrodriver), and standoff distance (*r*) is in units of metres.

These equations are not easily applied, especially on site, with a hand calculator. For simpler, practical use it is recommended that the tables and graphs in Attewell *et al.* (1992b) be adopted. The tables in particular are easily accessed, and, for a known standoff distance *r*, values of best fit, one-half standard deviation and one standard deviation are clearly given.

(There are immediately available, equivalent and reliable estimator equations for quarry blasting and similar operations. In such cases site measurements would need to be taken, and the results of those measurements compared with limiting vibration measurements proposed in national and other codes and standards.)

In quite general terms, end-bearing (H-section) piles would be expected to generate more vibration since they are required to enter stronger, stiffer ground. Consequently less of the input energy per blow goes into advancing the pile and more enters the ground in the form of vibration. It is often stated that bottom driven piles are usually easier to put in, and less noise is generated. However, it is not always certain that less vibration is transmitted, as is often claimed. Smaller displacement sheet piles generally require less energy to drive them, and should generate less vibration. However they are often driven close to existing structures. Martin (1980) states that vertical mode vibration is much greater than in the other component directions, but the evidence from the substantial body of data accumulated by researchers at Durham University indicates that horizontal radial mode

vibration is also significant but that transverse mode vibration is much less so except when measurements of vibration are made along the projected line of sheet piling when whip of the pile on entry can cause transverse motion. It is when structural response needs to be analysed in detail, and also when the effects of vibration on certain highly sensitive advanced manufacturing operations need to be assessed, that the directional components of vibration need to be taken fully into account.

With respect to the type of ground in which piling takes place, and as a general observation, piling into loose granular soils or fills can cause consequential settlement, while piling into dense soils may cause heave. The comments in Table SI 20.5 relate to the different types of hammer.

GROUND VIBRATION FROM BLASTING

Equation 5 above is most generally used not only in construction practice but also for assessing the effects of vibration from blasting operations (quarrying and civil engineering) upon neighbouring buildings. It is interesting to note the form of the equivalent Indian Standard 6922 (Indian Standards Institution, 1973):

$$v = K_1 (W^{2/3}/r)^{1.25} \quad\quad\quad\quad (13)$$

where, as before,
v is the ground particle velocity in mm/s,
K_1 is a constant, 880 for soils, weathered rock or soft rock, and 1400 for hard rock,
W is the explosive charge per delay (kg), and
r is the distance (m) from the blast point.

Table SI 20.5 Performance of different types of piling hammer

Hammer type	Comment
Diesel	Least controllable; energy available to the pile depends on the soil stiffness
Air	Generally pre-defined work rate; smaller hammers generate relatively low vibrations
Drop	Controlled by operator in terms of drop height, allowing the levels of vibration to be reduced if necessary
Hydraulic drop	Variable hammer weight and controllable drop height
Vibrodriver	Effective in essentially granular soils (SERF pilemaster is appropriate for essentially cohesive soils); not usually noted for excessive vibration, but the vibrations are continuous and periodic; vibrations attenuate rapidly with distance; an appropriate technique in suitable ground when buildings are 15-20m away from the driver.

Table SI 20.6   Some values of wave velocity c for different types of ground

Ground	Ground velocity, c, m/s	Range of c, m/s
Soils	1500	200 to 1800
Weathered and soft rocks	2500	1800 to 3200
Hard rock	5000	3200 to 7500

In contrast to the comment accompanying equation 5 it will be noted that in the case of this Indian Standard the site constants are defined at the outset (K = 880 or 1400; n = 1.25), although there is a note on monitoring and the use of 'pilot tests' 'to determine the maximum charge W that can be used in blasting operations to keep the ground particle velocity at the site of structures within the safe values noted below and for determining the accelerations for design'.

It is also noted that 'if the delay time' is equal to or exceeds $r/4c$, where c is the longitudinal wave velocity (celerity) in the ground in metres per second, then 'the ground motions are governed by the total charge weight in a single delay'. This statement is unclear since although it (correctly) implies that for long delays the weight of explosive per delay should be entered into equation 13 it tends to create some doubt as to the definition of W given in this equation. Ground velocity c is defined in Table SI 20.6.

For large charges of more than 100kg/delay, where the safety criteria noted below are violated and it is desired to design structures against the effects of blasts, the following equation is given for deducing the design acceleration in the horizontal direction:

$$(a/g) = K_2 W^{0.83} R^{-2} \quad ....(14)$$

where

a is the design acceleration in cm/s^2,
g is the acceleration due to gravity in cm/s^2,

K_2 is a constant (which may be taken as 4 for soil, weathered and soft rock and 6 for hard rock),
W is the charge per delay in kilograms, and
r is the distance of the structure from the blast point in metres.

STATISTICAL APPRAISAL OF VIBRATION DATA

In the overwhelming majority of cases, vibration estimation by one or other of the above empirical equations, or one or two site measurements, will suffice for comparison with recommended limits. There may be occasions, however, when a positive deduction may need to be drawn as to the probability (the risk) of a particular measurement being exceeded, assuming that all the other conditions (piling energy, explosive weight, type of ground, pile penetration, distance, and so on) remain the same. The basis of this analysis is that the vibration velocity used in a scaled distance predictive law is a mean value, and that it may be assumed, for any scaled distance, that the constitutive measurements which compile the empirical equation are distributed normally about the mean. Thus the vibration velocity distribution as a function of $W^{1/2}/r$, or $r/W^{1/2}$, may be fully described by a mean and a standard deviation.

If Z is the value of a standard normal variable, calculation of this value gives the cumulative probability up to Z from Table SI 20.7.

Table SI 20.7 Values of Z, the standard normal variable showing the cumulative probability up to Z

Z	0.00	0.01	0.02	0.03	0.04	0.05	0.06	0.07	0.08	0.09
0.0	.5000	.5040	.5080	.5120	.5160	.5199	.5239	.5279	.5319	.5359
0.1	.5398	.5438	.5478	.5517	.5557	.5596	.5636	.5675	.5714	.5753
0.2	.5793	.5832	.5871	.5910	.5948	.5987	.6026	.6064	.6103	.6141
0.3	.6179	.6217	.6255	.6293	.6331	.6368	.6406	.6443	.6480	.6517
0.4	.6554	.6591	.6628	.6664	.6700	.6736	.6772	.6808	.6844	.6879
0.5	.6915	.6950	.6985	.7019	.7054	.7088	.7123	.7157	.7190	.7224
0.6	.7257	.7291	.7324	.7357	.7389	.7422	.7454	.7486	.7517	.7549
0.7	.7580	.7611	.7642	.7673	.7704.	.7734	.7764	.7794	.7823	.7852
0.8	.7881	.7910	.7939	.7967	.7995	.8023	.8051	.8078	.8106	.8133
0.9	.8159	.8186	.8212	.8238	.8264	.8289	.8315	.8340	.8365	.8621
1.0	.8413	.8438	.8461	.8485	.8508	.8531	.8554	.8577	.8599	.8621
1.1	.8643	.8665	.8686	.8708	.8729	.8749	.8770	.8790	.8810	.8830
1.2	.8849	.8869	.8889	.8907	.8925	.8944	.8962	.8980	.8997	.9015
1.3	.9032	.9049	.9066	.9082	.9099	.9115	.9131	.9147	.9162	.9177
1.4	.9192	.9207	.9222	.9236	.9251	.9265	.9279	.9392	.9305	.9319
1.5	.9332	.9345	.9357	.9370	.9382	.9394	.9406	.9418	.9429	.9441
1.6	.9452	.9463	.9474	.9484	.9495	.9505	.9515	.9525	.9535	.9545
1.7	.9554	.9564	.9573	.9582	.9591	.9599	.9608	.9616	.9625	.9633
1.8	.9641	.9649	.9656	.9664	.9671	.9678	.9686	.9693	.9699	.9706
1.9	.9713	.9719	.9726	.9732	.9738	.9744	.9750	.9756	.9761	.9767
2.0	.9772	.9778	.9783	.9788	9793	.9798	.9803	.9808	.9812	.9817
2.1	9821	.9886	.9830	.9834	.9838	.9842	.9846	.9850	.9854	.9857
2.2	.9861	.9864	.9868	.9871	.9875	.9878	.9881	.9884	.9887	.9890
2.3	.9893	.9896	.9898	.9901	.9904	.9906	.9909	.9911	.9913	.9916
2.4	.9918	.9920	.9922	.9925	.9927	.9929	.9931	.9932	.9934	.9936
2.5	.9938	.9940	.9941	.9943	.9945	.9946	.9948	.9949	.9951	.9952
2.6	.9953	.9955	.9956	.9957	.9959	.9960	.9961	.9962	.9963	.9964
2.7	.9965	.9966	.9967	.9968	.9969	.9970	.9971	.9972	.9973	.9974
2.8	.9974	.9975	.9976	.9977	.9977	.9978	.9979	.9979	.9980	.9981
3.0	.9987	.9987	.9987	.9988	.9988	.9989	.9989	.9989	.9990	.9990
3.1	.9990	.9991	.9991	.9991	.9992	.9992	.9992	.9992	.9993	.9993
3.2	.9993	.9993	.9994	.9994	.9994	.9994	.9994	9995	.9995	.9995
3.3	.9995	.9995	.9995	.9996	.9996	.9996	.9996	.9996	.9996	.9997
3.4	.9997	.9997	.9997	.9997	.9997	.9997	.9997	.9997	.9997	.9998
3.5	.9998	.9998	.9998	.9998	.9998	.9998	.9998	.9998	.9998	.9998
3.6	.9998	.9998	.9999	.9999	.9999	.9999	.9999	.9999	.9999	.9999
3.7	.9999	.9999	.9999	.9999	.9999	.9999	.9999	.9999	.9999	.9999
3.8	.9999	.9999	.9999	.9999	.9999	.9999	.9999	.9999	.9999	.9999
3.9	1.0000									

The equation for Z is

$$Z = S/(M \times 10^{sd}) \dots\dots\dots (15)$$

where

S is the specified limiting vibration velocity value,

M is the median vibration velocity value,

sd is the standard deviation of the (assumed) normal distribution of data points, and

Z is the percentage probability of being

under the specified value. By definition, the percentage probability of being over the specified value is 100 - Z.

Equation 15 applies if the measured (median) value M is less than the specified value S. If M is greater than S then the following equation applies:

$$Z = M/(S \times 10^{sd}) \quad ....(16)$$

The percentage value obtained from this equation defines the probability of exceeding the specified value.

PARTICULAR SITE CONDITIONS WHERE PILING CAN CAUSE PROBLEMS

A check list of potentially sensitive site conditions can be useful to the project engineer.

(a) In all cases of old buildings, and particularly those having historical and architectural value, damage to which would arouse sensitivities and involve exceptional expense for their repair.
(b) In all cases, including those above, where there has been obvious movement, damage in general and cracking in particular.
(c) Buildings have old lath and plaster ceilings in poor condition.
(d) Roofed buildings where tiles have slipped.
(e) Buildings founded on a stiff surface crust above alluvial clays and silts.
(f) Near buildings, such as hospitals, old people's homes, night shift workers' homes, and solicitors' offices, which house 'sensitive' occupants.
(g) Near buildings with glass roofs and walls constructed with a high glass content. Some swimming pools may enter into this category.
(h) Near buildings housing sensitive equipment such as computers (*see* Boyle, 1990), precision grinding and milling machines, sensitive hospital equipment, photographic processing equipment, wafer stepper equipment for silicon chip manufacture, optical and electron microscopes, and continuous weighing machines. Small free-standing articles in houses, chandeliers, and even furniture can vibrate to the annoyance of inhabitants. Upper storey occupants and

household items will tend to experience amplified vibrations.
(i) When piles are being driven through a firm transmitting stratum on which neighbouring structures are founded. There can be wave guide effects, causing higher than expected levels of vibration to be transmitted for greater distances, when a band of stiff soil or rock is bounded by material having different acoustic impedance (ρc) characteristics.
(j) When piling through loose granular soils, particularly loose sands on which the neighbouring structures are founded. Repeated vibration can cause dynamic compaction settlement.
(k) When piling below the water table into sands that are prone to liquefaction.
(l) When piling through obstructions, of which boulders are perhaps the most common.
(m) Vibrodriving of piles and casings generally.
(n) When there is physical contact between the piling and the adjacent structure.

STRUCTURAL DAMAGE CRITERIA

Vibration measurements taken on the ground outside a building are not readily transferred to an assessment of structural vibration. There may be a general characteristic frequency with which a building as a body may resonate, but the interior elements of a building comprise a range of lengths and stiffnesses, each having a frequency in the incoming waveform with which it can resonate and thereby extract dynamic energy. Buildings may respond in a partial rigid mode when a characteristic dimension is small with respect to the wavelength of the incident vibration. There is a linear relation between peak radial vibration incident on a wall and the maximum transient strain in that wall. However, there is a dependence of structural strain upon structural length, larger structures being expected to show larger strains in response to a given incident vibration.

It must be recognised that vibration amplitude alone is not a direct determinant of building damage. A relatively small vibrational strain may trigger damage in buildings where structural elements are already at a critical level of pre-stress.

As noted below, there is no current British Standard which deals exclusively with ground and structural vibration, although this situation should be remedied in the near future. The ICE Conditions of Contract, 6th Edition (1991) Clause 29(2) ('All work shall be carried out without unreasonable noise or disturbance or other pollution') does not really state anything that is of practical help, and the following two sub-clauses do leave open to speculation just what is 'the unavoidable consequence of constructing and completing the Works' in the context of piling operations. Also, the DTp Specification for Road and Bridgeworks (1976) makes no mention of vibration. According to the ICE Model Procedures and Specifications for Piling (1978) tender documents are to include any conditions concerning limitations on noise and Clause 11.1 requires the contractor to minimise noise and disturbance. Under Section 61 of the Control of Pollution Act (1974) a consent to work agreement may be obtained by the project promoter (client) from the local authority in advance of a tender enquiry, a note to this effect and the local authority requirements being included in the tender (contract) documents so that the tenderers do not base their tenders on the use of unacceptable methods of work and plant. Alternatively, the contractor can be required by the project promoter to obtain the consent to work agreement before commencing work, a typical statement in the contract documentation being 'The contractor shall apply for all necessary permissions'. Local authorities are themselves able to set perceived tolerable vibration levels, but they do this from a background of imprecision because the case study evidence is limited and such data as is available is often of rather dubious quality. Limiting values for vibration will often be proposed by or on behalf of the promoter and accepted by the environmental health inspector when the consent to work agreement is being formulated. Proposal of or acceptance by a local authority of such limit values then places an obligation on that authority should vibration-related problems subsequently occur.

Some of the standards that have been applied by various national bodies to vibration are noted below. Reference may also be made to BS 5228: Part 4 (British Standards Institution, 1992a) for further information on consent to work agreements.

German Code

German Standard DIN 4150 Part 3 (1975) offers the reference values in Table SI 20.8 for the maximum resultant peak particle velocity in a building foundation. These values may be used for the assessment of the effects of short-term vibrations up to the level of which damage in the sense of a reduction in serviceability should not be expected.

Table SI 20.8 Reference values for peak particle velocity in a building foundation (after DIN 4150 1975)

Type of building	Reference value maximum v_{res} (mm/s)
Well reinforced buildings* with heavy structural members and well-reinforced skeleton structures in a state of conservation corresponding to generally accepted rules in structural engineering	30
Residential, commercial and other buildings of similar construction in a state of conservation corresponding to the generally accepted rules of structural engineering	8
Buildings not corresponding to those above and buildings protected as monuments	4

Note:
* The code states that for these buildings, particular attention should be paid to the real resistance to vibration loadings.

Table SI 20.9 Reference values for vibration in a foundation (after DIN 4150 1986)

Type of structure	Vibration velocity v_i (mm/s)			
	Foundation			Plane of floor of uppermost full storey
	At a frequency of			Frequency mixture
	<10Hz	10-50Hz	50-100Hz*	
Buildings used for commercial purposes, industrial buildings	20	20-40	40-50	40
Dwellings and buildings of similar design and/or use	5	5-15	15-20	15
Structures that, because of their particular sensitivity to vibration, do not correspond to those listed above and are of great intrinsic value (eg. buildings that are under a preservation order)	3	3- 8	8-10	8

* For frequencies above 100Hz, at least the values specified in this column shall be applied.

With respect to Table SI 20.8 there are several factors to consider.

1) The above reference values take no account of irregular settlements of the subsoil.
2) If maximum v_{res} is less than or equal to 2mm/s measured on a foundation, then no damage (such as cracking of plaster and glass) should be expected.
3) The above reference values refer to a few cycles of vibration per day over a frequency range of a few hertz (Hz) to 60Hz. Repeated pile vibrations would not satisfy this criterion. If the cycles are more frequent, then the above figures should be reduced by two-thirds. For frequencies greater than 60Hz, the code suggests using higher reference values.
4) If significant ceiling vibration in a building is noted, then the maximum permissible vertical vibration velocity v_v should be less than or equal to 20mm/s.

The code recommends actually measuring on the outer foundation or on external walls as close as possible to the ground on the side facing the source of vibration.

German Standard DIN 4150 Part 3 1986 changes some of the maximum resultant vibration velocity figures in the 1975 version of the code and takes account more explicitly of frequency. Most importantly, it abandons the maximum resultant vibration velocity as a criterion and recommends using the absolute maximum value of vibration in the x (horizontal), y (horizontal) and z (vertical) directions. This particular maximum vibration velocity value is designated v_i, and is shown in Table SI 20.9.

There are several further points to note.

1) The above guideline values relate to the foundation and in the plane of the floor of the uppermost full storey.
2) Fatigue effects are deemed not to be significant.
3) The vibration intervals create no resonance.
4) It does not necessarily follow that if the above values are exceeded damage will ensue. If the values are significantly exceeded, then further investigation is necessary.
5) If, in the case of short-term vibration, floors vibrate, then, where v does not exceed

20mm/s as measured in the z (vertical) direction at the point of maximum vibration velocity, which is usually at the centre of the floor, a reduction in the utility value of the floors is not to be expected.

6) Vibratory hammers and vibrators are not to be regarded as simply imposing short-term effects; resonance phenomena and symptoms of fatigue can occur in walls and floors of buildings.

Table SI 20.10 Swiss standard SN 640 312 1978 for ground vibration relating to different sources of disturbance

Structural type	Source of vibration	Frequency Hz	Guide values v_{max} resultant (mm/s)
I Reinforced concrete and steel construction (without plaster) such as industrial and commercial buildings; retaining walls; bridges; towers; above-ground pipelines. Underground structures such as caverns, galleries with or without concrete lining.	M	10-30 30-60	12 12-18 [a]
	S	10-60 60-90	30 30-40 [b]
II Buildings with foundation walls in masonry (brickwork, stonework) or concrete. Retaining walls of ashlar construction. Underground structures such as caverns, tunnels and galleries with masonry lining. Pipelines buried in soft ground (soil).	M	10-30 30-60	8 8-12 [a]
	S	10-60 60-90	18 18-25 [b]
III Buildings with foundations and basement floors of concrete construction, with wooden beam construction in upper floors; brickwork walls.	M	10-30 30-60	5 5-8 [a]
	S	10-60 60-90	12 12-18 [b]
IV Buildings which are especially sensitive or worthy of protection.	M	10-30 30-60	3 3- 5 [a]
	S	10-60 60-90	8 8-12 [b]

Notes:
M means machines, traffic, constructional equipment. S means blasting.
[a] Lower value applies at 30Hz and the upper one at 60Hz. Interpolate for intermediate frequencies.
[b] Lower value applies at 60Hz and upper value at 90Hz. Interpolate for intermediate frequencies.

7) The three axis vibration transducers shall be placed close to one another in the lowest storey of the building, either on the foundation of the outer wall, or in the recesses of the outer wall. For buildings having no basement the point of measurement shall lie not more than 0.5m above ground level and be preferably located on the side of the building facing the source of vibration. One of the directions of measurement shall be parallel to one of the side walls of the building.

8) A z-axis measurement shall be taken at the centre of the floors of a building.

9) Horizontal vibrations of up to 5mm/s, measured at an upper storey, should cause no damage in terms of a reduction in serviceability in those 'buildings used for commercial purposes, industrial buildings, and buildings of similar design' and in 'dwellings and buildings of similar design and/or use'. If the quoted vibration velocity figures are exceeded it does not follow that damage is inevitable.

International Organisation for Standardisation

ISO Draft Standard ISO/DIS4866, covering the measurement and evaluation of structural vibration, has been prepared.

Swiss Standard

SN 640 312:1978 gives the relations in Table SI 20.10 between different types of structures and guide values of vibration velocity. The figures relate to machine, traffic and constructional vibration (M) as well as to blasting (S).

Table SI 20.11 Swedish recommendation for limiting vibration velocity from blasting

Object	Limiting vibration (peak)		
	Displacement mm	Velocity mm/s	Acceleration m/s^2
Concrete bunker, steel reinforced		200	
High rise apartment block. Modern concrete or steel frame design	0.4	100	
Underground rock cavern roof, hard rock, span 15-18m		70-100	
Normal block of flats. Brick or equivalent walls	70		
Light, concrete building		35	
Swedish National Museum Building structure Sensitive exhibits		25	5
Computer centre Computer supports	0.1		2.5
Circuit break Control room			0.5-2

Swedish recommendations

The typical vibration limits in Table SI 20.11 are enforced for short-duration construction blasting when a building foundation is on hard rock.

Australian and Austrian Standards

Permissible peak particle velocity values of 25mm/s and 15mm/s, independent of frequency, may be taken for Australia and Austria, respectively.

Indian Standard

The structural damage threshold from underground blasting, calculated from equation 11, is given by the Indian Standards Institution (1973) as not to exceed the following safe ground particle velocities in:

soils, weathered or soft rock ...   50mm/s
hard rock ....   70mm/s.

It is also stated that 'where monitoring of ground particle velocity by means of suitable instruments is adopted as a means of vibration control, the peak ground particle velocity may not exceed the following:

soils, weathered or soft rock ...   70mm/s;
hard rock ....   100mm/s'.

It is noted in the Standard that the safe ground velocity values given above 'are lower than those

which may be intolerable to human beings' and that the 'values are appropriate for masonry and will be conservative for concrete of M 150 quality'.

'Soft rocks' are defined as 'shale, sandstone phyllites, laminated slates, mica schist, weathered hard rocks and other soft rock material'. 'Hard rocks' are defined as 'granite, basalt, quartzite, marble, crystalline schists, massive slates and other hard massive crystalline rocks'. 'Threshold damage' is defined as 'formation of new plaster cracks, widening of old cracks'. There is no definition of ground particle velocity, but it assumed that it is zero-to-peak and that the equations refer to the individual orthogonal motions rather than resultants. There is reference in the Standard to Langefors and Kihlström (1963) in the context of delayed action blasting. There is also a note to the effect that the frequency response for velocity measurement (the frequency response of the geophone) should be flat above 10Hz and in the case of acceleration measurement (the response of the accelerometer) should be flat in the range zero to 100Hz.

Australian Standard

Recommendations with respect to permissible ground vibration near to buildings, when blasting under water, and with respect to air blast overpressure are contained in the publication AS 2187, Part 2 1983 (Standards Association of Australia, 1983). These permissible levels of vibration are shown in Table SI 20.12.

Table SI 20.12 Australian recommendations for peak particle velocity related to structures (after Standards Association of Australia, 1983)

Type of building or structure	Recommended maximum peak particle velocity of vibration (resultant) (mm/s)
1. Historic buildings and monuments, and buildings of special value or significance	2
2. Houses and low-rise residential buildings; commercial buildings not included in item 3 below	10
3. Commercial and industrial buildings or structures of reinforced concrete or steel construction	25

Table SI 20.13 Vibration effects on different subjects: the parameters to measure and the ranges of sensitivity of apparatus to use, as in BS 5228 Part 4 (British Standards Institution, 1992a).

Examples	Subject area	Measurement parameter and ranges of sensitivity		
		Displacement mm	Velocity mm/s	Acceleration g
Laboratory facilities	Equipment and processes	$(0.25$ to $1)$ x 10^{-3} [0.1Hz to 30Hz]		$(0.1$ to $5)$ x 10^{-3} [30-200]Hz
Microelectronics facilities	Eqipment and processes		$(6$ to $400)$ x 10^{-3} [3Hz to 100Hz]	$(0.5$ to $8)$ x 10^{-3} [5Hz to 200Hz]
Precision machine tools	Equipment and processes	$(0.1$ to $1)$ x 10^{-3}		
Computers	Equipment and processes	$(35$ to $250)$ x 10^{-3}		0.1 to 0.25 r.m.s [up to 300Hz]
Microprocessors	Equipment and processes			0.1 to 1
Hospitals and dwellings	People		0.15 to 15 (vertical) [8Hz to 80Hz] 0.4 to 40 (horizontal) [2Hz to 80Hz)	0.5 to 50 (vertical r.m.s) [4Hz to 8Hz]
Offices	People		0.5 to 20 (vertical) [8Hz to 80Hz] 1 to 52 (horizontal) [2Hz to 80Hz]	$(1$ to $100)$ x 10^{-3} (vertical r.m.s) [4Hz to 8Hz]
Workshops	People		1 to 20 (vertical) [8Hz to 80Hz] 3.2 to 52 (horizontal) [2Hz to 80Hz)	$(4$ to $650)$ x 10^{-3} (vertical r.m.s) [4Hz to 8Hz]
Residential or commercial	Buildings		1 to 50	
Gas or water mains	Underground services	$(10$ to $400)$ x 10^{-3}	1 to 50	

British Standard

It is re-stated that there is currently no British Standard which deals *exclusively* with ground and structural vibration. The Control of Pollution Act 1974 and, in Northern Ireland, the Pollution Control and Local Government (Northern Ireland) Order 1978 SI 1049, which defines 'noise' as including 'vibration' (Section 73(1) of the 1974 Act and Article 53(1) of the 1978

Order), contain provisions for the abatement of nuisances caused by noise and vibration. British Standard BS 5228:1992 (British Standards Institution, 1992a), which has superseded BS 5228:1975, comprises four parts which deal with noise. Part 4: 'Noise control on construction and open sites' was withdrawn and replaced by a new Part 4: 'Code of practice for noise and vibration control applicable to piling operations'. This latter Part of the Standard was originally in draft form for public comment. Is is noted that the 1992 Part 4 Standard addresses civil engineering driven piling operations, but not vibration caused by blasting. The vibration limits proposed in the new British Standard draft for driven piling are given in Table SI 20.13.

From a consensus of all the national codes, and in the context of pile vibrations, it may be concluded that piling should not be conducted close enough to a building to induce peak particle velocities greater than 10mm/s if minor damage is to be avoided. The equivalent distance is between about 5m and 10m from the building. If structural strains are to be monitored, then, for guidance, it may be noted that brickwork has a tolerance to dynamic strain in excess of 100 microstrains. A special series of tests conducted on brickwork by Durham University showed that a maximum $2.3\mu\varepsilon/mm/s$ were induced by drophammer and that the least vibration of $0.43\mu\varepsilon/mm/s$ was induced by high frequency vibrodriver.

The following points need to be noted with respect to Table SI 20.12:

1. This table does not cover high-rise buildings, buildings with long-span floors, specialist structures such as reservoirs, dams and hospitals and buildings housing scientific equipment sensitive to vibration. These require special considerations which may necessitate the taking of additional measurements on the structure itself, to detect magnification of ground vibrations which might occur within the structure. Particular attention should be given to the response of suspended floors.
2. In a specific instance, where substantiated by careful investigation, a value of peak particle velocity other than that recommended in the table may be used.
3. The peak particle velocities in the table have been selected taking no consideration of human discomfort and the effect on sensitive

equipment within the building. In particular, the limits recommended for buildings Types 2 and 3 may cause complaints.

Advanced manufacturing facilities

As shown above, some advanced technological manufacturing operations, for example micro-electronics manufacture, demand the setting of very low limits for a virtually vibration-free environment. Semiconductor wafer steppers (wafer fabs) need to be isolated from general movement of the production plant operation and for any internal vibration, as from fans and so on, to be designed out. The frequency range of concern is about 5Hz to 100Hz, and one manufacturer quotes the upper limit specification in Table SI 20.14 for vertical vibration.

There are certain points concerning Table SI 20.13 that need to be considered:

1. Except where root mean square (r.m.s.) accelerations are quoted, all measurement ranges, whether displacement, velocity or acceleration, are in terms of zero-to-peak.
2. Typical ranges for equipment and processes vary considerably depending on the sensitivity of the equipment installed.
3. The above ranges depend upon the dominant frequency of vibration.
4. g is the acceleration due to gravity, i.e. $9.81m/s^2$.
5. The ranges noted above relate in the main to transient vibrations. But the frequencies (typically 25Hz to 40Hz) of vibratory pile drivers are usually much higher than the fundamental frequencies of buildings. Particular problems could arise during the running up and running down phases of the vibrodriver when the frequencies tend to approximate to the natural frequencies of building floor slabs and of loose or medium-loose soil upon which the building is constructed.

Table SI 20.14. Frequency and acceleration limits for semiconductor manufacture

frequency < 4Hz	acceleration < $0.4cm/s^2$
4Hz < frequency < 25Hz	log(acceleration) < 1.63 log(frequency) - 1.38
25Hz < frequency < 250Hz	acceleration < $8cm/s^2$

Table SI 20.15 Ground vibration criteria for sensitive manufacturing operations

Frequency	Vertical		Horizontal	
	Acceleration (cm/s²)	Velocity (μm/s)	Acceleration (cm/s²)	Velocity (μm/s)
0.5	0.62	1973	0.19	605
1	0.38	605	0.065	103
2	0.066	52.5	0.065	51.7
3	0.10	53	0.10	53
5	0.047	15	0.115	36.6
10	0.23	36.6	0.20	31.8
20	0.61	48.5	0.53	42.2
30	2.5	132.6	1.10	58.4
40	3.5	139.3	1.70	67.6
50	4.2	133.7	2.05	65.2
100		6.5	103.4	3.00

These acceleration values are listed as zero-to-peak. Frequencies over 250Hz have a negligible effect on the wafer stepper.

If the vibration is assumed to take the form of a simple sine wave, the above acceleration limits can then be transformed into equivalent vibration velocity limits. It is found, for example, that at 1Hz the velocity is 636μm/s, at 10Hz it is 283μm/s, at 20Hz it is 438μm/s, and at 30Hz it is 424μm/s. Thereafter the velocity declines: 100Hz: 127μm/s; 200Hz: 63μm/s; 250Hz: 51μm/s. These tolerable levels are generally higher than those suggested in the draft British Standard (Table SI 20.13 above) in which the quoted lower figure of 6μm/s would be very difficult to achieve in any working environment.

Another manufacturer shows more complex graphical criteria for both vertical and horizontal criteria. These criteria are expressed for some stepped frequencies in Table SI 20.15.

Several conclusions can be drawn from a comparison of these horizontal and vertical allowable levels of zero to peak vibration, and from these general levels in comparison with those quoted above. First, the allowable horizontal vibration velocity is significantly less than the allowable vertical vibration velocity at frequencies below 1Hz and above about 30Hz. Between these frequencies the allowable velocities are similar except that in the vicinity of about 5Hz the allowable horizontal velocity exceeds the vertical velocity. More notably, with the exception of the vertical velocities at

frequencies less than about 0.5Hz these levels up to about 100Hz are less than those quoted in the earlier example.

A further specification for microelectronics fabrication, this time by a design office, provides similar allowable vibration limits to those in the last example. Curves by Bolt, Beranek and Newman Inc (Gordon and Wilby, 1986) show an upper vibration limit of 25.4μm/s r.m.s. (36.3μm/s zero-peak) at 4Hz decreasing linearly to 20.32μm/s r.m.s. (29.0μm/s zero-peak) at 8Hz, maintaining this permissible velocity limit to 126Hz. These figures are below the threshold of human perception which is about 32μm/s r.m.s. (45.7μm/s zero-peak).

Since such manufacturing processes tend to continue round the clock, construction vibration above a specified level would either be prohibited, or loss of production yield (through wastage) would have to be tolerated (with compensation), or production would have to cease (again, presumably, with compensation payable by the contractor and as a charge to the client).

NEW BUILDING FOUNDATION DESIGN

In a ground vibration environment such design requires a knowledge of the dynamic properties of the soil on which the structural foundation will stand. In particular, the dynamic modulus of elasticity is required. One method of investigation uses the Indian Standard 5249

(1977). A pit is excavated and a concrete block cast at the bottom. The block and the soil together then form a mass-spring system having certain resonant frequencies which can be excited by the application of a suitable force. The numerical value of the frequency of such resonances is controlled by the block geometric and inertial properties and also by the soil properties. Since the resonance frequencies can be measured on site, and since the block properties are fixed and known, the soil properties can be found by calculation.

The block is excited by a sweeping unbalance shaker powered by an electric motor. The amount of imbalance may be varied to change the applied force - typically the unbalance could be set at about 5.7 kilogram millimetres, but must not exceed the ISO:5249(1977) limit. There are in fact resonant frequencies possible for the block-spring (soil) system, and they are identified by noting the changes in peak amplitude response as the shaker sweeps through a speed range. These resonances are (1) vertical z axis (pure translational); (2) torsional about z-z axis (pivotal point at the system centre of gravity); (3) upper rocking mode about x-x axis (combined translation in y and rotation in x-y plane, with pivotal point below system centre of gravity); (4) lower rocking mode about x-x axis (combined translation in y and rotation in x-y plane, with pivotal point above centre of gravity); (5) upper rocking about y-y axis (combined translation in x and rotation in z-x plane, with pivotal point below centre of gravity); (6) lower rocking about y-y axis (combined translation in x and rotation in z-x plane, with pivotal point above centre of gravity). For further information on application and analysis, reference should be made to the Indian Standard.

CONTROL OF POLLUTION ACT 1974

The Environmental Protection Act 1990 has re-enacted the provisions of the Control of Pollution Act 1974 except in respect of noise (and vibration). Under the Control of Pollution Act 1974, legal action can be taken under common law to restrict nuisance or to seek compensation for damage. Some of the more salient controls under this Act relevant to noise and vibration comprise:

Section 58 Summary proceedings by local authorities (Statutory Notice 58(1)
    Appeal to Magistrates' Court 58(3)
    High Court injunction (58(8)
Section 59 Summary proceedings by occupier of premises
    Magistrates' Court may act on a complaint 59(1).
Section 60 Control on noise (and vibration) on construction sites
    Types of works of which 60(1)(b) and (c) would create vibration
    Local authority can serve notice 60(2).
    Details of steps to be taken to minimise 'noise' 60(3)(b)
Section 61 Prior consent for work on a construction site
    Application by client 61(3)
    Need for sufficient information 61(4)
    Limitation and qualifications to agreement 61(5).

Under English law (applicable also to Wales), action takes the form of seeking an injunction to restrain a defendant from continuing a nuisance. Damages are also sometimes also sought. If proceedings are brought in a County Court, it is necessary for the plaintiff to claim damages as well as an injunction, otherwise the court has no jurisdiction. In Scotland, the corresponding remedy is an action for interdict and damages - actions to be brought in the Sheriff Court or in the Court of Session. It is not necessary in either case to claim damages as well as interdict.

Clients and contractors need also to be aware of the distinctions between two legal types of nuisance: public and private. A *public nuisance* is so widespread as to render it not reasonable for one person to take proceedings on his/her own responsibility to stop it. The community as a body must take on the responsibility. Such a nuisance can be a crime, a misdemeanour at common law and as such the subject of an indictment. Criminal prosecutions may be brought by the Attorney General or, for example, the Environmental Health Department of a local authority. Civil proceedings may be brought by the Attorney General alone, or at the instigation of the Environmental Health Department of the local authority or a private individual. This route applies to an individual only when the individual can show in civil proceedings that the nuisance is the cause of special damage over and above that sustained by the public at large.

An action for *private nuisance* is not necessarily based on compliance with conditions restricting noise and/or vibration detailed in a planning permission, or any provision under the Control of Pollution Act 1974. Further, it is not necessarily a sufficient defence to demonstrate that the best practical means have been taken to reduce or prevent noise or vibration, nor that the cause of the nuisance is the exercise of a business or trade in a reasonable or proper manner. The principal remedies are the seeking of an injunction or damages. An injunction will not generally be given where damages would be an adequate remedy.

HUMAN PERCEPTION CRITERIA

Some of the vibration limits outlined in Table SI 20.13 relate to the effect of vibration upon people. Very low levels of vibration (around $32\mu m/s$ r.m.s. velocity, equivalent to $45.7\mu m/s$ zero-peak velocity) - much lower than those capable of causing even minor architectural damage to property - can be felt by humans, generally in the frequency range 1-80Hz, and there is a psychological amplification of vibration when it is experienced in the home. Several authorities have formulated criteria (the Reiher-Meister scale (1931) being particularly well-known), and these have been reviewed by Broadhurst *et al.* (1989). As a practical example of the application of such a criterion, the Reiher-Meister scale sets perception and annoyance levels at vibration velocities of 0.3mm/s and 3mm/s, respectively. (This perception level is higher than that quoted above.) Such an annoyance level would typically occur at a distance of about 15m from a driven pile. For quick reference, and in terms of frequency-weighted root mean square (r.m.s.) *acceleration*, the following general criteria can be used:

0.315 to 0.63 m/s^2: 'a little uncomfortable'
>2 m/s^2: 'extremely uncomfortable'
around 10 m/s^2: 'damage to health'.

Perhaps the best known of the human environmental standards is the German Standard DIN 4150:1975 (Provisional) upon which the

Building Research Establishment (1983) based its recommendations.

German Standard

DIN 4150 uses a K or KB perception parameter as the guide value. It can be expressed in terms of either vibration displacement, velocity or acceleration. The K equations in the 1975 Standard differ from those in the earlier 1970 Standard which itself superseded the Dieckmann (1958) recommendations but did not differ greatly from them. KB (DIN 1975) is given as:

Particle acceleration:
$$KB = a\alpha \left\{ [1 + (f^2/f_0)]^{\frac{1}{2}} \right\} \quad \text{......................(17)}$$

Particle velocity:
$$KB = v\beta f \left\{ [1 + (f^2/f_0)]^{\frac{1}{2}} \right\} \quad \text{....................(18)}$$

Particle displacement:
$$KB = A\gamma f^2 \left\{ [1 + (f^2/f_0)]^{\frac{1}{2}} \right\} \quad \text{................(19)}$$

where
a = particle acceleration (m/s^2)
v = particle velocity (mm/s)
A = particle displacement (mm)
f = frequency (Hz)
f_0 = 5.6 Hz (reference frequency)
α, β, γ are constants which for peak values of a, v and A are:
α = 20.2m^{-1}s^{-2};
β = 0.13mm^{-1}s^{-2};
γ = 0.80mm^{-1}s^{-2}.

KB values, used above, are defined as shown in Table SI 20 16.

The following notes indicate how the KB values in Table SI 20.16 should be used in different environmental settings:

(a) Continuous vibrations and repeated vibrations with interruptions are those which occur continuously for longer than 2 hours.

(b) With vibrations, such as from pile driving, that take place only over a few days and during daylight hours, guide values up to twice those in column 4 of Table SI 20.16 may be used provided that damage limits are not exceeded.

(c) The guide values in the brackets should not be exceeded when vibrations are equal to or greater than 5Hz.

Table SI 20.16 *KB* values related to the environmental condition and to the nature of the vibrations

Construction area	Time	KB Guide Values	
	By day 0600 to 2200hr By night 2200 to 0600 hr	Continuous vibrations and those occurring repeatedly with interruptions	Rarely occurring vibrations
1. Residential weekend house, etc.	Day Night	0.2(0.15)	4.0
2. Urban residential and commercial	Day Night	0.3(0.20) 0.2	8.0 0.2
3. Commercial, including offices	Day Night	0.4 0.3	12.0 0.3
4. Industrial	Day Night	0.6 0.4	12.0 0.4
5. Special area according to type of usage and proportion of housing	Day Night	0.1 to 0.6 0.1 to 0.4	4 to 12 0.15 to 0.4

Table SI 20.17a Tolerable vibration velocity in humans (British Standards Institution, 1992b)

Frequency (Hz)	Vibration velocity, zero-to-peak (mm/s)	
	x, y axes	z axis
1.00	0.804	2.250
1.25	0.643	1.610
1.60	0.502	1.110
2.00	0.402	0.796
2.50	0.402	0.569
3.15	0.402	0.402
4.00	0.402	0.281
5.00	0.402	0.225
6.30	0.402	0.179
8.00	0.402	0.141
10.00	0.402	0.141
12.50	0.402	0.141
16.00	0.402	0.141
20.00	0.402	0.141
25.00	0.402	0.141
31.50	0.402	0.141
40.00	0.402	0.141
50.00	0.402	0.141
63.00	0.402	0.141
80.00	0.402	0.141

Table SI 20.17b Tolerable vibration acceleration in humans (British Standards Institution, 1992b)

Frequency (Hz)	Vibration acceleration, m/s² r.m.s.	
	x, y axes	z axis
1.00	3.57×10^{-3}	1.00×10^{-2}
1.25	3.57×10^{-3}	8.94×10^{-3}
1.60	3.57×10^{-3}	7.91×10^{-3}
2.00	3.57×10^{-3}	7.07×10^{-3}
2.50	4.46×10^{-3}	6.32×10^{-3}
3.15	5.63×10^{-3}	5.63×10^{-3}
4.00	7.14×10^{-3}	5.00×10^{-3}
5.00	8.93×10^{-3}	5.00×10^{-3}
6.30	1.13×10^{-2}	5.00×10^{-3}
8.00	1.43×10^{-2}	5.00×10^{-3}
10.00	1.79×10^{-2}	6.25×10^{-3}
12.50	2.23×10^{-2}	7.81×10^{-3}
16.00	2.86×10^{-2}	1.00×10^{-2}
20.00	3.57×10^{-2}	1.25×10^{-2}
25.00	4.46×10^{-2}	1.56×10^{-2}
31.50	5.63×10^{-2}	1.97×10^{-2}
40.00	7.14×10^{-2}	2.50×10^{-2}
50.00	8.93×10^{-2}	3.13×10^{-2}
63.00	1.13×10^{-1}	3.94×10^{-2}
80.00	1.43×10^{-1}	5.00×10^{-2}

British Standard

British Standard 6472 (British Standards Institution, 1992b) provides criteria for tolerable human exposure to vibration in the frequency range 1Hz to 80Hz. Criteria are expressed in the form of acceleration-frequency and velocity-frequency curves for three axes of posture - (x): back to chest (horizontal); (y): right side to left side (horizontal); (z): foot to head (vertical). There are base curves, also expressed in tabular data, and to which may be added multiplying factors related to particular locations and to whether the vibration occurs during a daytime period or at night. The base curve zero-to-peak vibration velocity and acceleration data are as shown in Tables SI 20.17a and SI 20.17b, respectively.

The multiplying factors that are applied to the vibration velocities in Tables SI 20.17a and SI 20.17b are given in Table SI 20.18, and those factors for continuous daytime vibration, given in the third column of Table SI 20.18, apply to a 16

hour exposure period. The estimated vibration dose value (eVDV) corresponding to a unity multiplying factor is then approximately $0.1 \text{m/s}^{1.75}$, that is:

$$eVDV = 1.4 \times a(\text{r.m.s.}) \times t^{¼}$$
$$= 1.4 \times 0.005 \times (16 \times 60 \times 60)^{¼} \quad ....(20)$$

The multiplying factors for night time apply to a 8 hour exposure period. The estimated vibration dose value corresponding to a unity multiplying factor is then approximately $0.091 \text{m/s}^{1.75}$, that is:

$$eVDV = 1.4 \times a(\text{r.m.s.}) \times t^{¼}$$
$$= 1.4 \times 0.005 \times (8 \times 60 \times 60)^{¼} \quad ....(21)$$

The r.m.s. acceleration corresponding to the vibration dose values varies according to the duration of exposure. Table SI 20.19 from BS 6472:1993 shows how the r.m.s acceleration corresponding to a low probability of adverse comments during the daytime (that is,

Table SI 20.18 Multiplying factors applied to the British Standard tolerable vibration velocities in humans shown in Tables SI 20.17a and SI 20.17b.

Place	Time	Multiplying Factors (*see* Notes 1 and 5)	
		Exposure to continuous vibration (16h day, 8h night (*see* Note 2)	Intermittent vibration excitation with up to 3 occurrences (*see* Note 8)
Critical working areas (e.g. some hospital operating theatres, precision laboratories see Notes 3 and 10)	Day	1	1
	Night	1	1
Residential	Day	2 to 4 (*see* Note 4)	60 to 90 (*see* Notes 4 and 9)
	Night	1.4	20
Office	Day	4	128 (*see* Note 6)
	Night	4	128
Workshops	Day	8 (*see* Note 7)	128 (*see* Notes 6 and 7)
	Night	8	128

Table SI 20.19 Frequency weighted r.m.s acceleration (m/s^2 r.m.s.) corresponding to a low probability of adverse comment

Place	Exposure periods				
	16h	1h	225s	14s	0.9s
Residential building daytime	0.01 to 0.02	0.02 to 0.04	0.04 to 0.08	0.08 to 0.16	0.16 to 0.32

multiplying factors of 2 to 4 in Table SI 20.18) varyies with exposure duration. BS 6472: 1992 also shows how vibration dose values may be used to assess the severity of impulsive and intermittent vibration (*see* BS 6481).

The total vibration dose value for the day is approximately given by

$$eVDV = 1.4 \times a(\text{r.m.s.}) \times t^{¼} \quad \text{......................} (22)$$

where
eVDV is the estimated vibration dose value (in m/s$^{1.75}$),
a(r.m.s.) is the r.m.s. value (in m/s^2), and
t is the total duration of vibration exposure (in seconds).

Reference needs to be made to the notes accompanying this part of the Standard.

The following notes apply to Table SI 20.18:

1. The table leads to magnitudes of vibration below which the probability of adverse comments is low (any acoustical noise caused by structural vibration is not considered).
2. Doubling of the suggested vibration magnitudes may result in adverse comment and this may increase significantly if the magnitudes are quadrupled (where available, dose/response curves may be consulted).
3. Magnitudes of vibration in hospital operating theatres and critical working places pertain to periods of time when operations are in

progress or critical work is being performed. At other times magnitudes as high as those for residences are satisfactory provided that there is due agreement and warning.

4. Within residential areas people exhibit wide variations of vibration tolerance. Specific values are dependent upon social and cultural factors, psychological attitudes and expected degrees of intrusion.

5. Vibration is to be measured at the point of entry to the entry to the subject. Where this is not possible then it is essential that transfer functions be evaluated.

6. The magnitudes for vibration in offices and workshop areas should not be increased without considering the possibility of significant disruption of working activity.

7. Vibration acting on operators of certain processes such as drop forges or crushers, which vibrate working places, may be in a separate category from the workshop areas considered in Table SI 20.17a (z-axis vibration velocity). The vibration magnitudes specified in relevant standards would then apply to the operators of the exciting processes.

8. Guidance is given in Appendix C of BS 6472:1992 on assessment of human response to vibration induced by blasting.

9. When short-term works such as piling, demolition and construction give rise to impulsive vibrations it should be borne in mind that undue restriction on vibration levels can significantly prolong these operations and result in greater annoyance. In certain circumstances higher magnitudes can be used.

10. In cases where sensitive equipment or delicate tasks impose more stringent criteria than human comfort, the corresponding more stringent values should be applied. Stipulation of such criteria is outside the scope of BS 6472:1992.

The Standard also addresses the matter of repeated exposures to vibration. Where vibration conditions are constant (or regularly repeated) throughout the day, only one representative period, in seconds (of duration t_1) need be measured. If the measured vibration dose value is VDV_1, the total vibration dose for the day (VDV_d) will then be given by

$$VDV_d = (t_d/t_1)^{¼} \times VDV_1 \quad ....(23)$$

where t_1 is the duration of exposure per day (s).

The Standard states that, if in a day, there is a total of N periods of various durations with vibration dose value VDV_n, the total vibration dose value for the day is given by summing the individual n VDVs:

$$VDV = (\Sigma VDV_n{}^4)^{¼} \quad ....(24)$$

In practice, it will be difficult for contractors to determine magnification factors to be applied throughout the numerous properties bordering a construction site, so the factors actually adopted may have to be assumed equal to those for the worst case if the recommendations in the Standard are to be rigorously applied. Also, there may well be a detrimental tendency to apply these standards to buildings.

Reference may also be made to BS 6841 (British Standards Institution, 1987).

International Organisation for Standardisation

ISO 2631 - 1978 (E) (International Organisation for Standardisation, 1978.01.15) addresses the problem of human exposure to vibration in very much the same way as does British Standard 6472:1984 except that vibration levels are here expressed in acceleration (m/s^2 root mean square) units rather than both the acceleration and the vibration velocity (mm/s zero-to-peak) units of the British Standard. As in the British Standard, the one-third octave band centre frequencies from 1Hz to 80Hz are used. Standing-up, sitting-down and also lying-down postures are accommodated in the Standard. Since this ISO Standard pre-dates BS 6472:1984, and since the elements of BS 6472:1984 have been described above, it is not described further here.

VIBRATION RELATIVE TO NORMAL EVERYDAY ACTIVITY IN BUILDINGS

The above vibration levels should be assessed in the context of normal environmental vibrations, as measured by New (1986). It will be quickly realised that it is unrealistic to impose limits that are lower than the ambient vibrations experienced during normal building usage unless the ambient vibrations can be isolated out from the sensitive person or object receiving them. In any case, many externally imposed vibrations are of a temporary nature, and higher levels can be tolerated over a shorter period. Typical normal vibration levels are shown in Table SI 20.20.

PROPOSED ACTION TO BE TAKEN ON A CONSTRUCTION SITE

There should be detailed pre-piling and/or pre-blasting structural surveys carried out jointly by the project engineer and a qualified building surveyor in order to establish:

* Position of the building in relation to the civils work
* Form of the property construction
* Age of the building
* State of repair of the building
* Presence of serious existing defects
* The need for any temporary support, such as shoring or canopies, so that this work can be included in the relevant tender document
* The need for structural monitoring during construction

These factors should always receive careful consideration during the preliminary design stage of the works in case the route and positioning of the works can be so arranged as to minimise detrimental effects on property and people. At the same time there should also be an appraisal of alternative construction methods if significant environmental problems are thought likely to arise from the implementation of current

methods. For example, in the case of deep excavation (say, construction of a storm sewage overflow chamber) close to property, a design incorporating trench sheeting and a lighter-duty hammer may be specified rather than resorting to sheet piling. In another instance there could be a clause in the construction contract that specifies use of continuous flight auger piles if driven piles cause vibration greater than a specified level.

Property schedules should be prepared by a building surveyor at least three months before construction is planned to take place. The schedules would include locations and measurements (lengths and widths) of all pre-existing structural cracks, together with colour photographic records of them. Demec points or cover glass slips should be cemented either side of, or across, the cracks, respectively. There should be a level and plumb survey, with special attention being given to the condition of the damp course. The level survey can be performed by the Resident Engineer's staff from a temporary bench mark well away from the influence of the construction work being performed. Detailed attention should be paid to the following items:

* Walls for tilt and bulge
* Plasterwork, particularly at door frames
* Window panes (steel framed and timber framed)
* Any loose and broken roof tiles
* Chimney pots
* Brick chimneys
* Leaking flashings

Table SI 20.20 Typical vibration velocity levels experienced in a normal environment (after New, 1986)

Source	Resultant peak particle velocity (mm/s)		
	Modern steel framed office	Modern masonry building	Old dwelling house (thick lime mortar masonry)
Normal footfalls	0.02-0.2	0.05-0.5	0.02-0.3
Foot stamping	0.2-0.5	0.3-3.0	0.15-0.7
Door slams	10-15	11-17	3-7
Percussive drilling	5-25	10-20	10-15

Note:
The transducers were located at varying positions on walls, usually within 1m to 4m from the source of the vibration.

*Rainwater pipes and gutterings
*Renderings to exterior walls
*Loose and free-standing fitments and ornaments.

Copies of the property schedules should be given to the property owners and the inhabitants.

Where the investigation provides evidence of damage, the building and any cracking therein should be monitored for a period of three months before construction work begins. If there is continuing movement prior to construction, then the property owner must be informed and this movement taken account of when any liability for reinstatements is under review. There is some advantage in the client being generous as far as reinstatements are concerned, but an agreement should be entered into whereby the cost of reinstatement is in full and final settlement of the matter and that responsibility for any further damage will not fall on the participating parties to the particular construction contract.

During piling or blasting, monitoring - an essential condition of a consent to work agreement - of vibration will be needed. A clause may be inserted in the contract documentation that this work will be done on a day-to-day basis by the contractor, but the environmental health inspector will also take his own measurements. Construction work conducted within specified vibration limits is at the risk of the Employer, whereas work above those limits is at the risk of the Contractor. However, it is the responsibility of the Resident Engineer and his staff to ensure that the vibration is monitored at all times, that correct records are kept, and that the specified vibration limits are adhered to.

Because of the effects of vibration upon people the public relations aspect of a Resident Engineer's work can become very important in quelling people's fears and thereby reducing the possibility of legal restrictions being placed on the work. The public must be made aware of the work to be done, the reason for it, that blasting, piling, and/or extraction will be necessary, and they are given information on the controls, times, advance warnings, together with a contact name, address and telephone number. Complaints in the first instance should be directed to the Resident Engineer.

In the first instance public contact should be achieved at the pre-contract stage through advance press release and circulars distributed to households likely to be affected. Political representatives should be informed by personal letter. Responsibility for dissemination of this information should lie with the project engineer.

Approximately one month before work begins on site, the Resident Engineer (Supervisor) should contact, by an information leaflet and personal letter, those property owners and occupiers of property likely to be affected by the works. Any complaint during construction must be responded to immediately, with a visit by the Resident Engineer to the complainant in order to determine its precise nature. If the complaint concerns alleged damage to property, then the Resident Engineer should refer the matter to the surveying department of the organisation responsible for the civils works, or to an independent firm, as the case may be, and also to the Public Relations Department of the client organisation. Complaints of vibration are often made by people really wanting reassurance as to the well-being of their property. Concern can very often be reduced by taking specific measurements at the property (ideally using an analogue trace read-out), also explaining recommended limits and indicating procedures that will be taken by the client should those limits be exceeded, and then demonstrating that the measured vibration can also sometimes be approached or exceeded by internal household vibrations (banging doors, jumping on floors, etc.). Ornaments on glass shelves in display cabinets often rattle during normal household activity and will be particularly prone to rattling when vibrated from outside the house. It would be advisable to seek permission to pack such ornaments carefully into boxes before work begins.

It is very important that the occurrences of blasting, piling and/or pile extraction be fully documented. When and where any damage is caused, or is alleged to have been caused, the records will assist in demonstrating liability or otherwise, and can be used in conjunction with measurements taken at, in, or in the vicinity of the property. Full documentation serves also to demonstrate to all concerned that care is being taken with the monitoring.

SOME POINTS OF POLICY IN RESPECT OF PILING AND PILE EXTRACTION

In all cases where piling and/or extraction is known to be necessary, or thought to be possible, whether as part of the permanent works or for temporary ground support, a consent to work agreement in accordance with the 1974 Control of Pollution Act should be obtained by the client from the local environmental health inspector in advance of tender enquiry. A note to the effect that such a consent to work agreement has been entered into should be included in the contract documents, and full details of the agreement should be given in the documents and itemised in the bill of quantities for contractor pricing.

A joint consultative procedure may exist between the client, if a local or public authority, and the statutory undertakers, in which case this procedure should be followed.

The permitted work periods for piling and/or pile extraction should be clearly stated in the Specification section dealing with this particular item. In general, the specified working hours would typically be from 08.00hrs until 17.30hrs Monday to Friday, inclusive. There would be a cumulative maximum period of 6 hours per day for driving or extraction of piles. Any exceptions to this rule should be identified during consultations between the contracting parties and the local authority, and be included in the contract documents.

Every person, organisation or company approached and/or consulted in connection with the works should be informed of the proposed hours of working and of proposed limiting noise and vibration levels.

BLASTING

Human perception

The fourth column in Table SI 20.18 above gives the multiplying factors to be used in order to specify vibration magnitudes that are tolerable to humans.

For a given vibration magnitude an individual's tolerance to perceived vibration decreases as the number of events increases. For three or more blasts in a 16 hour day, a factor F can be derived:

$$F = 1.7N^{\frac{1}{2}}T^{-d} \quad \dots \dots \dots \dots \dots \dots \dots \dots \dots \dots \dots \dots \dots \dots (25)$$

where

N is the number of events in a 16 hour day (and for $D > 3$),
T is the duration of events in seconds, and
d is zero for T less than 1 second.
For $T > 1$ sec., $d = 0.32$ for wooden floors;
$d = 1.22$ for concrete floors.

A blasting event is defined in the Standard as a vibration exceeding 0.5×10^{-3}m/s velocity (zero to peak) or background vibration level, whichever is the greatest. Duration is defined as the period of time in seconds that this level is exceeded. This relationship does not apply when values lower than those given by the factors for continuous vibration in Table SI 20.18 result.

There are useful applications of these recommendations in the BS 6472:1992.

Some possible outline clauses for inclusion in contract documentation

'The Contractor shall limit his charges and so design his blast such that the peak particle velocity does not exceed the values given in Table(...) according to the particular environmental conditions pertaining at the site.

'Vibration measurements and the reporting of those measurements shall be the responsibility of the Contractor.

'The location for the vibration measurements shall be at ground level immediately adjacent to the building, service or other item which is nearest to the face being blasted.

'The vibration measurements shall be made using an instrument approved by the Engineer, calibrated immediately before contract use and issued with a calibration certificate if the Engineer so requires, and capable of measuring along three orthogonal axes. So when the instrument is positioned, one of these axes shall be horizontal and parallel to the centre line of the excavation and another shall be vertical. This requirement may be varied with permission from or direction from the Engineer's Representative. The Contractor shall provide concrete plinths or other means of instrument support if and as required by the manufacturer's instructions or those of the Engineer's Representative. Any such supports shall be removed when no longer required. An adequate and agreed period of notice shall be given by the Contractor to the Engineer's Representative before firing any charge so that blasting arrangements can be inspected and instrument reading witnessed if the Engineer so requires.

'Before beginning any stage of the works that involves blasting operations, the Contractor shall submit his proposals in writing to the Engineer. These proposals must include the type of explosive and detonator, charge weights, delay sequences, firing patterns, and the likely variations in these proposals in relation to locations along the route of the works which are considered to be critical with respect to blasting operations. "Locations" in this context means both the geology at the tunnel face and structures (both above ground and below ground).

'In addition to the requirements of the Statutory Regulations, the shotfirer shall be a competent person, at least 21 years of age, who is experienced in the work and who has received adequate instruction as to the dangers connected therewith and the precautions to be observed. The shotfirer must be approved by the Engineer and appointed in writing by the Contractor. Explosives shall not be handled or used except by or under the direct supervision of the appointed shotfirer. The shotfirer must keep a legible record of the shot hole configuration in sketch form on pro-forma sheets, the number of shots fired, their time of firing, type and weight of explosives used, and the type and weight of detonators used for each and every location. In all cases copies of the shot firing records and vibration readings in an agreed form shall be supplied to the Engineer's Representative before the end of every shift on which shots were fired.

'Charges shall be fired so as not to interfere with or be a danger or nuisance to residents or property or operations carried out therein. At all times when firing is to take place the Contractor shall provide a means to prevent flying debris. Shafts and similar bounded excavations shall be protected by solid timber covers, fastened together and weighted down. Open excavations shall be fully and effectively covered by blast mats and heavy netting. The Contractor shall also provide sufficient watchmen, pickets, notices, warning sirens, fencing and/or other measures as

are deemed to be necessary by the police for the protection of persons and property.'

Additional safety measures to be taken when blasting in gaseous strata

Introductory note

Gaseous means inflammable or noxious. Inflammable gas is usually associated with tunnelling in Coal Measures strata, but it may have deeper seated sources or may be conducted into other strata by jointing systems. It may also be associated with vegetation decay. It is necessary to carefully identify such a potential hazard at the project site investigation stage.

Possible specification clauses

'During the Contract, shaft sinking and tunnelling/heading operations will be carried out through (Coal Measures) strata (and possibly through abandoned coal workings) which will yield or are likely to yield inflammable (and toxic) gas. The Contractor shall allow for the following requirements which are the minimum acceptable for the safe execution of the works.

'Before work commences, a competent person shall be nominated by the Contractor, approved by the Engineer and appointed in writing to ascertain thereafter the condition of all areas of the works affected by inflammable or noxious gases.

'Each shaft sinking and tunnel/heading shall be ventilated at all times by a fan of adequate capacity to dilute and render harmless all inflammable or noxious gases and to provide air containing a sufficiency of oxygen. The fan shall be situated at a minimum of 5 metres from the edge of the shaft top and be provided with sufficient ventilation ducting to within a maximum distance of 5 metres from the base of the shaft sinking or tunnel face, as the case may be. The fan shall be capable of delivering or exhausting at the end of the air duct a minimum quantity of 15 cubic metres of air per minute per square metre of area of the face to be ventilated. The use of compressed air equipment for ventilation will be strictly prohibited. Other ventilation requirements are outlined below.

'Each shaft sinking and heading/tunnel shall be provided with at least one safety lamp and an approved automatic inflammable gas detector. Continuous monitoring for inflammable gases and oxygen contents shall be instituted, with monitors set as near as possible to the most hazardous areas. Regular tests shall also be undertaken for carbon dioxide content.

'All safety lamps and gas detectors shall be expertly and regularly maintained. Each safety lamp shall be issued to and used only by a competent person, at least 21 years of age, who has received adequate instructions as to the dangers connected therewith and the precautions to be observed. The contractor must keep written records of such instruction.

'Lighting and all other electrical equipment shall be either intrinsically safe or flameproof of a type approved by the Health and Safety Commission. Smoking and naked lights will be prohibited in the tunnels/heading and associated shafts, and on the surface within 5 metres of the shaft edge and any exhaust fan outlet.

'If the amount of inflammable gas (methane) in the general body of the air or in cavities in any shaft sinking or tunnel/heading exceeds 1.25 percent by volume, the electrical supply shall be cut off from the surface from all apparatus other than the telephone or any electrical safety fan. No shots shall be charged or fired. If the amount of inflammable gas in the main body of the air exceeds 2 percent by volume, all persons in the affected areas shall be withdrawn immediately. The nominated competent person to ascertain the conditions of the area(s) affected and the measures taken to render safe the affected area(s) shall be approved by the Engineer, such approval not relieving the Contractor of his obligations under the Contract. Personnel will only be allowed to re-enter the area(s) after safe conditions to the satisfaction of the Engineer have been restored.

'Items have been included in the Bills of Quantities to enable the Contractor to allow for the provisions herein made.

'If the Engineer considers that the conditions within the shaft sinking or tunnel/heading require further restrictions, then he may apply relevant sections of the Mines and Quarries Act 1954 and any subsequent Regulations.

'Only *Permitted* explosives shall be used for blasting initiated by means of electric detonators approved for use with Permitted explosives. All other shot firing equipment shall be approved by

the Health and Safety Commission for use in coal mines. The shotfirer should hold a current Mining Qualification Board (MQB) shotfirer's certificate or be competent to the satisfaction of the Engineer, and shall carry out tests for gas before and after blasting. Records of gas tests shall be kept and results submitted to the Engineer at the end of every shift on which shots are fired.

Further Ventilation Requirements

'In addition to the above requirements, the following shall apply in order to render the shaft or tunnel/heading ventilation satisfactory.

(a) The amount of carbon dioxide in the general body of the air shall not exceed 0.5 percent by volume.

(b) The amount of oxygen in the general body of the air shall not be less than 20 percent by volume.

(c) The amount of respirable dust in the general body of the air shall not be more than 3 milligrams per cubic metre.

(d) The amount of respirable quartz in the general body of the air shall not be more than 0.25 milligrams per cubic metre. (Respirable particles of less than 7μm in size create the problem.)

(e) The amount of any other substance must not be more than the threshold limit values currently in use by the Health and Safety Commission. In the case of an inflammable gas, this value must not exceed 25 percent of the lower explosive limit of the mixture by volume in air.

'When the Contractor proposes to use a tunnelling machine for the excavation of rock strata, ventilation shall be by an exhausting system fitted with an effective dust filter.'

Further reference on this subject may be made to the Control of Substances Hazardous to Health (COSHH) Regulations (1988) and to Kirby and Morris (1990).

EFFECT OF VIBRATION ON FRESH CONCRETE

Akins and Dixon (1979) suggest that vibration

Table SI 20.21 Permissible vibration velocity levels for concrete

Concrete age	Permissible particle velocity (mm/s zero to peak)
0 to 4 hours 4hrs to 24 hrs 1 day to 7 days	No limit 5, but preferably no vibration 50

during the initial 2 to 4 hours after mixing and placing can actually increase concrete strength by up to 35%. The critical time follows the initial set and before acquisition of appreciable strength. Suggested limits for vibration are given in Table SI 20.21.

Piles may therefore be driven close to newly cast-*in situ* piles up to about 4 hours after casting, or after a few days, but not in between. It would be sensible for contract documents to include a specification clause which draws attention to the possibility that the ultimate strength of 'green' concrete may be affected by the close proximity of construction activities which cause significant vibration. The clause would thus require the contractor to have regard to this during his programming of the work, and would perhaps draw express reference to his method statement.

NOISE

Sound exists over a very wide range of frequencies. Young people are able to detect frequencies of between 20Hz and 20kHz. Frequencies above 20kHz are in the ultrasound region and frequencies below 20Hz are in the infrasound region.

Sound intensity is usually presented as a sound level using a logarithmic unit, the decibel (dB), as discussed above. A sound level change of 1dB can just be detected by the human ear. A sound level increase of 10dB anywhere within the range of hearing is perceived by the ear as a doubling in loudness, and correspondingly a drop of 10dB is perceived as a halving in loudness.

A suitable sound-measuring instrument duplicates the ear variable sensitivity to sound of different frequencies. This can be achieved by building into the instrument a filter having a similar frequency response to that of the ear. This filter is called an A-weighting filter because it conforms with the internationally standardised A-weighting curves. Measurements obtained with this filter are termed A-weighted sound level measurements and the unit is the dB. The sound from sources of noise, particularly from construction operations, often fluctuates substantially during a given period of measurement. An average value can be measured, the equivalent sound pressure level $L_{Aeq,T}$ being the equivalent continuous sound level which would deliver the same sound energy as the actual A-weighted fluctuating sound measured in the same time period T. Another weighting is L_{10}, the A-weighted sound level exceeded for 10% of the measurement period.

In order to determine the composition of a sound it is necessary to determine the sound level at each frequency individually. Sound level values are stated in octave bands. The audible frequency range is divided into ten octave bands having centre frequencies and bandwidth according to international standards. The centre frequencies of each consecutive octave band are twice the centre frequency of the previous one, and the upper frequency of each octave band is twice the lower frequency. Reference to the octave bands is usually by their centre frequencies, with, for example, the 500Hz octave band covering a range from 354Hz to 707Hz. The centre frequency is the geometrical average of the upper and lower frequencies, so

$$\text{centre frequency} = (\text{lower frequency} \times \text{upper frequency})^{\frac{1}{2}} = 500\text{Hz}.$$

Conversely,

$$\text{lower frequency limit} = 2^{-\frac{1}{2}} \times \text{centre frequency},$$

and

$$\text{upper frequency limit} = 2^{\frac{1}{2}} \times \text{centre frequency}.$$

Levels of noise arising from more than one

source are not added directly because they are logarithmic quantities, but the total level does exceed that attributable to one source alone. Two equally intense sound sources produce a sound level which is 3dB higher than one alone, and ten sources produce a 10dB higher sound level. Sound attenuates with distance travelled, in free air from a point source by 6dB for each doubling of the distance. Attenuation inside a building or confined workplace is less than this because of reverberation effects from walls and ceilings, and linings in the case of tunnels and shafts.

In addition to the frequency weightings there are also time weightings, the choice usually being between three standardised weightings or dampings. The 'S' weighting has a high damping, giving a slow display movement and having an effective averaging time of 1 second. The 'F' weighting has low instrument damping, so giving a more rapid display movement with an effective averaging time of approximately 0.125 second.

The 'I' weighting has a very fast rising time constant and a very slow falling time constant. The intention is to present a value which represents how loud the human ear judges a short duration sound, and is so directed at assessing annoyance rather than the risk of damage to hearing. Further to the three weightings above, some sound meters are also able to measure the actual peak sound pressure level of a short duration sound even as short as 20 microseconds. This is targeted at the risk of damage to hearing.

The British Noise at Work Regulations (1989) require that action be taken where exposure to noise is likely to reach or exceed any of the three 'action' levels that are given in Table SI 20.22. Table SI 20.23 shows some likely noise exposure levels in $L_{EP,d}$ terms and Table SI 20. 23 relates some typical dB levels to the actual environmental conditions that produce the noise. For further information the book by Adams and McManus (1994) may be consulted.

Table SI 20.22 Noise at Work Regulations (1989)

Action required where $L_{EP,d}$ is likely to be: (*refer to* Note 1 below)	Below 85 dB(A)	85 dB(A) First AL	90 db(A) Second AL
EMPLOYER'S DUTIES			*See* Note 2
General Duty to Reduce Risk Risk of hearing damage to be reduced to the lowest level reasonably practicable (Regulation 6)	•	•	•
Assessment of Noise Exposure Noise assessments to be made by a Competent Person (Regulation 4)		•	•
Record of assessments to be kept until a new one is made (Regulation 5)		•	•
Noise Reduction Reduce exposure to noise as far as is reasonably practicable by means other than ear protectors (Regulation 7)			•
Provision of Information to Workers Provide adequate information, instruction and training about risks to hearing, what employees should do to minimise risk, how they can obtain ear protectors if they are exposed to between 85 and 90 dB(A), and their obligations under the Regulations (Regulation 11)		•	•
Mark ear protection zones with notices, so far as is reasonably practicable (Regulation 9)			•

(Table SI 20.22 is continued on the next page)

Table SI 20.22 Noise at Work Regulations (1989) (Continued)

Action required where $L_{EP,d}$ is likely to be: (*refer to* Note 1 below)	Below 85 dB(A)	85 dB(A) First AL	90 db(A) Second AL

Ear Protectors
Ensure that, as far as is practicable that protectors are:
  *provided to employees who ask for them
  (Regulation 8(1))
  *provided to all exposed (Regulation 8(2))
  *maintained and repaired (Regulation 10(1)(b))
  *used by all exposed (Regulation 10(1)(a))
		•
		
		•
		
			•
			•
			•

Ensure so far as is reasonably practicable that all who go into a marked ear protection zone use ear protectors (Regulation 9(1)(b))			*See* Note 3 •

Maintenance and Use of Equipment
Ensure, as far as is practicable, that:
  *all equipment provided under the Regulations is used, except for the ear protectors provided between 85 dB(A) and 90 dB(a) (Regulation 10(1)(a))
  *all equipment is maintained (Regulation 10(1)(b))
		•
		•
			•
			•

EMPLOYEE'S DUTIES			

Use of Equipment
So far as is practicable:
  *use ear protectors (Regulation (10)(2))
  *use any other protective equipment (Regulation (10)(2))
  *report any defects that are discovered to his or her employer (Regulation (10)(2))
		
		•
		•
			•
			•
			•

MACHINE MAKER'S AND SUPPLIER'S DUTIES			
Provision of Information Provide information on the noise likely to be generated (Regulation 12)		•	•

Notes on Table SI 20.22:
Subscript '*EP*' denotes 'effectively perceived'.
1. The dB(A) action levels are values of daily personal exposure to noise ($L_{EP,d}$).
2. All the actions indicated at 90 dB(A) are also required to be implemented where the peak sound pressure is at or above 200 Pa (140 dB re 20μPa).
3. This requirement applies to all who enter the zones, even if they do not stay long enough to receive an exposure of 90 dB(A)$L_{EP,d}$.

Table SI 20.23 Likely noise exposure (after the Health and Safety Executive, 1993)

Activity		Likely noise exposure $L_{EP,d}$	
		Average	Range
Agent (up to 50% day on site)		<80	
Asphalt paving		<85	
Blasting		100+	
Bricklayer		83	81-85
Carpenter		92	86-96
Concrete	chipping/drilling	85+	
	floor finishing	85	
	grinding	85+	
Concrete worker		89	
Crushing	mill worker	85+	
Driver	crawler tractor	85+	
	dumper	85+	
	excavator	<85	
	grader	85+	
	loader	<85	
	roller	85+	
	wheeled loader	89	
	wheeled tractor	<85	
Engineer	supervising pour	96	
	surveying	<80	
Foreman	supervising workers	80	
Formwork setter		92	89-93
Ganger	concrete pour	93	92-93
	general work	94	
Guniting		85+	
Labourer	concrete pour	97	95-98
	digging/scabbling	100	
	general work	84	
	shovelling hard core	94	
	shuttering	91	
M&E installer			
	general	89	82-96
	small work	84	78-89
Piling operator		85+	
Piling worker		100+	
Reinforcement worker			
	building site	86	82-89
	bending yard	84	77-87
Sandblasting		85+	

Table SI 20.24 Some typical dB levels

Sound level in dB	Environmental conditions	Sound pressure in μbars
140		
134	Threshold of pain	1000
130		
	Pneumatic concrete breaker	
120		
114	Loud car horn (at 1m)	100
110		
	Inside old aeroplane	
100		
94	Inside a metro (underground) train	10
90		
	Inside bus	
80		
74	Average traffic on a street corner	1
70		
	Conversational speech	
60		
54	Typical company office	0.1
50		
	Living room in a suburban area	
40		
34	In a library	0.01
30		
	Bedroom at night	
20		
14	Broadcasting studio	0.001
10		
0	Threshold of hearing	0.0002

The dB levels in Table SI 20.24 can be assessed in the context of noise logging by the Building Research Station (*BRE News of Construction Research*, April 1994) which indicated that, from a sample of 1000 homes, 56% were exposed to a noise level of >55dB(A). This is the level recommended by the World Health Organisation which should not be exceeded if significant community annoyance is to be prevented. The measured percentage rose to 63% at night, when the recommended WHO level is 45dB(A). Road traffic is the main source of noise, being heard outside over 95% of the properties.

REFERENCES

Adams, M. and McManus, F. (1994) *Noise and Noise Law*, Wiley Chancery Law, Chichester, England. (ISBN 0471 93708 8.)

Akins, K.P. and Dixon, D.E. (1979) Concrete structures and construction vibrations, *Symp. on Vibrations of Concrete Structures*, American Concrete Institute SP-60, 213-247.

Attewell, P.B. (1986a) Estimation of ground vibration caused by driven piling in soil. In: *Contributions on the Influence of Earthwork Construction on Structures* (Proc. XII Int. Conf. Soil Mech. Found. Engrg, San Francisco, Calif., Aug. 11-16 1985), D. Reséndiz and M.P. Romo (eds), 1-10.

Attewell, P.B. (1986b) Noise and vibration in civil engineering, *Mun. Engr.*, **3**, 139-158.

Attewell, P.B. and Farmer, I.W. (1973) Attenuation of ground vibration from pile driving, *Ground Engrg*, 6(4), 26-29.

Attewell, P.B. and Ramana, Y.V. (1966) Wave attenuation and internal friction as functions of frequency in rocks, *Geophysics*, **31**, 1049-1056.

Attewell, P.B., Selby, A.R. and O'Donnell, L. (1992a) Estimation of ground vibration from driven piling based on statistical analyses of recorded data, *J. Geotech. and Geol. Engrg*, 10(1), 41-59.

Attewell, P.B., Selby, A.R. and O'Donnell, L. (1992b) Tables and graphs for the estimation of ground vibration from driven piling operations, *J. Geotech. and Geol. Engrg* 10(1), 61-87.

Attewell, P.B., Selby, A.R. and Uromeihy, A. (1989) Appraisal of ground vibration from civil engineering construction, *Int. J. Min. Geol. Engrg*, 7, 183-208.

Boyle, S. (1990) The effects of piling operations in the vicinity of computing systems, *Ground Engineering*, 23(5), 23-27.

British Standards Institution (1975) *Code of Practice for Noise Control on Construction and Demolition Sites*, BS 5228, HMSO, London (under review).

British Standards Institution (1984) *Guide to the Evaluation of Human Exposure to Vibration in Buildings (1 Hz to 80 Hz)*, BS 6472: 1984, London.

British Standards Institution (1987) *British Standard Guide to Measurement and Evaluation of Human Exposure to Whole Body Mechanical Vibration and Repeated Shock*, London, BS 6841: 1987, London.

British Standards Institution (1992a) *Noise Control on Construction and Open Sites: Part 4. Code of Practice for Noise and Vibration Control Applicable to Piling Operations*, BS 5228: 1992, London 64pp. ISBN 0 580 20381 6.

British Standards Institution (1992b) *Guide to Evaluation of Human Exposure to Vibration in Buildings (1Hz to 80Hz)*, BS 6472: 1992, London, 18pp. ISBN 0 580 19963 0.

Broadhurst, K.A., Wilton, T.J. and Higgins, J.P. (1989) Review of current standards and recommendations for vibration and noise, *Trans. Inst. Min. and Metall.*, **85**, A210-A213.

Building Research Establishment (1983) *Vibrations: Building and Human Response*, Digest No. 278, HMSO, London.

Control of Pollution Act (1974) Part III, HMSO, London.

Control of Substances Hazardous to Health Regulations (1988) amended by SI 1990/2026, SI 1991/2431, and SI 1992/2382.

Dieckmann, D. (1958) A study of the influence of vibration on man, *Ergonomics*, 1(4), 347-355.

Environmental Pollution Act (1990), HMSO, London.

German Standard (Deutsche Norm) (1975) *Vibrations in Buildings; Principles, Predetermination and Measurement of the Amplitude of Oscillations*, DIN 4150 Part 1 with Part 2 and Part 3, DK 534.83: 534.647:699.84.

German Standard (Deutsche Norm) (1986) *Structural Vibration in Buildings. Effects on Structures*, DIN 4150 Part 3, UDC 534.83: 534.647:699.84.

Gordon, C.J. and Wilby, J.F. (1986) *The Control of Vibration in Buildings with Special Reference to the Design of Microelectronics Manufacturing and Research Facilities*, BBN Laboratories Report, California, USA.

Health and Safety Executive (1989) *The Control of Substances Hazardous to Health Regulations*, Statutory Instrument 1988 No. 1657, HMSO, London, 43pp.

Health and Safety Executive (1993) *Noise in Construction: Further Guidance on the Noise at Work Regulations 1989*, Pamphlet published by the HSE and available free from HSE Books, PO Box 1999, Sudbury, Suffolk CO10 6FS.

Indian Standards Institution (1973) *Criteria for Safety and Design of Structures Subject to Underground Blasts* (First reprint April 1982), IS: 6922 - 1973.

Indian Standards Institution (1977) *Method of Test for Determination of Dynamic Properties of Soil* (First revision), IS: 5249 - 1977.

Institution of Civil Engineers (1991) *ICE Conditions of Contract and Forms of Tender, Agreement and Bond for Use in Connection with Works of Civil Engineering Construction*, 6th Edition, Institution of Civil Engineers, Association of Consulting Engineers, and the

Federation of Civil Engineering Contractors, Thomas Telford Ltd, London, 54pp.

International Organisation for Standardisation (1978(E)) *Guide for the Evaluation of Human Exposure to Whole Body Vibration*, Second Edition, ISO 2631.

International Organisation for Standardisation (1987) Draft: *Shock and Vibration Sensitive Equipment. Methods of Measurement and Reporting Data of Shock and Vibration Effects in Buildings*, ISO 8569.

International Organisation for Standardisation (1989) *Mechanical Vibration and Shock-Measurement and Evaluation of Vibration Effects on Buildings - Guidelines for the Use of Basic Standard Methods*, ISO 4866.

Kirby, C.E. and Morris, W. (1990) A response to COSHH, *Trans. Inst. Min. Metall.*, **99**, A138-A146.

Langefors, S.L. and Kihlström, B. (1963) *Modern Techniques of Rock Blasting*, John Wiley and Sons, New York.

Martin, D.J. (1980) *Ground Vibrations from Impact Pile Driving during Road Construction*, Transport and Road Research Laboratory Report No. 544, Crowthorne, Berkshire, England.

New B.M. (1986) *Ground Vibration caused by Civil Engineering Works*, Transport and Road Research Laboratory Report 53, Crowthorne, Berks.

Noise at Work Regulations 1989.

Reiher, H. and Meister, F.J. (1931) Die Empfindlichkeit der Menschen gegen Erschütterungen, *Forschung auf dem Gebiet der Ingenieurwesens*, **2**(II), 381-386.

Richart, F.E. and Woods, R.D. (1987) Vibrations. In: *Ground Engineer's Reference Book*, F.G. Bell (ed), Butterworths, London, 17/4-17/13.

Selby, A.R. and Swift, J.S. (1989) Recording and processing ground vibrations caused by pile driving, *Proc. Int. AMSE Conf. on 'Signals and Systems'*, Brighton, England, **6**, 101-113.

Standards Association of Australia (1983) *Explosives - Storage, Transport and Use known as the SAA Explosives Code Part 2 Use of Explosives*, AS 2187, Part 2-1983, Standards House, 80 Arthur St., North Sydney, NSW, Australia, 32pp.

Swiss Standard (1978) *Effects of Vibrations on Buildings*, SN 640 312:1989.

Uromeihy, A. (1990) Ground vibration measurements with special reference to pile driving, unpublished PhD thesis, University of Durham, England.

Vuolio, R. (1990) Blast vibration: threshold values and vibration control, PhD thesis, Finnish Academy of Technology, Helsinki, Finland. (Acta Polytechnica Scandinavica, Civil Engineering Construction Series No. 95, 146pp.)

BIBLIOGRAPHY

Ashley, C. and Parkes, D.B. (1976) Blasting in urban areas, *Tunnels and Tunnelling*, Sept. 1976, 60-67.

British Standards Institution (1982) *Guide to the Selection and Use of Elastomeric Bearings for Vibration Isolation of Buildings*, BS 6177:1982, London.

Das, B.M. (1983) *Fundamentals of Soil Dynamics*, Elsevier, Amsterdam.

Health and Safety Executive (1986) *Noise from Portable Breakers*, IAC L21.

Health and Safety Executive (1989) *Noise at Work Guides. Noise Guide no. 1: Legal Duties of Employers to Prevent Damage to Hearing; Noise Guide no. 2: Legal Duties of Designers, Manufacturers, Importers and Suppliers to Prevent Damage to Hearing. The Noise at Work Regulations 1989*. (ISBN 011 885430 5.)

Health and Safety Executive (1990) *Noise at Work: Noise Assessment, Information and Control. Noise Guides 3 to 8 HS(G)56.* (ISBN 011 885430 5.)

Health and Safety Executive (1992) *Introducing the Noise at Work Regulations: A Brief Guide to the Requirements for Controlling Noise at Work*, IND(G)75L.

Land Compensation Act 1973, HM Stationery Office, London.

New, B.M. (1982) Vibrations caused by underground construction, *Proc. Tunnelling '82*, London, 217-229.

Pekeris, C.L. and Lifson, H. (1957) Motion of the surface of a uniform elastic half space produced by a buried pulse, *J. Acoust. Soc. Amer.*, **29**, 1233-1238.

Wiss, J.F. (1981) Construction vibrations: state-of-the-art, *J. Geotech. Engrg Div., ASCE*, **107**(GT2), Feb. '81, 167-181.

21

SEWER AND WATER MAIN LININGS

New construction is usually of circular cross-section in contrast to much of the older Victorian 'egg-shape' cross-section. The circular section lends itself to easier construction; the egg-shape, with smaller radius at invert, encourages higher gravity flow velocities, and better cleansing, at low volume (dry weather) flow rates. More modern techniques of lining are usually by reinforced concrete segments for the primary lining, with a cast-*in-situ* concrete permanent lining. Older sewers were lined in high quality durable brick, typically of Accrington character, and a mortar having lower durability. Sometimes for new sewer tunnels a patent 'one-pass' type of primary lining (for example, Charcon or C.V. Buchan p.c. concrete segments with hydrophilic sealing glands), which allows a secondary lining to be dispensed with, might be used. Grouted-up Lytag will often be used for void infilling at the extrados. These linings will not have the long-term integrity of a two-stage lining, particularly if the soil and groundwater conditions become aggressive during the lifetime of the lining. Lining segments of trapezoidal shape are particularly suitable for adjustment of line and level, and for use when tunnelling round curves. As one example, this type of lining, with a Hydrotite gasket, was used at Southport on North West Water's Coastal Water's Interceptor Project (modified IChemE Green Book project) in association with a Lovat M131 TBM used in both open-face (conveyor belt and compressed air) and EPB (with screw) modes. The 15mm taper on the trapezoidal liner assisted the building of 600m radius curves. The rings themselves had dowels on the circle joints and bolts on the cross joints, with caulking grooves cast on the segments. In another example, the C.V. Buchan trapezoidal segments used in the Barking Reach combined cycle gas turbine power station cooling water tunnels were sealed with a C E Heinke EPDM sealing gasket.

Many of the older sewers are now well beyond their 'design lives' of 100 years and are in need of extensive repair at best or, at worst, are in a state of total collapse. Although many old interceptor sewers have been or will be bypassed by new sewers, some of the old sewers cannot be duplicated for economic or other reasons. If they have collapsed they must be re-built using high quality bricks and modern mortars, often comprising a resin base. Some sewers must be re-lined entirely to ensure their watertightness and their long-term resistance to aggressive chemicals and abrasion, but the labour costs involved in bricklaying would usually rule this method out of consideration. One method is to use glass-reinforced plastic, pinned or expanded by internal air pressure to the old lining. The 'Alphacrete' technique, claimed to be less expensive, involves the use of weldmesh and 'chicken wire' fixed to fabric-backed polypropylene or high density polypropylene. Grout is then sprayed on and trowelled up in man-entry tunnels to form a ferrocement lining behind the permanent shuttering.

More generally for a water main, a single-pass, wedge block (expanded) lining might include 10 No. pre-cast concrete segments plus a key segment. Insertion of the key re-compresses the ground behind the lining so that it can assist the lining to withstand an internal water pressure. However, construction of this type of lining requires the leaving of a 1m gap between the back of the tunnelling machine (or shield) and the last erected ring of support. Good self-support capability of the ground is therefore required. (London Clay, for example, offers such suitable support.) If the ground cannot stand unsupported while an expanded lining is erected then it is better to adopt a bolted primary lining together with a secondary lining.

New steel fibre-reinforced tunnel segments have been used by Miller Construction on the baggage handling tunnel driven at Heathrow Airport in London under the New Engineering Contract conditions of contract (*see* Section 2.8 in the main text, Part One of the book). The segments were developed in conjunction with Crendon Tunnel Linings.

BIBLIOGRAPHY

Tiedemann, H.R., Parker, H.W. and Hansmire, W.H. (1987) Precast concrete segmental linings - practical considerations, *Proc. Conf. Rapid Excav. and Tunneling*, New Orleans, USA; Littleton: *Soc. Mining Engrs*, **1**, 87-95.

Ward, W.H. (1970) *Yielding of the Ground and the Structural Behaviour of Linings of Different Flexibility in a Tunnel in London Clay*, Building Research Station Current Paper 34/70, Garston, Watford, Hertfordshire.

INDEX

Printed in the USA/Agawam, MA
December 15, 2011